AF389383

Movement Biomechanics and Motor Control

Movement Biomechanics and Motor Control

Special Issue Editor

Carlo Albino Frigo

MDPI • Basel • Beijing • Wuhan • Barcelona • Belgrade • Manchester • Tokyo • Cluj • Tianjin

Special Issue Editor
Carlo Albino Frigo
Politecnico di Milano
Italy

Editorial Office
MDPI
St. Alban-Anlage 66
4052 Basel, Switzerland

This is a reprint of articles from the Special Issue published online in the open access journal *Applied Sciences* (ISSN 2076-3417) (available at: https://www.mdpi.com/journal/applsci/special_issues/Movement_Biomechanics_Control).

For citation purposes, cite each article independently as indicated on the article page online and as indicated below:

LastName, A.A.; LastName, B.B.; LastName, C.C. Article Title. *Journal Name* **Year**, *Article Number*, Page Range.

ISBN 978-3-03936-266-0 (Hbk)
ISBN 978-3-03936-267-7 (PDF)

Contents

About the Special Issue Editor

Carlo Albino Frigo Associate Professor at the Department of Electronics, Information, and Bioengineering of Politecnico di Milano, teacher at the Biomedical Engineering degree, he has been President of the Italian Society of Clinical Movement Analysis, Vice President of ISPO-Italy (International Society for Prosthetics and Orthotics), committee member of ESMAC (European Society for Movement Analysis in Adults and Children), invited speaker at the Honorary Baumann Lecture at ESMAC-2012 congress, member of several scientific societies and editorial boards, and participated in several European projects and educational courses. His main interests are in the areas of movement analysis, motor rehabilitation, musculoskeletal modelling, biomechanics, and motor control.

Editorial

Special Issue: Movement Biomechanics and Motor Control

Carlo Albino Frigo

Department of Electronics, Information and Bioengineering, Politecnico di Milano, I-20133 Milano, Italy; carlo.frigo@polimi.it

Received: 30 April 2020; Accepted: 4 May 2020; Published: 6 May 2020

When Applied Science invited me to organize a Special Issue on Movement Biomechanics and Motor Control, more than one year ago, I was surprised, but also flattered by the invitation. Actually, Movement Biomechanics and Motor Control was the name of my laboratory at Politecnico di Milano. The name comes from the intense activity we developed in this field many years ago, mainly thanks to my great friend Prof. Paolo Crenna, who passed away prematurely. I started contacting old friends, Prof. Marco Schieppati in particular, who suggested a number of outstanding researchers to me and accepted to work on the excellent review that you can see published in the present issue [1]. I was impressed by the warm responsiveness of most authors who accepted to contribute with their original papers. As I supposed, some of the contacted persons already had a plan of publication, but some of them accepted to change it in favor of this Special Issue. This was a sign of the interest of this topic, which is really multifaceted and still has wide areas that are worth investigating. This special issue provides a demonstration of this. The papers collected deal with anticipatory postural adjustments [2] and anticipatory locomotor adjustments [3], strategies to tackle obstacles [2–4], the effects of weight unloading on gait [5], posture control in special populations: ataxic children [6] and obese subjects [7], surface perturbation during posture [8], effects of sensory information and feedback on postural control [9,10], elderly behavior during a motor-motor double task [11], upper limb control [12,13], different aspects related to running: ankle joint dynamic stiffness [14] and fatigue [15]. There is also a flash on an ecologic condition where a pedestrian has to program its strategy to cross a road in between two moving vehicles [16], and a study on visual-manual control in monkeys [17].

While selecting the contributions I tried to comply with the main inspiration of this special issue, that was not biomechanics alone (an extremely wide area), nor motor control alone (extremely wide as well), but the integration between biomechanical and motor control aspects. Actually, all of the collected papers investigate motor control with a basis in more or less complicated movement biomechanics methodologies and use biomechanical concepts, so the purpose seems to have been fulfilled.

I am quite sure that such a wide spectrum of research studies offers a useful overview of the present interests and future perspectives in this area.

I must thank all the authors that contributed to realize this high quality editorial initiative, and I hope this collection will be useful and stimulating for future studies and applications.

Funding: This research received no external funding.

Conflicts of Interest: The author declares no conflict of interest.

References

1. Ghai, S.; Nardone, A.; Schieppati, M. Human Balance in Response to Continuous, Predictable Translations of the Support Base: Integration of Sensory Information, Adaptation to Perturbations, and the Effect of Age, Neuropathy and Parkinson's Disease. *Appl. Sci.* **2019**, *9*, 5310. [CrossRef]
2. Artico, R.; Fourcade, P.; Teyssèdre, C.; Caderby, T.; Delafontaine, A.; Yiou, E. Influence of Swing-Foot Strike Pattern on Balance Control Mechanisms during Gait Initiation over an Obstacle to Be Cleared. *Appl. Sci.* **2019**, *10*, 244. [CrossRef]
3. Fiset, F.; McFadyen, J.B. The Switching of Trailing Limb Anticipatory Locomotor Adjustments is Uninfluenced by what the Leading Limb Does, but General Time Constraints Remain. *Appl. Sci.* **2020**, *10*, 2256. [CrossRef]
4. Baggen, R.J.; van Dieën, J.H.; Van Roie, E.; Verschueren, S.M.; Giarmatzis, G.; Delecluse, C.; Dominici, N. Age-Related Differences in Muscle Synergy Organization during Step Ascent at Different Heights and Directions. *Appl. Sci.* **2020**, *10*, 1987. [CrossRef]
5. Kabbaligere, R.; Layne, S.C. Adaptation in Gait to Body-Weight Unloading. *Appl. Sci.* **2019**, *9*, 4494. [CrossRef]
6. Farinelli, V.; Palmisano, C.; Marchese, S.M.; Strano, C.M.M.; D'Arrigo, S.; Pantaleoni, C.; Ardissone, A.; Nardocci, N.; Esposti, R.; Cavallari, P. Postural Control in Children with Cerebellar Ataxia. *Appl. Sci.* **2020**, *10*, 1606. [CrossRef]
7. Cimolin, V.; Cau, N.; Galli, M.; Capodaglio, P. Balance Control in Obese Subjects during Quiet Stance: A State-of-the Art. *Appl. Sci.* **2020**, *10*, 1842. [CrossRef]
8. Ramli, F.N.M.; Dzahir, A.M.M.; Yamamoto, S.-I. Estimation of Transition Frequency during Continuous Translation Surface Perturbation. *Appl. Sci.* **2019**, *9*, 4891. [CrossRef]
9. Sotirakis, H.; Hatzitaki, V.; Munoz-Martel, V.; Mademli, L.; Arampatzis, A. Center of Pressure Feedback Modulates the Entrainment of Voluntary Sway to the Motion of a Visual Target. *Appl. Sci.* **2019**, *9*, 3952. [CrossRef]
10. Misiaszek, E.J.; Chodan, D.C.S.; McMahon, J.A.; Fenrich, K.K. Influence of Pairing Startling Acoustic Stimuli with Postural Responses Induced by Light Touch Displacement. *Appl. Sci.* **2020**, *10*, 382. [CrossRef]
11. Lee, Y.-J.; Liang, N.J.; Wen, Y.-T. Characteristics of Postural Muscle Activity in Response to A Motor-Motor Task in Elderly. *Appl. Sci.* **2019**, *9*, 4319. [CrossRef]
12. Choi, W.; Lee, J.; Li, L. Analysis of Three-Dimensional Circular Tracking Movements Based on Temporo-Spatial Parameters in Polar Coordinates. *Appl. Sci.* **2020**, *10*, 621. [CrossRef]
13. Fehse, U.; Schmitz, G.; Hartwig, D.; Ghai, S.; Brock, H.; Effenberg, O.A. Auditory Coding of Reaching Space. *Appl. Sci.* **2020**, *10*, 429. [CrossRef]
14. Garofolini, A.; Taylor, S.; Mclaughlin, P.; Mickle, J.K.; Frigo, A.C. Ankle Joint Dynamic Stiffness in Long-Distance Runners: Effect of Foot Strike and Shoes Features. *Appl. Sci.* **2019**, *9*, 4100. [CrossRef]
15. Luo, Z.; Zhang, X.; Wang, J.; Yang, Y.; Xu, Y.; Fu, W. Changes in Ground Reaction Forces, Joint Mechanics, and Stiffness during Treadmill Running to Fatigue. *Appl. Sci.* **2019**, *9*, 5493. [CrossRef]
16. Kim, H.S.; Kim, W.J.; Chung, H.-C.; Choi, G.-J.; Choi, M. Behavioral Dynamics of Pedestrians Crossing between Two Moving Vehicles. *Appl. Sci.* **2020**, *10*, 859. [CrossRef]
17. Sakazume, Y.; Furubayashi, S.; Miyashita, E. Functional Roles of Saccades for a Hand Movement. *Appl. Sci.* **2020**, *10*, 3066. [CrossRef]

 *applied sciences*

Review

Human Balance in Response to Continuous, Predictable Translations of the Support Base: Integration of Sensory Information, Adaptation to Perturbations, and the Effect of Age, Neuropathy and Parkinson's Disease

Shashank Ghai [1,2], Antonio Nardone [3] and Marco Schieppati [4,*]

[1]   School of Physical & Occupational Therapy, McGill University, Montréal, QC H3G 1Y5, Canada; shashank.ghai@mail.mcgill.ca

[2]   Feil & Oberfeld Research Centre of the Jewish Rehabilitation Hospital: Centre for Interdisciplinary Research of Greater Montreal (CRIR), Laval, QC H7V 1R2, Canada

[3]   Istituti Clinici Scientifici Maugeri IRCCS, Pavia, and Department of Clinical-Surgical, Diagnostic and Pediatric Sciences, University of Pavia, 27100 Pavia, Italy; antonio.nardone@icsmaugeri.it

[4]   Istituti Clinici Scientifici Maugeri IRCCS, Via Maugeri 4, 27100 Pavia, Italy

*   Correspondence: marco.schieppati@icsmaugeri.it

Received: 13 November 2019; Accepted: 2 December 2019; Published: 5 December 2019

Featured Application: A strong point of balance challenging experiments is that, in addition to representing a paradigm for understanding the control of dynamic equilibrium and the processes of sensory integration in a complex motor task, they are useful in quantifying sensory and motor impairments in patients with balance problems of central and peripheral origin. Moreover, these protocols can be easily and successfully adapted to balance training for rehabilitation purposes.

Abstract: This short narrative review article moves from early papers that described the behaviour of healthy subjects balancing on a motorized platform continuously translating in the antero-posterior direction. Research from the laboratories of two of the authors and related investigations on dynamic balancing behaviour are briefly summarized. More recent findings challenging time-honoured views are considered, such as the statement that vision plays a head-in-space stabilizing role. The time interval to integrate vision or its withdrawal in the balancing pattern is mentioned as well. Similarities and differences between ageing subjects and patients with peripheral or central disorders are concisely reported. The muscle activities recorded during the translation cycles suggest that vision and amplitude changes of the anticipatory postural activities play a predominant role in controlling dynamic balance during prolonged administration of the predictable perturbation. The potential of this paradigm for rehabilitation of balance problems is discussed.

Keywords: dynamic balance; motorized platform; antero-posterior translation; reflex responses; anticipatory adjustments; vision; ageing; neuropathy; Parkinson's disease

---

1. Introduction

Balance is one of the 4000 most commonly used words, according to the Collins dictionary [1] (admittedly, possibly boosted by its usage in economics). We would sustain here this trend by proposing a survey of the balancing behaviour when humans are forced to keep their equilibrium on a flat support base that continuously and predictably moves back and forth in the antero-posterior direction, i.e., along the sagittal axis of the standing subjects (Figure 1).

The antero-posterior (A-P) platform translation challenges body balance differently than the medio-lateral (M-L) translation. In the former case, both legs behave equally, in a symmetrical way: the feet act as a solid interface between body and platform because of their length and the efficacy of the action of the plantar- and dorsi-flexor foot muscles, and the distance between the feet may not be critical to the balancing task. Conversely, in the medio-lateral perturbations, the legs and pelvis configure a parallelepiped shape when standing on the platform [2], where both legs are equally displaced with respect to the vertical plane (in the same sense, both legs to the right or to the left). This compels reciprocal relaxation, contraction of the agonist neck, trunk, pelvis (mainly), and lower limb muscles of both sides of the body in order to minimize displacement in the frontal plane. Further, the medio-lateral perturbations imply a rotational torque component, albeit minimal, because the centre of mass (CoM) of the body lies somewhat ahead of the centre of the ankle joints during normal standing, and the inertia of the former and displacement of the latter would create an asymmetrical compensation.

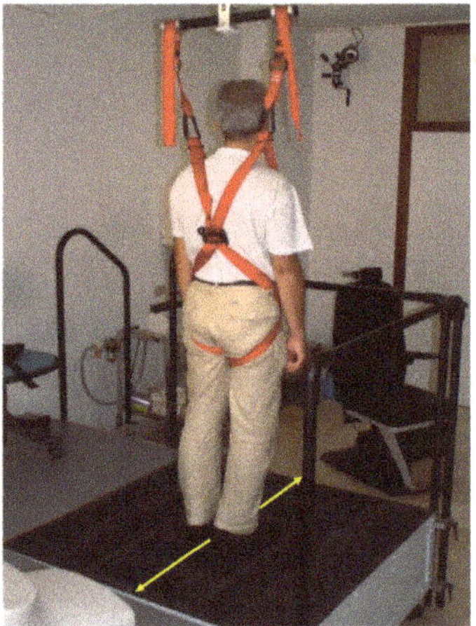

Figure 1. This shows a platform that can produce movements of the supporting base in various directions in the horizontal space according to different driving functions.

While many studies on balance control have addressed perturbations in the sagittal plane (such as A-P impulsive and continuous translations), less attention has been devoted to the balancing behaviour in response to M-L translations, and no prolonged perturbations have been administered, yet. Hence, this short review will not consider the findings obtained with perturbations in the frontal plane. This article will not address the mechanisms responsible for the maintenance of the erect stance under static conditions, either, even if during quiet stance, antero-posterior sway of the body occurs and can be easily detected [3]. Besides, this article will not deal with the automatic postural activities accompanying gait, be it simple walking along a linear path or more demanding walking as when we move along curved trajectories (e.g., [4,5]). As a further stipulation, we will not deal here with another type of perturbation, which has been popular in the past, consisting of the toe-up or toe-down tilt of the supporting platform. This perturbation challenges balance to a minor extent compared to the translations, because the platform is most often set to rotate (has the centre of rotation) around the ankle joint, thereby producing reflex responses in the stretched plantar- or dorsi-flexor muscles, but small displacements of the CoM with respect to the support base (see [6]). The main aim of that experimental condition has been indeed to study the stretch reflexes and their modulation under various sensory and environmental conditions [7] or in the case of neurological disorders [8]. A good review article on the broader issue of the control of perturbed balance is available [9].

2. The Continuous Predictable Balance Perturbations Administered by the Antero-Posterior Translation of the Platform

The peculiar challenge to our ability to remain upright, when we are standing in the critical situation represented by the continuous predictable back and forth displacement of the support base, prompts our balance capacities aimed at keeping the body's CoM over the moving support base. The brain controls our postural muscles accordingly, by activating short and long latency reflex responses, by preparing anticipatory activities once the features of the perturbation become known, and by putting in place corrective activities when the reactions are not perfectly calibrated. The sensory input from the periphery (proprioceptors, including the vestibular system, and exteroceptors, including vision) plays a crucial role [10]. The muscle spindles of the postural muscles (see [11]) and the vestibular system and vision would be paramount, since cutaneous receptors of the foot sole seem to contribute little to the control of stance [12,13], probably through modulation of the stretch reflex excitability [14]. The vestibular system would play a head stabilizing role [15,16] in keeping with [17], who showed that head displacement on the translating platform in chronic vestibular patients with eyes closed (EC) was larger than the 95th percentile of healthy subjects.

In addition, where the sequence of support-base perturbations goes on for any longer, the brain's computational effort and the muscle forces diminish over time, featuring a clear-cut adaptation phenomenon (see [18]). However, in many such experimental conditions, the analysis of the muscle activities and the body's mechanics have been limited to the epochs (the perturbation cycles) recorded in the period when the steady-state is reached (the term steady-state is reductive, owing to the high intra- and inter-subject variability of the balancing behaviour; see [19]). These limitations notwithstanding, though, this protocol has been rich with interesting conclusions about the strategies put in place for maintaining equilibrium and about the processes of the integration of proprioceptive and visual inputs.

Just before the turn of the century, two groups independently exploited this paradigm. Buchanan and Horak [20] observed the effects of the frequency of a sizable antero-posterior translation of the support platform on the movements of the body segments, with and without vision (eyes open (EO)). They found that at low translation frequencies (<0.3 Hz), subjects rode the platform showing no major joint motion, and the role of vision hardly affected this behaviour. At higher translation frequencies (<0.5 Hz), behaviour was characterized by large amplitude motions of the head and trunk when vision was not available, whereas vision strongly reduced these antero-posterior oscillations to the point that the head moved less than the platform (a "fixed-in-space" head behaviour). Corna et al. [21] observed the same behaviour, whereby subjects behaved as a non-rigid, noninverted pendulum and stabilised head in space with vision. Without vision, the head oscillated more than the platform, again at low translation frequencies. Therefore, the balancing strategy during this type of continuous perturbation shifts from a pendulum to an inverted pendulum type, passing from active head-and-trunk control with vision to maximal body compliance to the translation pattern with EC.

3. Muscle Activities

Muscle stretch (mainly of the postural muscles of the leg) certainly occurs during this condition. The feet are moved backwards with respect to the CoM of the body during the backward translation phase, producing ankle dorsiflexion and a stretch of the triceps surae. The forward translation produces an ankle plantarflexion and a stretch of the pretibial muscles. In passing, the intrinsic foot muscles are not unrelated to the reactions, and they play their own role, scarcely addressed so far [22,23]. Short and long latency leg reflexes are elicited in the leg muscles by the displacement of the body segments [24]. Trunk and neck muscles contribute prominent balance correcting effects [25–27] as they also do under similar perturbation conditions, seated [28].

Notably, the continuous, predictable translations of the support base elicit proactive strategies [29–31]. The related muscle activities are aimed at counteracting the balance perturbations elicited by the platform displacement reversal at the dead points of the platform translation cycles.

These activities are not connected to proprioceptive reflexes triggered by muscle stretch, but configure anticipatory postural adjustments. Figure 2 shows an example of a recording of the tibialis anterior activity during the first two cycles of a continuous perturbation at a 0.2 Hz frequency and the trace of the platform antero-posterior translation. The first cycle elicits a brisk EMG response due to the muscle stretch induced by the forward displacement of the platform. Before the turn-around point preceding the successive translation cycle, EMG activity not related to muscle stretch appears, to be followed by a stretch reflex response more modest than that occurring during the first cycle. The anticipatory activities appear when the leg muscles are not stretched and are appropriate for counteracting the inertia of the body at the time of the turning points of the platform. Anticipatory activities may not be optimally tuned to the complex combination of active and passive body movements from the beginning of the perturbation sequence and may be adjusted as the perturbation proceeds [32,33].

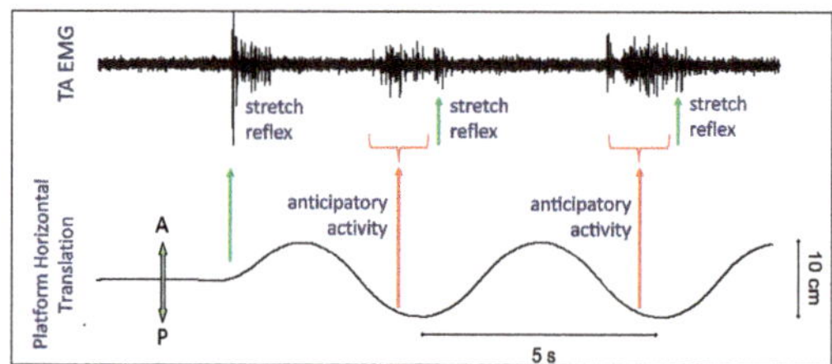

Figure 2. Top trace shows the tibialis anterior (TA) EMG at the beginning of a series of perturbations delivered with eyes open at a 0.2 Hz frequency. The bottom trace is the platform antero-posterior translation.

4. Vision

During quiet stance, unsteadiness in the dark is a sign of impaired proprioception (see [34] for a good review article, with historic hints). For instance, closing the eyes increased the body's sway in all the subjects, as detected by instrumented platforms [35]. The increase in sway can be inconspicuous in healthy subjects, whereas in patients with neuropathy or deficits of the vestibular system, the postural imbalance when standing or walking with EC can be remarkable [36–38]. An investigation based on continuous random platform tilts [39] suggested that vision modulates the vestibular noise, thereby reducing the threshold for position detection. When healthy subjects stand with EC on the antero-posterior translating platform, the balancing behaviour consisting of large head antero-posterior displacements (compared to those of the platform and head with EO) is obvious [40]. The large head displacements when balancing on that platform with EC occur therefore in the absence of any proprioceptive malfunction. This leads to two easy conclusions: (1) vision does increase stability when standing, both under quiet conditions and when balancing on the translating platform; (2) standing on the translating platform is quite different from a quiet stance in that, even with full somato-sensation, unsteadiness becomes an issue without vision. It is perhaps appropriate here to mention that healthy subjects never fall and rarely take a compensatory step during the trials, which commonly comprise numerous perturbation cycles [41].

The translating platform is not just a means for evaluating the role of vision in stance by enhancing the likelihood to detect clear-cut differences between eyes open/eyes closed conditions. It is an interesting protocol to address the role of vision and its disorders in the control of dynamic balance as well. It has been shown that the head antero-posterior oscillations, normally limited with EO, increase steadily with the severity of poor vision. When subjects balance on the mobile platform wearing test lenses to modify visual acuity from clear vision to blurred vision, head stabilization in space progressively worsens with the decrease in visual acuity and becomes similar to the EC behaviour when vision is severely blurred [42]. Thus, a simple light/dark dichotomy hypothesis for the visual control of dynamic stance is not supported, because the balancing pattern changes gradually from head-fixed-in-space to large head oscillation as a function of visual acuity. This suggests that the visual input is integrated as is the general somaesthetic feedback and that we have no capacity of extrapolating simple information from the fixed environment from an indistinct image of it. It is notable that, in this case, head stability gradually diminishes progressively with visual acuity deterioration even when all balancing trials are performed within a short period of time and within the same laboratory, i.e., when subjects are well accustomed to both the environment and the continuous perturbation protocol.

Hence, clear vision is a powerful player in the control of balance, particularly under critical dynamic conditions. However, what is the actual purpose of vision under these conditions? Is it necessary for minimizing body segments' oscillations? Two questions deserve an answer. (1) Are there conditions in which we can do without vision because other sensory receptors firing under unstable conditions come to the rescue? (2) Are supra-postural visual tasks able to modulate the stabilizing effect of vision? The former question receives an explicit answer: no. We have already mentioned that reduced vision is not enough to stabilize body oscillations on the platform. Moreover, blind people, including congenitally blind subjects (therefore, people used to moving in their environment and counteracting postural perturbations of various natures for years without vision), do not sway less than sighted people when administered the continuous perturbations of balance on the periodically moving platform. Their head oscillations are quite superimposed to those of the sighted people balancing with EC [43]. Hence, they have not learned to better exploit their somato-sensation or vestibular input in order to enhance stability, whereas they might have done so. On the contrary, normal subjects need not learn to exploit proprioception because it is so easy for them to open their eyes under critical conditions. Therefore, one is compelled to ask whether stability is an obligation for our body under the critical balancing condition and if vision is the means for achieving stability. The lesson from the blind subjects is that they can easily tolerate ample head and body oscillations, even if, in principle and if needed, they should be able to diminish their oscillations thanks to the proprioception and the vestibular input. The latter question receives an explicit answer: yes. As a matter of fact, under certain conditions, sighted subjects can tolerate ample head and body oscillations, when balancing on the translating platform, despite having normal vision, for instance while reading a text. In this case, as the text moves with the moving platform to which it is fixed as an integral whole, the head moves as much as the platform, and the distance between the eyes and the text is kept constant [44]. Therefore, these findings suggest that head stabilization in space can be revoked by the brain to enhance the performance of a non-postural task. Head stabilization is not necessarily produced by vision to obtain a dependable spatial reference. Not only that: if proprioception and the labyrinth do produce head stabilizing effects, their role would be anyway contingent and subject to the supra-postural task.

5. Playing with Vision

As said above, the differences in head oscillation eyes open/eyes closed are easily quantified. To summarize, oscillations are smaller than those of the platform with vision and larger without vision. One question had been raised some time ago. Given these clear-cut differences, it should be easy to investigate the phenomena occurring at the passage vision/no-vision (or vice versa) while subjects were balancing on the translating platform. Beyond curiosity, such an experiment can give hints as to the time necessary to integrate vision into the balance control [45]. As a matter of fact, closing the eyes in the middle of a series of perturbations soon produces a shift from small to large head oscillations, and the balancing pattern becomes definitely equal to that with EC. The same is true for the opposite change. The time period for integration of vision (the delay to reduction in head oscillation) is between 1.0 and 1.5 s, shorter than that for vision withdrawal (1.5–2.5 s). Of course, the periodic nature of the perturbations and the length of the translation cycles can affect the statistical estimate of the "true" delay. This time interval comprises the time to integrate the available sensory inputs, i.e., to change from an allocentric reference with EO to a proprioceptive reference with EC and vice versa and to adjust the calibration of the motor activity to reach the best motor control with the new sensory set. With a different technique, the delay in body sway on changing visual conditions has been calculated during quiet stance later on [46]. Interestingly, the delays observed during the platform continuous translations on addition or withdrawal of vision are broadly similar to those measured during quiet stance. This shows that the time to embodiment (or to disappearance) of the balance stabilizing effect of vision does not depend on the balancing condition at hand, is not at all negligible in duration, and may be critical in the prompt adaptation to changing conditions [47].

6. Aging

Gait perturbation paradigms are effective to improve reactive responses during walking in healthy elderly people [48]. Balance recovery is less prompt and effective in the oldest elderly people, in particular when sensory information is manipulated, and more so when more than one system is challenged [49,50]. However, the capacity to maintain quiet upright stance does not appear to be markedly altered in normal elderly subjects when estimated under static [51,52] and dynamic conditions by delivering isolated rotation tilt of the support basis triggering stretch responses [53]. This can be related on the one hand to the relatively easy task of standing quietly required by static force-platform measurements and on the other on the limited modifications with age in the latency and amplitude of the reflex responses [53–55].

It has been suggested, based on pseudorandom tilts of the support base, that older subjects rely more on proprioception rather than visual and vestibular cues for dynamic balance control [56]. When healthy elderly subjects are administered the periodic antero-posterior translations of the support base, they can perform even better than their young peers, provided vision is available and the frequency of oscillation is low (Figure 3, control). The smaller head oscillation is accompanied by higher flexibility of their body, as indicated by a low cross-correlation value between feet and head. With EC and high perturbation frequency, though, their oscillation is much larger than for the young subjects, and they assume a stiffer state [57]. Interestingly, when elderly persons rely on anticipatory adjustments during continuous perturbations, their behaviour is again similar to that of the young subjects, but they are more destabilized when perturbations are unanticipated (externally triggered) [58]. Van Ooteghem et al. [59] showed that older adults maintained the capacity to learn adaptive postural response minimization of instability similarly to young adults. However, despite practice, they still maintained a stiff attitude. Others have found, with continuous multidirectional perturbations, that elderly people do not behave in a strongly different way than the young subjects, but muscle weakness and postural asymmetries do represent a cause for enhanced risk of falling [49]. Anxiety is an additional problem in elderly, as seen by their increased sense of instability when balancing on a platform translating fore and aft in unpredictable mode [60].

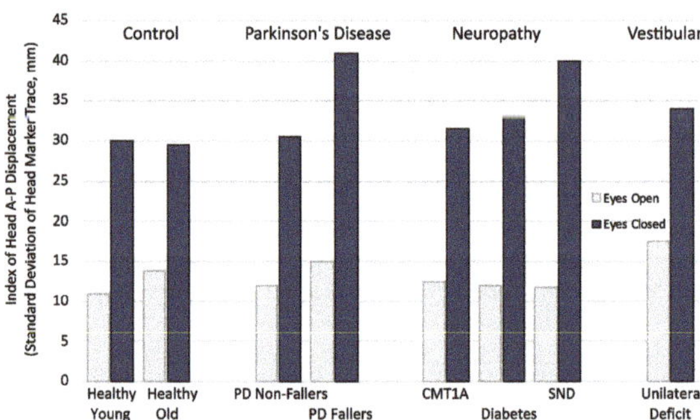

Figure 3. Mean indexes of head antero-posterior (A-P) displacement while standing on a platform moving continuously back and forth 6 cm in the A-P direction at a 0.2 Hz frequency with both eyes open and eyes closed. Despite data from small numbers of subjects per group contributing to the column values, overall, vision stabilizes the head compared to no-vision. CMT1A, Charcot-Marie-Tooth type 1A; SND, sensory neurone disease.

7. Neuropathy

How do neuropathic patients behave when they are exposed to such a critical balancing condition? These patients feature loss of sensation from the receptors innervated by the abnormal sensory fibres. The lesion can have several origins, from occupational exposure [61], metabolic, viral, toxic (chemotherapy), immune, and genetic nature, hypovitaminosis, and aging [62]. Diabetic neuropathy is a common example of sensory neuropathy. Small and large sensory fibres, or both, can be affected. The nerve fibres can show demyelination or axon degeneration [63]. Of course, in many cases including diabetes, neuropathies can affect both sensory and motor fibres [64], and in some cases, pure motor neuropathies can be observed.

It is known that body sway during quiet stance is increased in neuropathic patients [65,66] and that standing on foam with EC can detect subclinical neuropathies in older adults [67]. Therefore, neuropathic patients should experience severe instability when standing on the translating platform, all the more so when vision is not available. However, when neuropathic patients balance on the continuously back and forth translating platform, most of them show a surprisingly good performance (Figure 3, CMT1A, diabetes), in spite of a clear-cut unsteadiness during quiet stance [68]. This has been shown to occur in patients affected by diabetic sensorimotor neuropathy and Charcot–Marie–Tooth type 1A (CMT1A) neuropathy [69]. There are three reasons that can explain this behaviour. First, the proprioceptive input that drives the balancing behaviour originates not only in the more distal parts of the lower limbs (typically more affected by the fibre degeneration than other body parts), but likely from many different segments of the body as well, and the comprehensive afferent input would be more than enough for controlling balance. Second, at least as far as the Charcot–Marie–Tooth type 1A condition is concerned (the disease mainly affects the largest myelinated fibres; see [70]), the relatively good balancing capacity may depend on the integrity of the smaller (type II) afferent fibres originating from the muscle spindles. Third, the results are obtained under steady-state conditions, i.e., under conditions where subjects have learned to adapt optimally to the continuous perturbing stimulus by top-down control of the coordinated body movements in order not to stumble or fall as a consequence of the successive perturbations.

Conversely, patients affected by a rare form of pure sensory neuropathy with a lesion to both the peripheral and central axons of large nerve fibre tract (ganglionopathy or sensory neurone disease) ((Figure 3, SND) show severe unsteadiness under both static and dynamic conditions. When standing on the moving platform, the oscillation of their head is not larger than in healthy subjects with EO, but becomes abnormally large with EC. In a sense, the findings confirm the modest role of the large sensory afferent fibres in this task seen in Charcot–Marie–Tooth disease type 1A. However, removal of vision is more troublesome [71]. In these SND patients, the degeneration of the central branch of the afferent fibres would be responsible for the deterioration of equilibrium, possibly through abnormal reactivity of the brain cortices and circuits that elaborate and integrate the visual feedback for the control of balance [72].

Attempts have been made at understand the actual role of the somatosensory input in the balancing behaviour induced by the continuous perturbations of stance. A typical approach has been to vibrate postural muscles, since it is known that muscle-tendon vibration selectively activates the muscle spindle receptors, in particular those originating from the annulo-spiral endings of the spindle [73,74]. The vibration of either the Achilles tendon or the tendon of the tibialis anterior muscle in subjects standing on the translating platform produced only minor disequilibrium with EO [75]. On the one hand, these findings together with those from the pathological model mentioned above confirm that equilibrium under critical conditions is not severely affected by a continuous proprioceptive disturbance and that, more generally, the Ia muscle afferent input and the induced stretch reflex modulation are not essential in the control of stance under continuous perturbations. On the other hand, they suggest that anticipation rather than feedback would be the main mechanism by which the central nervous system manages this challenging situation. In the mentioned paper [75], however, abnormal head control did emerge with EC when the postural leg muscles were vibrated. This would be connected with abnormal cortical processing of the perturbation related sensory inflow because the unnatural discharge originating in the vibrated spindles would lead to gating of the proprioceptive volley reaching the somatosensory cortex [76].

8. Parkinson's Disease

If the balancing behaviour is affected not only by vision or by somato-sensation, but is also dependent on the supraspinal integration of these afferent inputs into a coherent motor response [8,77–79], balance performance should be considerably degraded in Parkinson's disease (PD) patients, particularly when the support base translates continuously. Quiet stance is affected in these patients, but in several of them, body sway is comprised within the normal values of age matched controls [52,65]. More than that, in partial attenuation of the statements about their proprioceptive deficits, the capacity of the patients to score their instability consciously is not affected, showing that they are able to collect and integrate the sensory inputs from the periphery and convert them into an explicit evaluation of their sway [65,80]. When patients with Parkinson´s disease are standing on the moving platform continuously delivering the cyclic perturbations, their trunk and head do not oscillate more than age matched healthy subjects, regardless of vision availability. Again, their preserved reflex responses and the capacity to anticipate the successive perturbation cycles [81] may help them to react properly to the postural challenge [82]. It should be considered that medicated patients maintain the capacity of improving automatic postural responses with practice [83]. However, patients characterized as fallers based on their medical history do show larger head oscillations compared to non-fallers when they close their eyes [84]. Further, these fallers show smaller cross-correlation and larger time delays between head and platform motion. Interestingly, in that cohort, head displacement increased with levodopa dose, suggesting the possibility for medication to worsen balancing capacity while improving the score of the Unified Parkinson's Disease Rating Scale.

As has been done in healthy young subjects, the translating platform has been exploited to answer the question of whether patients with PD are more sluggish in their ability to adapt promptly in the balancing behaviour to sudden changes in visual conditions [85]. Patients change body kinematics and EMG patterns to those appropriate for the new visual condition as normal subjects do when balancing on the moving platform. However, patients are slower in changing their behaviour (i.e., shifting from head-fixed-in-space with EO to body riding the platform with EC), in particular for the change EC to EO, as if adding a new important sensory inflow would represent an integration problem. These findings should be considered in the framework of the integration deficits in PD [86].

9. Adaptation to the Repeated Perturbation Cycles

The findings from the studies reported in the preceding sections were obtained mostly under steady-state conditions, i.e., when the initial uncontrolled reactions to the onset of the platform translation have vanished. Usually, the first few translation cycles are omitted from the analysis. Yet, these initial cycles and associated reactions and their changes as a function of the successive cycles can tell something about the way subjects comply with the condition and, in perspective, whether the problems in balancing behaviour affect more the transient phase or the steady-state behaviour or how pathologies affect these initial reactions. Adaptation implies a gradual shift from a reactive response to the abrupt onset of the perturbation series and from conscious control of the balancing behaviour to a more automatic processing, including modulation of the reflex responses triggered by the continuous stimulation (see [87] for the neck responses in seated subjects). Hence, understanding the adaptation phenomenon adds to our capacity of better exploiting this challenging perturbation procedure in order to understand the mechanisms and processes whereby the appropriate balancing behaviour is achieved and maintained in order to produce a more suitable behaviour.

The strength of the reflex response triggered by a certain ankle rotation during a series of horizontal translations is modulated appropriately by its repetition [88]. Even more so, the very same responses elicited by the translating platform itself should be liable to modulation by the continuous perturbation cycles. By using a sequence of discrete impulsive perturbations, Keshner et al. [89] had already suggested that adaptation was due to a generalized habituation in the postural control system. Others have shown that event related cortical potentials connected to the postural stimulus diminish over time with the successive cycles [90]. Of course, at the onset of the sequence of perturbations, a combined postural and startle response ensues [91,92].

Taken together, the studies show that adaptation occurs when a subject's balance is challenged by repeated perturbations. Under these circumstances, the sensorimotor processes underlying the adaptation to the repeated perturbations would require the collection of information about the amplitude velocity and frequency of the displacement of the support surface and of the instantaneous and continuously varying position of the body's CoM relative to the platform. This information would improve the performance by reducing redundant muscle activity (the initial co-contraction of the antagonist leg muscles) and enhancing the intersegmental coordination [33,41]. The reduction in the muscle activity concerns both the reflex responses and the anticipatory activities, a sign of coordinated modulation of responses occurring at different times of the perturbation cycle. Interestingly, the reduced muscle bursts of activity do not modify their time relationship with the perturbation cycle [93]. This indicates that the adaptation modulates so nicely with the neuronal excitability and muscle activity that it need not modify the time windows of their activation to comply with the changes in the body's mechanics. As a consequence, the head back-and-forth oscillation is diminished as well [41,94]. The speed of the adaptation process is susceptible to fatigue, even if the basic task of controlling the CoM of the body is substantially maintained [95–97].

10. Rehabilitation

Predictable balance perturbations have been successfully employed for rehabilitation of balance in selected patient groups [98] including vestibular patients [17], who may poorly behave on the platform when not compensated [99], and cerebellar patients, of both degenerative and vascular origin [100]. In all groups, such a training improved balance, allowing recommending the continuous translation as an effective and measurable way of rehabilitation. To our knowledge, clinical trials of balance rehabilitation in neuropathic patients based on this protocol have not been exploited, yet. However, these patients can improve balance control following multisensory training [101], strongly suggesting that continuous perturbations with and without vision could enhance their compensation capacities. Patients with Parkinson's disease show minor adaptation problems that can contribute to their balance dysfunction [8,85,102]. A balance training with a continuous, predictable translation of the support base has improved their balance and coordination capacities as much as training based on standardized exercises [103]. Interestingly, gait speed was also improved in both cases. Overall, the continuous and predictable translations of a support base onto which standing patients would be trained to enhance their reflex and balance correcting responses and to calibrate their anticipatory adjustments seem to be useful, yet still not exploited for all the potential advantages.

11. Conclusions and Perspectives

This short narrative review summarized some findings of studies on dynamic balance control. We briefly considered the potential of this protocol for both basic and translational research. It disclosed interesting features of balance control in healthy subjects and patients with balance problems. Much can still be done. Quantification of the variability of the balancing behaviour has not been addressed in depth. We are still missing normative values for the cross-correlation and time lag between moving body segments, which could likely be exploited as markers for initial balance problems. The effect of fatigue and adaptation while the training is administered needs further inquiry as well. Importantly, this protocol has proven helpful in rehabilitating balance in various cohorts of patients. Again, the rehabilitating potential can be further exploited, not only in patients with balance problems, but also in healthy older subjects, for which the risk of falling owing to stumbling or slipping is not uncommon.

Author Contributions: Conceptualization, S.G. and M.S.; validation, M.S.; writing, original draft preparation, M.S.; review and editing, S.G. and A.N.

Funding: This research received no external funding.

Acknowledgments: Several studies mentioned in this review were supported by the Ricerca Finalizzata Funding scheme of the Ministry of Health and by the Progetti di Ricerca di Interesse Nazionale Funding scheme of the Ministry of Education and Research, Italy, years 1999 to 2011. S.G. is currently funded by FRQS Postdoctoral fellowship.

References

1. Collins Dictionary. Available online: https://www.collinsdictionary.com/dictionary/english/balance (accessed on 4 October 2019).

2. Rietdyk, S.; Patla, A.E.; Winter, D.A.; Ishac, M.G.; Little, C.E. Balance recovery from medio-lateral perturbations of the upper body during standing. *J. Biomech.* **1999**, *32*, 1149–1158. [CrossRef]

3. Balasubramaniam, R.; Wing, A.M. The dynamics of standing balance. *Trends Cogn. Sci.* **2002**, *6*, 531–536. [CrossRef]

4. Guglielmetti, S.; Nardone, A.; De Nunzio, A.M.; Godi, M.; Schieppati, M. Walking along circular trajectories in Parkinson's disease. *Mov. Disord.* **2009**, *24*, 598–604. [CrossRef] [PubMed]

5. Godi, M.; Giardini, M.; Schieppati, M. Walking along curved trajectories. Changes with age and Parkinson's Disease. Hints to rehabilitation. *Front. Neurol.* **2019**, *10*, 532. [CrossRef] [PubMed]

6. Di Fabio, R.P.; Badke, M.B.; Duncan, P.W. Adapting human postural reflexes following localized cerebrovascular lesion: Analysis of bilateral long latency responses. *Brain Res.* **1986**, *363*, 257–264. [CrossRef]

7. Bove, M.; Nardone, A.; Schieppati, M. Effects of leg muscle tendon vibration on group Ia and group II reflex responses to stance perturbation in humans. *J. Physiol.* **2003**, *550*, 617–630. [CrossRef]

8. Schieppati, M.; Nardone, A. Free and supported stance in Parkinson's disease. The effect of posture and 'postural set' on leg muscle responses to perturbation, and its relation to the severity of the disease. *Brain* **1991**, *114*, 1227–1244. [CrossRef]

9. Rogers, M.W.; Mille, M.L. Balance perturbations. *Handb. Clin. Neurol.* **2018**, *159*, 85–105. [CrossRef]

10. Allum, J.H.; Honegger, F. Synergies and strategies underlying normal and vestibular deficient control of balance: Implication for neuroprosthetic control. *Prog. Brain Res.* **1993**, *97*, 331–348. [CrossRef]

11. Schieppati, M.; Nardone, A. Group II spindle afferent fibers in humans: Their possible role in the reflex control of stance. *Prog. Brain Res.* **1999**, *123*, 461–472. [CrossRef]

12. Meyer, P.F.; Oddsson, L.I.; De Luca, C.J. The role of plantar cutaneous sensation in unperturbed stance. *Exp. Brain Res.* **2004**, *156*, 505–512. [CrossRef] [PubMed]

13. Ferguson, O.W.; Polskaia, N.; Tokuno, C.D. The effects of foot cooling on postural muscle responses to an unexpected loss of balance. *Hum. Mov. Sci.* **2017**, *54*, 240–247. [CrossRef] [PubMed]

14. Abbruzzese, M.; Rubino, V.; Schieppati, M. Task-dependent effects evoked by foot muscle afferents on leg muscle activity in humans. *Electroencephalogr. Clin. Neurophysiol.* **1996**, *101*, 339–348. [CrossRef]

15. Horak, F.B.; Buchanan, J.; Creath, R.; Jeka, J. Vestibulospinal control of posture. *Adv. Exp. Med. Biol.* **2002**, *508*, 139–345. [CrossRef]

16. Goodworth, A.D.; Peterka, R.J. Influence of bilateral vestibular loss on spinal stabilization in humans. *J. Neurophysiol.* **2010**, *103*, 1978–1987. [CrossRef]

17. Corna, S.; Nardone, A.; Prestinari, A.; Galante, M.; Grasso, M.; Schieppati, M. Comparison of Cawthorne-Cooksey exercises and sinusoidal support surface translations to improve balance in patients with unilateral vestibular deficit. *Arch. Phys. Med. Rehabil.* **2003**, *84*, 1173–1184. [CrossRef]

18. Miall, R.C.; Wolpert, D.M. Forward models for physiological motor control. *Neural Netw.* **1996**, *9*, 1265–1279. [CrossRef]

19. Schieppati, M.; Giordano, A.; Nardone, A. Variability in a dynamic postural task attests ample flexibility in balance control mechanisms. *Exp. Brain Res.* **2002**, *144*, 200–210. [CrossRef]

20. Buchanan, J.J.; Horak, F.B. Emergence of postural patterns as a function of vision and translation frequency. *J. Neurophysiol.* **1999**, *81*, 2325–2339. [CrossRef]

21. Corna, S.; Tarantola, J.; Nardone, A.; Giordano, A.; Schieppati, M. Standing on continuously moving platform: Is body inertia counteracted or exploited? *Exp. Brain Res.* **1999**, *124*, 331–341. [CrossRef]

22. Schieppati, M.; Nardone, A.; Siliotto, R.; Grasso, M. Early and late stretch responses of human foot muscles induced by perturbation of stance. *Exp. Brain Res.* **1995**, *105*, 411–422. [CrossRef] [PubMed]

23. Kelly, L.A.; Kuitunen, S.; Racinais, S.; Cresswell, A.G. Recruitment of the plantar intrinsic foot muscles with increasing postural demand. *Clin. Biomech.* **2012**, *27*, 46–51. [CrossRef] [PubMed]

24. Nardone, A.; Giordano, A.; Corrà, T.; Schieppati, M. Responses of leg muscles in humans displaced while standing. Effects of types of perturbation and of postural set. *Brain* **1990**, *113*, 65–84. [CrossRef]

25. Carpenter, M.G.; Allum, J.H.; Honegger, F. Directional sensitivity of stretch reflexes and balance corrections for normal subjects in the roll and pitch planes. *Exp. Brain Res.* **1999**, *129*, 93–113. [CrossRef]

26. Granata, K.P.; Slota, G.P.; Bennett, B.C. Paraspinal muscle reflex dynamics. *J. Biomech.* **2004**, *37*, 241–247. [CrossRef]

27. Boudreau, S.A.; Falla, D. Chronic neck pain alters muscle activation patterns to sudden movements. *Exp. Brain Res.* **2014**, *232*, 2011–2020. [CrossRef]

28. Van Drunen, P.; Koumans, Y.; van der Helm, F.C.; van Dieën, J.H.; Happee, R. Modulation of intrinsic and reflexive contributions to low-back stabilization due to vision, task instruction, and perturbation bandwidth. *Exp. Brain Res.* **2015**, *233*, 735–749. [CrossRef]

29. Massion, J. Movement, posture and equilibrium: Interaction and coordination. *Prog. Neurobiol.* **1992**, *38*, 35–56. [CrossRef]

30. Aruin, A.S.; Forrest, W.R.; Latash, M.L. Anticipatory postural adjustments in conditions of postural instability. *Electroencephalogr. Clin. Neurophysiol.* **1998**, *109*, 350–359. [CrossRef]

31. Bouisset, S.; Do, M.C. Posture, dynamic stability, and voluntary movement. *Neurophysiol. Clin.* **2008**, *38*, 345–362. [CrossRef]

32. Aruin, A.S. Enhancing Anticipatory Postural Adjustments: A Novel Approach to Balance Rehabilitation. *J. Nov. Physiother.* **2016**, *6*, e144. [CrossRef] [PubMed]

33. Sozzi, S.; Nardone, A.; Schieppati, M. Calibration of the Leg Muscle Responses Elicited by Predictable Perturbations of Stance and the Effect of Vision. *Front. Hum. Neurosci.* **2016**, *10*, 419. [CrossRef] [PubMed]

34. Khasnis, A.; Gokula, R.M. Romberg's test. *J. Postgrad. Med.* **2003**, *49*, 169. [PubMed]

35. Redfern, M.S.; Yardley, L.; Bronstein, A.M. Visual influences on balance. *J. Anxiety Disord.* **2001**, *15*, 81–94. [CrossRef]

36. Paulus, W.; Straube, A.; Brandt, T.H. Visual postural performance after loss of somatosensory and vestibular function. *J. Neurol. Neurosurg. Psychiatry* **1987**, *50*, 1542–1545. [CrossRef] [PubMed]

37. Nardone, A.; Schieppati, M. Balance control under static and dynamic conditions in patients with peripheral neuropathy. *G Ital. Med. Lav. Ergon.* **2007**, *29*, 101–104. [CrossRef]

38. Nardone, A.; Schieppati, M. The role of instrumental assessment of balance in clinical decision making. *Eur. J. Phys. Rehabil. Med.* **2010**, *46*, 221–237.

39. Assländer, L.; Hettich, G.; Mergner, T. Visual contribution to human standing balance during support surface tilts. *Hum. Mov. Sci.* **2015**, *41*, 147–164. [CrossRef]

40. Ghai, S.; Ghai, I.; Effenberg, A.O. Effects of dual tasks and dual-task training on postural stability: A systematic review and meta-analysis. *Clin. Interv. Aging* **2017**, *12*, 557. [CrossRef]

41. Schmid, M.; Bottaro, A.; Sozzi, S.; Schieppati, M. Adaptation to continuous perturbation of balance: Progressive reduction of postural muscle activity with invariant or increasing oscillations of the center of mass depending on perturbation frequency and vision conditions. *Hum. Mov. Sci.* **2011**, *30*, 262–278. [CrossRef]

42. Schmid, M.; Casabianca, L.; Bottaro, A.; Schieppati, M. Graded changes in balancing behavior as a function of visual acuity. *Neuroscience* **2008**, *153*, 1079–1091. [CrossRef] [PubMed]

43. Schmid, M.; Nardone, A.; De Nunzio, A.M.; Schmid, M.; Schieppati, M. Equilibrium during static and dynamic tasks in blind subjects: No evidence of cross-modal plasticity. *Brain* **2007**, *130*, 2097–2107. [CrossRef] [PubMed]

44. Sozzi, S.; Nardone, A.; Schieppati, M. Vision Does Not Necessarily Stabilize the Head in Space During Continuous Postural Perturbations. *Front. Neurol.* **2019**, *10*, 748. [CrossRef] [PubMed]

45. De Nunzio, A.M.; Schieppati, M. Time to reconfigure balancing behaviour in man: Changing visual condition while riding a continuously moving platform. *Exp. Brain Res.* **2007**, *178*, 18–36. [CrossRef]

46. Sozzi, S.; Monti, A.; De Nunzio, A.M.; Do, M.C.; Schieppati, M. Sensori-motor integration during stance: Time adaptation of control mechanisms on adding or removing vision. *Hum. Mov. Sci.* **2011**, *30*, 172–189. [CrossRef]

47. Honeine, J.L.; Crisafulli, O.; Sozzi, S.; Schieppati, M. Processing time of addition or withdrawal of single or combined balance-stabilizing haptic and visual information. *J. Neurophysiol.* **2015**, *114*, 3097–3110. [CrossRef]

48. McCrum, C.; Gerards, M.H.G.; Karamanidis, K.; Zijlstra, W.; Meijer, K. A systematic review of gait perturbation paradigms for improving reactive stepping responses and falls risk among healthy older adults. *Eur. Rev. Aging Phys. Act.* **2017**, *14*, 3. [CrossRef]

49. Tsai, Y.C.; Hsieh, L.F.; Yang, S. Age-related changes in posture response under a continuous and unexpected perturbation. *J. Biomech.* **2014**, *47*, 482–490. [CrossRef]

50. Alcock, L.; O'Brien, T.D.; Vanicek, N. Association between somatosensory, visual and vestibular contributions to postural control, reactive balance capacity and healthy ageing in older women. *Health Care Women Int.* **2018**, *39*, 1366–1380. [CrossRef]

51. Schieppati, M.; Grasso, M.; Siliotto, R.; Nardone, A. Effect of age, chronic diseases and parkinsonism on postural control. In *Sensorimotor Impairment in the Elderly*; Stelmach, G.E., Hömberg, V., Eds.; Kluwer Academic Publishers: Dordrecht, The Netherlands, 1993; pp. 355–373.

52. Schieppati, M.; Hugon, M.; Grasso, M.; Nardone, A.; Galante, M. The limits of equilibrium in young and elderly normal subjects and in parkinsonians. *Electroencephalogr. Clin. Neurophysiol.* **1994**, *93*, 286–298. [CrossRef]

53. Nardone, A.; Siliotto, R.; Grasso, M.; Schieppati, M. Influence of aging on leg muscle reflex responses to stance perturbation. *Arch. Phys. Med. Rehabil.* **1995**, *76*, 158–165. [CrossRef]

54. Studenski, S.; Duncan, P.W.; Chandler, J. Postural responses and effector factors in persons with unexplained falls: Results and methodologic issues. *J. Am. Geriatr. Soc.* **1991**, *39*, 229–234. [CrossRef]

55. Woollacott, M.H.; Shumway-Cook, A.; Nashner, L.M. Aging and posture control: Changes in sensory organization and muscular coordination. *Int. J. Aging Human Dev.* **1986**, *23*, 97–114. [CrossRef] [PubMed]

56. Wiesmeier, I.K.; Dalin, D.; Maurer, C. Elderly use proprioception rather than visual and vestibular cues for postural motor control. *Front. Aging Neurosci.* **2015**, *7*, 97. [CrossRef] [PubMed]

57. Nardone, A.; Grasso, M.; Tarantola, J.; Corna, S.; Schieppati, M. Postural coordination in elderly subjects standing on a periodically moving platform. *Arch. Phys. Med. Rehabil.* **2000**, *81*, 1217–1223. [CrossRef] [PubMed]

58. Bugnariu, N.; Sveistrup, H. Age-related changes in postural responses to externally- and self-triggered continuous perturbations. *Arch. Gerontol. Geriatr.* **2006**, *42*, 73–89. [CrossRef]

59. Van Ooteghem, K.; Frank, J.S.; Horak, F.B. Practice-related improvements in posture control differ between young and older adults exposed to continuous, variable amplitude oscillations of the support surface. *Exp. Brain Res.* **2009**, *199*, 185–193. [CrossRef]

60. Castro, P.; Kaski, D.; Schieppati, M.; Furman, M.; Arshad, Q.; Bronstein, A. Subjective stability perception is related to postural anxiety in older subjects. *Gait Posture* **2019**, *68*, 538–544. [CrossRef]

61. Trivedi, S.; Pandit, A.; Ganguly, G.; Das, S.K. Epidemiology of Peripheral Neuropathy: An Indian Perspective. *Ann. Indian Acad. Neurol.* **2017**, *20*, 173–184. [CrossRef]

62. Brisset, M.; Nicolas, G. Peripheral neuropathies and aging. *Geriatr. Psychol. Neuropsychiatr. Vieil* **2018**, *16*, 409–413. [CrossRef]

63. Feldman, E.L.; Callaghan, B.C.; Pop-Busui, R.; Zochodne, D.W.; Wright, D.E.; Bennett, D.L.; Bril, V.; Russell, J.W.; Viswanathan, V. Diabetic neuropathy. *Nat. Rev. Dis. Primers* **2019**, *5*, 41. [CrossRef]

64. Vinik, A.I. Diabetic Sensory and Motor Neuropathy. *N. Engl. J. Med.* **2016**, *374*, 1455–1464. [CrossRef] [PubMed]

65. Schieppati, M.; Tacchini, E.; Nardone, A.; Tarantola, J.; Corna, S. Subjective perception of body sway. *J. Neurol. Neurosurg. Psychiatry* **1999**, *66*, 313–322. [CrossRef] [PubMed]

66. Lencioni, T.; Rabuffetti, M.; Piscosquito, G.; Pareyson, D.; Aiello, A.; Di Sipio, E.; Padua, L.; Stra, F.; Ferrarin, M. Postural stabilization and balance assessment in Charcot-Marie-Tooth 1A subjects. *Gait Posture* **2014**, *40*, 481–486. [CrossRef]

67. Anson, E.; Bigelow, R.T.; Swenor, B.; Deshpande, N.; Studenski, S.; Jeka, J.J.; Agrawal, Y. Loss of Peripheral Sensory Function Explains Much of the Increase in Postural Sway in Healthy Older Adults. *Front. Aging Neurosci.* **2017**, *9*, 202. [CrossRef]

68. Nardone, A.; Grasso, M.; Schieppati, M. Balance control in peripheral neuropathy: Are patients equally unstable under static and dynamic conditions? *Gait Posture* **2006**, *23*, 364–373. [CrossRef]

69. Nardone, A.; Tarantola, J.; Miscio, G.; Pisano, F.; Schenone, A.; Schieppati, M. Loss of large-diameter spindle afferent fibres is not detrimental to the control of body sway during upright stance: Evidence from neuropathy. *Exp. Brain Res.* **2000**, *135*, 155–162. [CrossRef]

70. Kars, H.J.; Hijmans, J.M.; Geertzen, J.H.; Zijlstra, W. The effect of reduced somatosensation on standing balance: A systematic review. *J. Diabetes Sci. Technol.* **2009**, *3*, 931–943. [CrossRef]

71. Nardone, A.; Galante, M.; Pareyson, D.; Schieppati, M. Balance control in Sensory Neuron Disease. *Clin. Neurophysiol.* **2007**, *118*, 538–550. [CrossRef]

72. Whitlock, J.R. Posterior parietal cortex. *Curr. Biol.* **2017**, *27*, R691–R695. [CrossRef]

73. Burke, D.; Hagbarth, K.E.; Löfstedt, L.; Wallin, B.G. The responses of human muscle spindle endings to vibration during isometric contraction. *J. Physiol.* **1976**, *261*, 695–711. [CrossRef]

74. Burke, D.; Hagbarth, K.E.; Löfstedt, L.; Wallin, B.G. The responses of human muscle spindle endings to vibration of non-contracting muscles. *J. Physiol.* **1976**, *261*, 673–693. [CrossRef]

75. De Nunzio, A.M.; Nardone, A.; Schieppati, M. Head stabilization on a continuously oscillating platform: The effect of a proprioceptive disturbance on the balancing strategy. *Exp. Brain Res.* **2005**, *165*, 261–272. [CrossRef]

76. Staines, W.R.; McIlroy, W.E.; Brooke, J.D. Cortical representation of whole-body movement is modulated by proprioceptive discharge in humans. *Exp. Brain Res.* **2001**, *138*, 235–242. [CrossRef]

77. Konczak, J.; Corcos, D.M.; Horak, F.; Poizner, H.; Shapiro, M.; Tuite, P.; Volkmann, J.; Maschke, M. Proprioception and motor control in Parkinson's disease. *J. Mot. Behav.* **2009**, *41*, 543–552. [CrossRef]

78. Schoneburg, B.; Mancini, M.; Horak, F.; Nutt, J.G. Framework for understanding balance dysfunction in Parkinson's disease. *Mov. Disord.* **2013**, *28*, 1474–1482. [CrossRef]

79. Hwang, S.; Agada, P.; Grill, S.; Kiemel, T.; Jeka, J.J. A central processing sensory deficit with Parkinson's disease. *Exp. Brain Res.* **2016**, *234*, 2369–2379. [CrossRef]

80. Ghai, S.; Ghai, I.; Schmitz, G.; Effenberg, A.O. Effect of rhythmic auditory cueing on parkinsonian gait: A systematic review and meta-analysis. *Sci. Rep.* **2018**, *8*, 506. [CrossRef]

81. Boonstra, T.A.; van Vugt, J.P.; van der Kooij, H.; Bloem, B.R. Balance asymmetry in Parkinson's disease and its contribution to freezing of gait. *PLoS ONE* **2014**, *9*, e102493. [CrossRef]

82. Weissblueth, E.; Schwartz, M.; Hocherman, S. Adaptation to cyclic stance perturbations in Parkinson's disease depends on postural demands. *Parkinsonism Relat. Disord.* **2008**, *14*, 489–494. [CrossRef]

83. Van Ooteghem, K.; Frank, J.S.; Horak, F.B. Postural motor learning in Parkinson's disease: The effect of practice on continuous compensatory postural regulation. *Gait Posture* **2017**, *57*, 299–304. [CrossRef] [PubMed]

84. Nardone, A.; Schieppati, M. Balance in Parkinson's disease under static and dynamic conditions. *Mov. Disord.* **2006**, *21*, 1515–1520. [CrossRef] [PubMed]

85. De Nunzio, A.M.; Nardone, A.; Schieppati, M. The control of equilibrium in Parkinson's disease patients: Delayed adaptation of balancing strategy to shifts in sensory set during a dynamic task. *Brain Res. Bull.* **2007**, *74*, 258–270. [CrossRef] [PubMed]

86. Jacobs, J.V.; Horak, F.B. Abnormal proprioceptive-motor integration contributes to hypometric postural responses of subjects with Parkinson's disease. *Neuroscience* **2006**, *141*, 999–1009. [CrossRef]

87. Blouin, J.S.; Descarreaux, M.; Bélanger-Gravel, A.; Simoneau, M.; Teasdale, N. Attenuation of human neck muscle activity following repeated imposed trunk-forward linear acceleration. *Exp. Brain Res.* **2003**, *150*, 458–464. [CrossRef] [PubMed]

88. Nardone, A.; Pasetti, C.; Schieppati, M. Spinal and supraspinal stretch responses of postural muscles in early Parkinsonian patients. *Exp. Neurol.* **2012**, *237*, 407–417. [CrossRef] [PubMed]

89. Keshner, E.A.; Allum, J.H.; Pfaltz, C.R. Postural coactivation and adaptation in the sway stabilizing responses of normals and patients with bilateral vestibular deficit. *Exp. Brain Res.* **1987**, *69*, 77–92. [CrossRef]

90. Fujiwara, K.; Maeda, K.; Irei, M.; Mammadova, A.; Kiyota, N. Changes in event-related potentials associated with postural adaptation during floor oscillation. *Neuroscience* **2012**, *213*, 122–132. [CrossRef]

91. Siegmund, G.P.; Blouin, J.S.; Inglis, J.T. Does startle explain the exaggerated first response to a transient perturbation? *Exerc. Sport Sci. Rev.* **2008**, *36*, 76–82. [CrossRef]

92. Allum, J.H.; Tang, K.S.; Carpenter, M.G.; Oude Nijhuis, L.B.; Bloem, B.R. Review of first trial responses in balance control: Influence of vestibular loss and Parkinson's disease. *Hum. Mov. Sci.* **2011**, *30*, 279–295. [CrossRef]

93. Schmid, M.; Sozzi, S. Temporal features of postural adaptation strategy to prolonged and repeatable balance perturbation. *Neurosci. Lett.* **2016**, *628*, 110–115. [CrossRef] [PubMed]

94. Kennedy, A.; Bugnariu, N.; Guevel, A.; Sveistrup, H. Adaptation of the feedforward postural response to repeated continuous postural perturbations. *Neurosci. Med.* **2013**, *4*, 45–49. [CrossRef]

95. Kennedy, A.; Guevel, A.; Sveistrup, H. Impact of ankle muscle fatigue and recovery on the anticipatory postural adjustments to externally initiated perturbations in dynamic postural control. *Exp. Brain Res.* **2012**, *223*, 553–562. [CrossRef] [PubMed]

96. Joseph Jilk, D.; Safavynia, S.A.; Ting, L.H. Contribution of vision to postural behaviors during continuous support-surface translations. *Exp. Brain Res.* **2014**, *232*, 169–180. [CrossRef]

97. Welch, T.D.; Ting, L.H. Mechanisms of motor adaptation in reactive balance control. *PLoS ONE* **2014**, *9*, e96440. [CrossRef]

98. Nardone, A.; Godi, M.; Artuso, A.; Schieppati, M. Balance rehabilitation by moving platform and exercises in patients with neuropathy or vestibular deficit. *Arch. Phys. Med. Rehabil.* **2010**, *91*, 1869–1877. [CrossRef]

99. Buchanan, J.J.; Horak, F.B. Vestibular loss disrupts control of head and trunk on a sinusoidally moving platform. *J. Vestib. Res.* **2002**, *11*, 371–389.

100. Nardone, A.; Turcato, A.M.; Schieppati, M. Effects of balance and gait rehabilitation in cerebellar disease of vascular or degenerative origin. *Restor. Neurol. Neurosci.* **2014**, *32*, 233–245. [CrossRef]

101. Missaoui, B.; Thoumie, P. Balance training in ataxic neuropathies. Effects on balance and gait parameters. *Gait Posture* **2013**, *38*, 471–476. [CrossRef]

102. Nanhoe-Mahabier, W.; Allum, J.H.; Overeem, S.; Borm, G.F.; Oude Nijhuis, L.B.; Bloem, B.R. First trial reactions and habituation rates over successive balance perturbations in Parkinson's disease. *Neuroscience* **2012**, *217*, 123–129. [CrossRef]

103. Giardini, M.; Nardone, A.; Godi, M.; Guglielmetti, S.; Arcolin, I.; Pisano, F.; Schieppati, M. Instrumental or Physical-Exercise Rehabilitation of Balance Improves Both Balance and Gait in Parkinson's Disease. *Neural Plast.* **2018**, *7*, 5614242. [CrossRef] [PubMed]

 *applied sciences*

Article

Influence of Swing-Foot Strike Pattern on Balance Control Mechanisms during Gait Initiation over an Obstacle to Be Cleared

Romain Artico [1,2,3], Paul Fourcade [1,2], Claudine Teyssèdre [1,2], Teddy Caderby [4], Arnaud Delafontaine [1,2] and Eric Yiou [1,2,*]

[1] CIAMS, Université Paris-Saclay, 91404 Orsay, France; romain.artico@u-psud.fr (R.A.); paul.fourcade@u-psud.fr (P.F.); claudine.Teyssedre@u-psud.fr (C.T.); arnaud.delafontaine@u-psud.fr (A.D.)
[2] CIAMS, Université d'Orléans, 45067 Orléans, France
[3] Ecole Nationale de Kinésithérapie et Rééducation, 75012 Saint Maurice, France
[4] IRISSE Laboratory, Université de la Réunion, 97430 Le Tampon, Ile de la Réunion, France; teddy.caderby@univ-reunion.fr
[*] Correspondence: eric.yiou@u-psud.fr; Tel.: +33-169-153-159; Fax: +33-169-156-222

Received: 29 October 2019; Accepted: 20 December 2019; Published: 28 December 2019

Abstract: Gait initiation (GI) over an obstacle to be cleared is a functional task that is highly challenging for the balance control system. Two swing-foot strike patterns were identified during this task—the rearfoot strike (RFS), where the heel strikes the ground first, and the forefoot strike (FFS), where the toe strikes the ground first. This study investigated the effect of the swing-foot strike pattern on the postural organisation of GI over an obstacle to be cleared. Participants performed a series of obstacle clearance tasks with the instruction to strike the ground with either an FFS or RFS pattern. Results showed that anticipatory postural adjustments in the frontal plane were smaller in FFS than in RFS, while stability was increased in FFS. The vertical braking of the centre of mass (COM) during GI swing phase was attenuated in FFS compared to RFS, leading to greater downward centre of mass velocity at foot contact in FFS. In addition, the collision forces acting on the foot were smaller in FFS than in RFS, as were the slope of these forces and the slope of the C7 vertebra acceleration at foot contact. Overall, these results suggest an interdependent relationship between balance control mechanisms and foot strike pattern for optimal stability control.

Keywords: anticipatory postural adjustments; balance control; foot strike pattern; gait initiation; obstacle clearance; biomechanics

---

1. Introduction

Like all terrestrial species, humans move in a gravity field that permanently attracts them towards the centre of the earth. Therefore, balance control is a key component of daily motor tasks. Gait initiation (GI), which corresponds to the transient phase between quiet standing and level walking, is a classical paradigm for investigating how balance is controlled during a functional whole-body movement [1,2]. Gait initiation can be broken down into a postural phase (corresponding to anticipatory postural adjustments (APAs)), followed by a swing phase that ends at the time of swing foot contact with the support surface. These APAs are characterised by a centre of pressure (COP) shift backwards towards the heels, and laterally towards the forthcoming swing leg. The backward centre of pressure shift reflects a strategy to generate the initial propulsive forces necessary to reach the intended centre of mass velocity at the end of gait initiation [3–6]. The mediolateral (ML) centre of pressure shift reflects a strategy to propel the centre of mass above (or beneath) the forthcoming stance foot in order to maintain ML stability during the swing phase of gait initiation [7–11].

Besides ML APAs, a major mechanism involved in stability control is the braking of the centre of mass fall during the swing phase of gait initiation. Previous lines of research have shown that braking is an active phenomenon that occurs in anticipation of the swing foot's collision with the ground, and involves the activation of the soleus of the stance leg [12–15]. Anticipatory centre of mass braking attenuates the vertical ground reaction forces (GRFs) at the time of foot collision (hereafter called collision forces), thus facilitating balance control and reducing the risk of injury linked to the transmission of this impact throughout the musculoskeletal system.

These collision forces are also known to depend on the swing-foot strike pattern at the time of collision [16–18]. At least two main swing-foot strike patterns have been identified in the literature—rearfoot strike (RFS), where the swing heel strikes the ground first, and forefoot strike (FFS), where the swing toe strikes the ground first [16]. Because studies have reported that participants systematically used an RFS pattern, foot strike patterns have so far not been considered in the literature on gait initiation. However, a recent study showed that when gait was initiated with the goal of clearing an obstacle longer than 30% of body height, a change from an RFS to an FFS pattern was observed in most trials [10]. Based on previous research on the effect of the foot strike pattern on collision forces during stepping down while walking [17,18] or running [16,19–22], it was proposed that this change in foot strike pattern reflected a strategy directed at attenuating collision forces.

It is surprising that the authors of these important lines of research, focusing on foot strike pattern effects, did not consider that the magnitude of the anticipatory braking might have changed between the RFS and FFS patterns. Indeed, such a change could at least partly account for the differences in the collision forces observed, with a larger anticipatory centre of mass brake being potentially responsible for greater attenuation of the collision forces. More generally, a brief literature review shows that the balance control mechanisms described above have been investigated by independent lines of research (for a recent review see [2]). Thus, it remains largely unknown whether these mechanisms (i.e., ML APAs, force braking and foot strike pattern) are independently controlled by the central nervous system (CNS) or are interdependent. Gait initiation with an obstacle to be cleared is a natural task that offers the opportunity to test the interdependence between these mechanisms since, as stated above, both foot strategies are naturally used to land on the support surface.

In the present article, it is expected that the ML APAs and the vertical centre of mass braking mechanisms are programmed according to foot strike patterns (FFS or RFS). Specifically, it is reasoned that the need to actively brake the centre of mass fall during the step–swing phase becomes less imperative in FFS if the FFS pattern is more effective than the RFS pattern to attenuate the collision forces, as repeatedly shown in the literature [16,19–22]. Because it has an energetic and attentional cost [12], the CNS may consequently choose to attenuate this active braking for economical purposes. If so, the duration of the swing phase of gait initiation can be expected to be shortened for the FFS pattern, since the centre of mass would fall more rapidly.

Of particular interest, recent studies reported that the amplitude of the ML APAs associated with gait initiation over an obstacle was programmed according to this temporal parameter (i.e., the duration of the swing phase), and was lower when this duration decreased with a lower obstacle [10,11]. In the present study, changes in ML APAs amplitude can therefore be expected because of the changes in the swing-foot strike pattern.

Hence, the present study aimed to investigate the effect of changing the swing-foot strike pattern (FFS vs. RFS) on the postural organisation of gait initiation over an obstacle. The hypothesis of an interdependent relationship between balance control mechanisms for optimal stability control was raised. Specifically, and for the reasons detailed above, it was expected that both active braking of the centre of mass fall and the ML APAs associated with gait initiation over an obstacle to be cleared would be attenuated in the FFS compared to the RFS pattern, but with no consequent alteration in ML stability or increase in collision forces.

2. Materials and Methods

2.1. Participants

In all, 13 young healthy subjects (eight males and five females, age (mean standard ± 1 deviation), 28.7 ± 1.5 years; height, 178.1 ± 7.2 cm; weight, 69 ± 8.3 kg) participated in the experiment. All were free of any known neuromuscular disorders. They provided written informed consent after being informed of the nature and purpose of the experiments, which were approved by the local ethics committee of University Paris-Sud, Paris-Saclay (EA 4532, 10 September 2019). The study complied with the standards set by the Declaration of Helsinki.

2.2. Experimental Protocol

The participants were instructed to initiate gait at maximal velocity with their preferred limb while clearing an obstacle, and to continue walking at the same velocity to the end of a five-meter track (Figure 1). The task was self-triggered, that is, the participants initiated gait only when they felt ready. As in our previous study [11], the obstacle height was 10% of the subject's height and its distance from the subject was 30% of the subject's height. Two conditions of gait initiation were performed. Under the first condition, the participants were instructed to strike the ground with the heel first after stepping over the obstacle (rearfoot strike (RFS) condition). Under the second condition, they were instructed to strike the ground with the forefoot first (forefoot strike (FFS) condition). We showed in our previous study [11] that participants used both the FFS and the RFS patterns under the condition where the obstacle had the same distance/height features as in the present study. In other words, no clear preferred foot strike pattern could be detected, suggesting that participants of the present experiment were not much constrained by the instruction on the foot strike pattern.

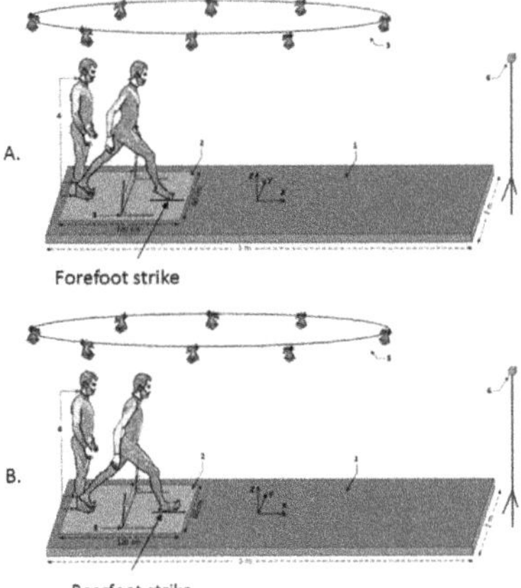

Figure 1. Schematic illustration of the experimental setup. (**A**) Gait initiation with a forefoot strike (FFS) pattern. (**B**) Gait initiation with a rearfoot strike (RFS) pattern. 1, walkway; 2, force plate; 3, obstacle; 4, reflective marker; 5, Vicon camera; 6, visual target.

The order of the conditions was randomised to avoid rank effects. For each condition, the subjects were allowed two familiarisation trials, after which five trials were recorded. In the initial posture, the participants stood upright barefoot, with their feet hip-width apart, their arms hanging loosely alongside their body and their body weight evenly distributed between their legs.

Gait was initiated on a force plate (600 × 1200 mm, AMTI, Watertown, MA, USA) positioned at the beginning of the track. The force plate was embedded in the track and was long enough so that the participant's swing foot always landed on it at the end of gait initiation. The obstacle consisted of a lightweight wooden rod (length, 65 cm; diameter, 1 cm) resting on two adjustable upright stands (Figure 1). The participant's toes served as the reference point for positioning the obstacle. Reflective skin markers (9 mm in diameter) were placed bilaterally at the hallux (toe markers), head of the fifth metatarsal (metatarsal markers) and posterior calcaneus (heel markers). In 10 subjects, a marker was also placed at the top spine level (C7) to investigate the transmission of the collision forces from the swing foot to the upper part of the musculoskeletal system. An additional marker was positioned at the middle of the obstacle to determine the safety distance, that is, the distance between the swing foot and the obstacle at the time of clearance. A V8i VICON eight-camera (Mcam2) motion capture system (Oxford Metrics Ltd., Oxford, UK) with a 64-channel analogue card was used to collect both the position of these markers and the force plate data at a sampling rate of 500 Hz. Kinematic and force plate data were acquired and synchronised using the Vicon acquisition software (Workstation, Version 5.1, Oxford, UK).

2.3. Data Analysis

After data collection, kinematic and force plate data were processed offline using a custom-made program written in MATLAB™ (Version 5.3 (R11), The MathWorks Inc., Natick, MA, USA). Data were low-pass filtered using a Butterworth filter with a 15 Hz cutoff frequency for kinematic data [23] and a 10 Hz cutoff frequency for kinetic data [24]. The ML and anteroposterior (AP) coordinates of the centre of pressure (yP and xP, respectively) were computed from force plate data in accordance with the instructions from the manufacturer (AMTI) as follows:

$$yP = \frac{Mx + Fy \times dz}{Fz}, \tag{1}$$

$$xP = \frac{-My + Fx \times dz}{Fz}, \tag{2}$$

where Mx and My are the moment around the AP and ML axes, respectively; Fy, Fx and Fz are the ML, AP and vertical ground reaction forces, respectively; and dz is the distance between the surface of the force plate and its origin.

Instantaneous acceleration of the centre of mass along the AP, ML and vertical axes was determined from the ground reaction force according to Newton's second law. The centre of mass velocity and displacement were computed by the successive numerical integration of centre of mass acceleration using integration constants equal to zero, namely, initial velocity and displacement null [3].

Gait initiation onset (t_0) and swing-foot contact were determined from force plate data. Since gait initiation onset did not necessarily occur simultaneously in the ML and AP axes, two t_0 times were estimated, one for each axis. T_{0y} and t_{0x} corresponded, respectively, to the instants when the ML and AP centre of pressure trace exceeded 2.5 standard deviations from its baseline value, respectively [24]. The one occurring first was considered t_0. Swing-foot contact corresponded to the instant when the ML and the AP centre of pressure trace shifted abruptly towards the swing-leg side and forward, respectively. Swing heel-off and swing toe-off were determined from VICON data. They corresponded to the instants when the swing heel marker's vertical position and the swing toe marker's anterior position exceeded their position in the initial static posture by 3 mm.

2.4. Dependant Variables

Initial posture, APAs and swing foot lift. Gait initiation was divided into three phases, namely APAs (from t_0 to swing heel-off), swing foot lift (from heel-off to toe-off) and swing phase (from swing toe-off to swing foot contact) (Figure 2). The ML and AP centre of mass position in the initial upright posture was estimated by averaging the centre of pressure position during a 250 ms period, lasting from t_0 minus 1500 ms to t_0 minus 1250 ms. No dynamic phenomena were detected during this period. APAs amplitude was characterised by the peaks of the backward and ML centre of pressure shift obtained during the APAs time window. The duration of APAs along the ML and AP axes was computed separately, as the t_0 times for these two axes did not necessarily occur simultaneously as stated above. The centre of mass velocity and displacement along the ML and AP axes were determined at swing toe-off and foot contact.

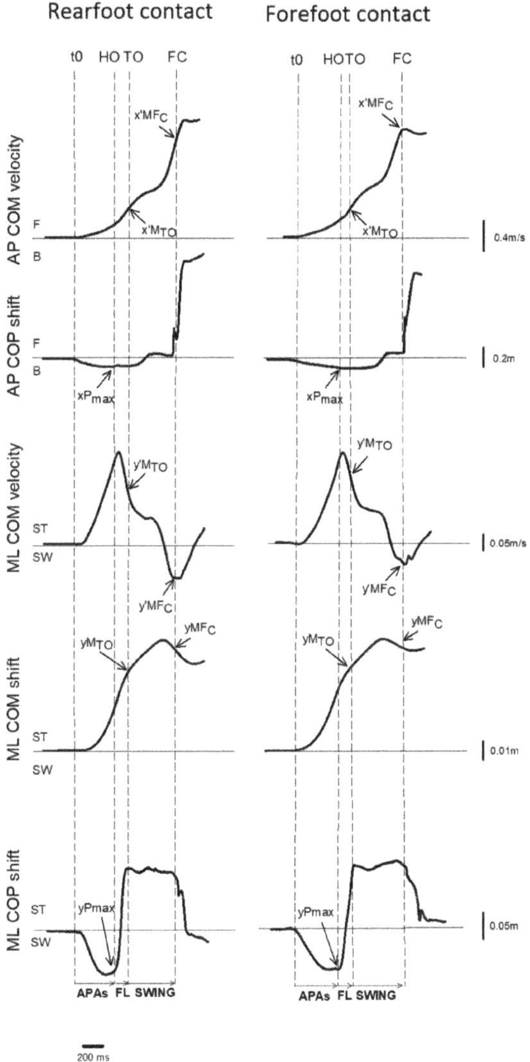

Figure 2. Typical biomechanical traces and representation of the main experimental variables (horizontal plane). Traces are from one representative subject initiating gait in the (**left**) rearfoot strike condition and the (**right**) forefoot strike (FFS) condition. Anteroposterior (AP) direction. x'M$_{TO}$, x'M$_{FC}$: centre of mass (COM) velocity at toe-off and foot contact; xPmax: peak of centre of pressure (COP) displacement during anticipatory postural adjustments (APAs); F: forward; B: backward. Mediolateral (ML) direction. y'M$_{TO}$, y'M$_{FC}$: COM velocity at toe-off and foot contact; yM$_{TO}$, yM_{FC}: COM displacement at toe-off and foot contact; yPmax: peak of COP displacement during APAs; ST: stance limb; SW: swing limb. Vertical dashed lines: t_0 onset variation of biomechanical traces; HO: swing heel-off; TO: swing toe-off; FC: swing foot contact. Horizontal arrows. APAs: anticipatory postural adjustments; FL: swing foot lift; SWING: swing phase of gait initiation.

Motor performance. Motor performance was quantified with the AP centre of mass velocity at foot contact, step length and swing phase duration. Step length corresponded to the distance covered by the heel marker of the swing leg from the initial posture to foot contact.

Safety distance. The vertical distance between the obstacle and the marker of the swing/rear heel, and between the obstacle and the marker of the swing/rear toe was measured at the instant these markers passed over the obstacle. The smaller of these two vertical distances corresponded to the safety distance.

ML dynamic stability. ML dynamic stability at foot contact (hereafter called ML stability) was quantified by the ML boundary of the base of support (BOS_{ymax}) and an adaptation of the margin of stability (MOS) [25]. The BOS_y was measured as the distance at foot contact between the marker on the fifth metatarsal of the stance foot and the heel marker of the swing foot in the case of a rearfoot strike, or the marker on the fifth metatarsal of the swing foot in the case of a forefoot strike. The MOS corresponded to the difference between BOS_{ymax} and the ML position of the extrapolated centre of mass at swing foot contact ($YcoM_{FC}$), that is, $MOS = BOS_{ymax} - YcoM_{FC}$. Based on the study of Hof et al. [25], the ML position of the extrapolated centre of mass at foot contact ($YcoM_{FC}$) was calculated as follows:

$$Y_{CO}M_{FC} = yM_{FC} + \frac{y'M_{FC}}{\omega_0}, \tag{3}$$

where yM_{FC} and $y'M_{FC}$ are, respectively, the ML centre of mass position and velocity at foot contact, and ω_0 is the eigenfrequency of the body modelled as an inverted pendulum calculated as $\omega_0 = \sqrt{\frac{g}{l}}$, where $g = 9.81$ m/s^2 is gravitational acceleration and l is the length of the inverted pendulum, which in this study was estimated to 57.5% of body height [26]. This quantity was held constant during the gait initiation. ML stability at foot contact is preserved on the condition that $YcoM_{FC}$ is within BOS_{ymax}, which corresponds to a positive MOS.

Centre of mass braking. The braking index introduced by Welter and colleagues provides evidence that the centre of mass does not merely fall under the force of gravity but that the CNS prepares for foot contact by decreasing the centre of mass vertical velocity to achieve a soft landing [15]. It was calculated according to the formula of Chong et al. [12] as follows:

$$\text{Braking index} = (V_{MIN} - V_{FC})/V_{MIN}, \tag{4}$$

where V_{MIN} is the minimum vertical centre of mass velocity occurring between mid- to late swing, and V_{FC} is the vertical centre of mass velocity at foot contact (Figure 3).

This index indicates the amount of change (i.e., braking) in the centre of mass vertical velocity at foot contact relative to its minimal value. If the centre of mass vertical velocity is actively decreased before foot contact, the braking index value is increasing.

Collision forces. The combined effect of the foot strike pattern (RFS vs. FFS) and vertical force braking on the collision forces and on the transmission of these forces through the body was assessed using variables extracted from force plate and VICON data, respectively. The variables obtained from the force plate were the peak and slope of the vertical ground reaction force trace occurring immediately after foot contact [16] (Figure 3). The variable obtained from VICON data was the slope (computed at foot contact) of the vertical acceleration of the marker placed at the C7 spine level.

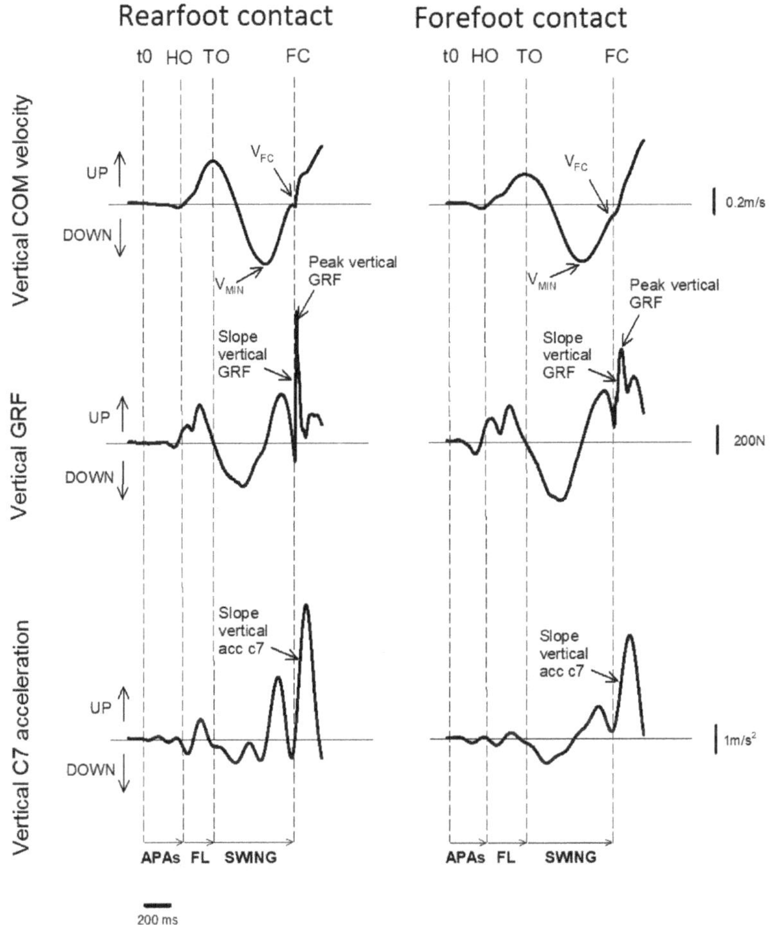

Figure 3. Typical biomechanical traces and representation of the main experimental variables (vertical direction). Traces are from one representative subject initiating gait in the (**left**) rearfoot strike condition and the (**right**) forefoot strike condition. COM, GRF, acc: centre of mass, ground reaction forces, and acceleration, respectively; V_{MIN}, V_{FC}: peak of negative vertical COM velocity and COM velocity at foot contact (FC). Vertical dashed lines: t_0 onset variation of biomechanical traces; HO: swing heel-off; TO: swing toe-off; FC: swing foot contact. Horizontal arrows. APAs: anticipatory postural adjustments; FL: swing foot lift; SWING: swing phase of gait initiation.

2.5. Statistics

Mean values and standard errors were calculated for each variable under each condition. The normality of data was checked using the Kolmogorov–Smirnov test and the homogeneity of variances was assessed using Bartlett's test. Paired *t*-tests were used to compare mean values. The alpha level was set at 0.05.

3. Results

3.1. Biomechanical Traces

The time course of the biomechanical traces was generally similar under both experimental conditions (see Figures 2 and 3). Swing heel-off was systematically preceded by postural dynamics that corresponded to APAs. During these APAs, the centre of pressure displacement reached a peak value in a backward direction and towards the swing leg side, while the centre of mass velocity was directed forwards and towards the stance leg side. The ML centre of mass velocity trace reached a first peak value towards the stance leg side at around heel-off. This trace then fell towards the swing leg side and a second peak value towards this side was reached a few milliseconds after foot contact. The ML centre of mass shift trace was bell-shaped and reached a peak value towards the stance leg side in the late part of the swing phase. The AP centre of mass velocity increased progressively until it reached a peak value a few milliseconds after swing foot contact, while the centre of mass was continuously shifted forward (see Figure 2).

The time course of the vertical ground reaction force and the centre of mass vertical velocity was also generally similar under both experimental conditions (see Figure 3). Following foot-off, the centre of mass was accelerated downwards (negative velocity indicates downward centre of mass movement) and then reversed. In fact, during single support, the centre of mass velocity shows a V shape indicating that the centre of mass fall was braked.

3.2. Initial Posture, APAs and Swing Foot Lift

Statistical analysis showed that there was no significant effect of the swing-foot strike pattern on the initial position of the centre of mass along the ML and the AP direction. Additionally, there was no significant effect of the swing-foot strike pattern on the peak of AP centre of pressure shift and AP/ML APAs duration (Figure 4). By contrast, the peak of the ML centre of pressure shift was significantly smaller under the FFS than under the RFS condition ($t_{12} = -2.43$, $p = 0.032$) and foot lift duration was significantly longer under the FFS than under the RFS condition ($t_{12} = 5.22$, $p < 0.001$). Foot lift duration was 88 ± 8 ms under the RFS condition and 105 ± 7 ms under the FFS condition. Finally, the ML centre of mass velocity at toe-off was significantly lower under the FFS than under the RFS condition ($t_{12} = -3.25$, $p = 0.006$), whereas the AP centre of mass velocity at toe-off was significantly higher ($t_{12} = 2.46$, $p = 0.003$).

3.3. Safety Distance

The safety distance values for the swing foot and for the rear foot were 13.6 ± 0.9 cm and 14.4 ± 1.4 cm, respectively, under the RFS condition, and were 13.7 ± 0.9 cm and 15.0 ± 1.5 cm, respectively, under the FFS condition, with no significant difference across conditions.

3.4. Motor Performance

The swing phase duration was significantly shorter ($t_{12} = -5.01$, $p < 0.001$) and the AP centre of mass velocity at foot contact was significantly smaller ($t_{12} = -4.64$, $p < 0.001$) under the FFS than under the RFS condition. By contrast, step length did not differ between the two conditions. Step length was 79.2 ± 4.0 cm under the RFS condition and 77.8 ± 4.5 cm under the FFS condition.

3.5. ML Dynamic Stability

The ML margin of stability (MOS) at foot contact was significantly greater under the FFS than under the RFS condition ($t_{12} = 6.12$, $p < 0.001$), showing that ML dynamic stability at foot contact was increased under the FFS compared to the RFS condition (Figure 5). The results further show that the ML size of the base of support (BOS_{ymax}) was larger under the FFS than under the RFS condition ($t_{12} = 4.77$, $p < 0.001$), and that the ML centre of mass velocity at foot contact was lower ($t_{12} = 5.75$,

$p < 0.001$). By contrast, the ML centre of mass position at foot contact was not significantly different between the two conditions.

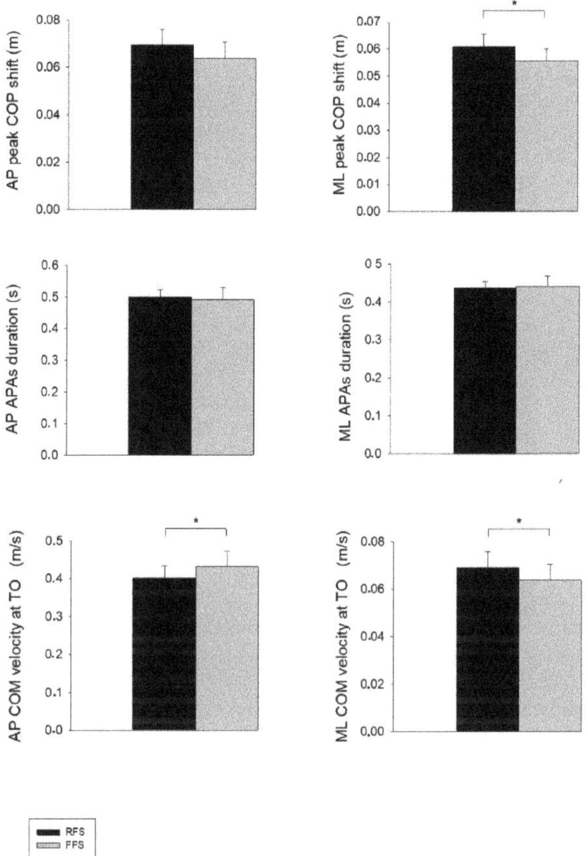

Figure 4. Effects of swing-foot strike pattern on selected APA parameters. Reported are mean values (all participants together) + 1 SE. APAs: anticipatory postural adjustments; TO: toe-off; COP: centre of pressure; COM: centre of mass; RFS: rearfoot strike condition; FFS: forefoot strike condition. Bars at the left side: mediolateral (ML) direction. Bars at the right side: anteroposterior (AP) direction. * Indicates a significant difference between bars.

3.6. Centre of Mass Braking

The braking index was significantly smaller under the FFS than under the RFS condition ($t_{12} = -2.44$, $p = 0.03$), that is, the downward vertical centre of mass velocity was braked less before foot contact under the FFS condition (Figure 6). As a consequence, the vertical centre of mass velocity at foot contact was greater under the FFS than under the RFS condition ($t_{12} = -2.19$, $p = 0.049$), and the swing phase was shorter (as reported above).

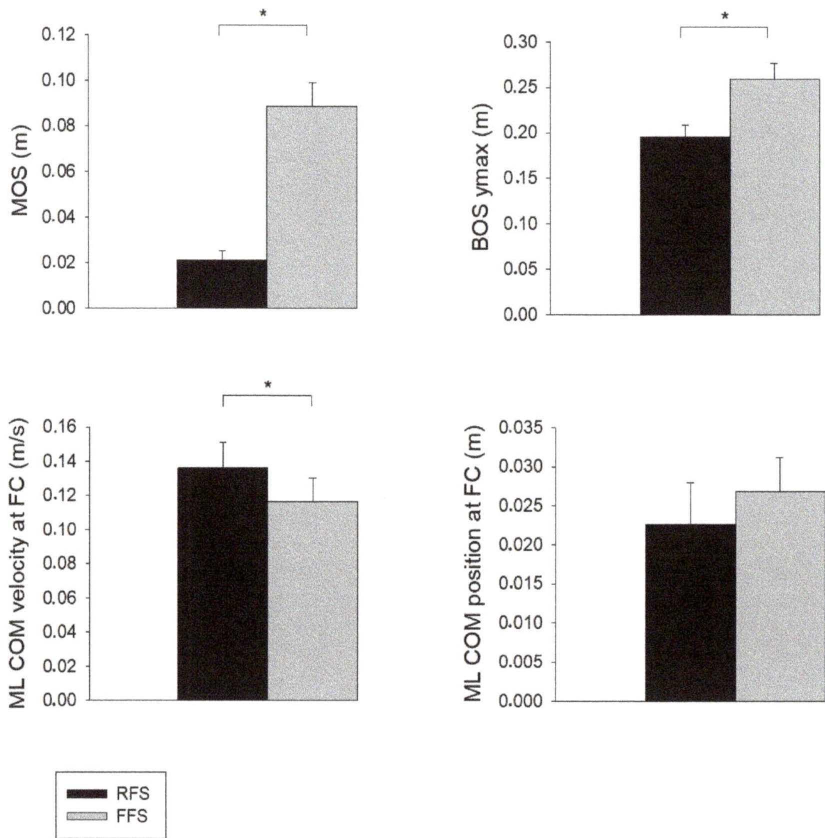

Figure 5. Effects of swing-foot strike pattern on selected stability parameters along the mediolateral (ML) direction. Reported are mean values (all participants together) + 1 SE. BOSymax: base of support size; MOS: margin of stability; COM: centre of mass; RFS: rearfoot strike condition; FC: swing foot contact; FFS: forefoot strike condition. * Indicates a significant difference between bars.

3.7. Collision Forces

The peak value of the vertical ground reaction forces, which occurred shortly after foot contact and reflected the collision forces acting at the foot, was significantly smaller under the FFS than under the RFS condition (t_{12} = −5.61, p < 0.001), as was the slope of these forces (t_{12} = −5.11, p < 0.001; see Figure 6). In addition, the slope of the vertical acceleration (computed at foot contact) of the marker located at the C7 spine level was significantly smaller under the FFS than under the RFS condition (t_{12} = −2.32, p = 0.045).

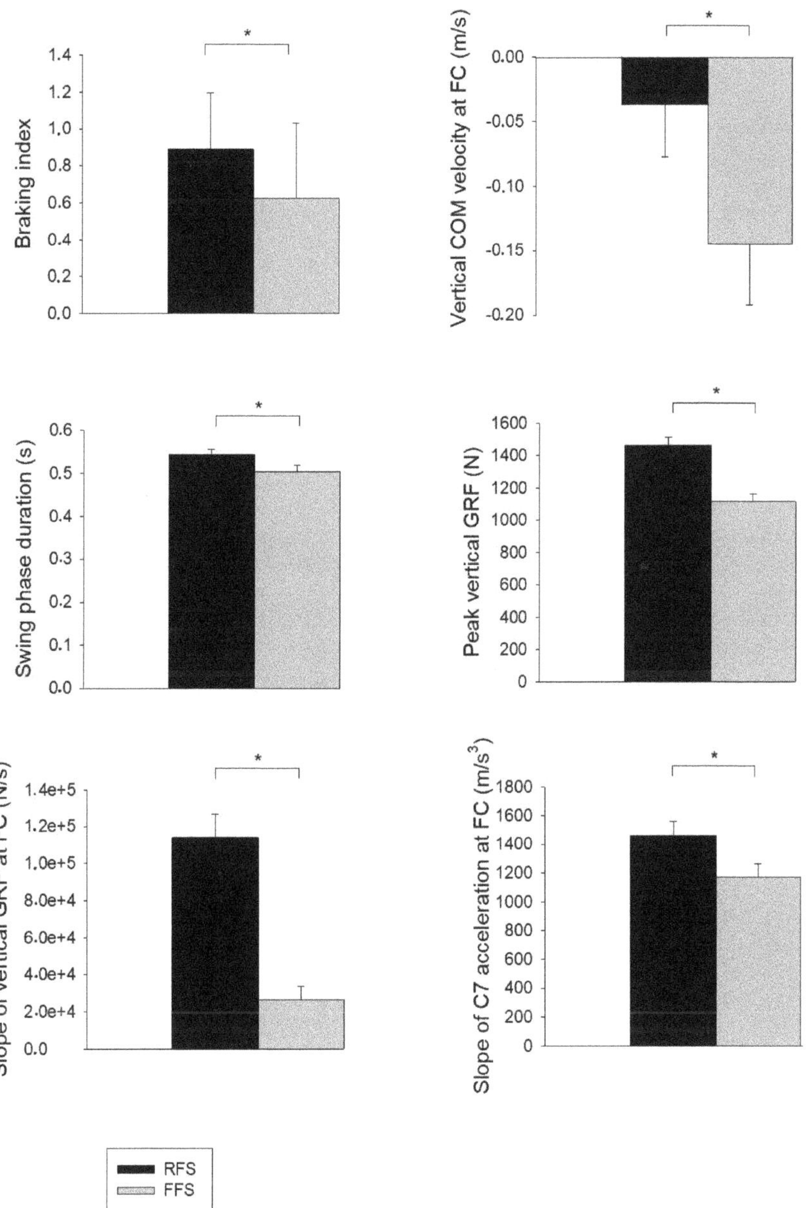

Figure 6. Effects of swing-foot strike pattern on collision force parameters, vertical force braking and swing duration. Reported are mean values (all participants together) ± 1 SE. GRF: ground reaction forces; COM: centre of mass; C7: 7th cervical vertebrae; FC: foot contact; RFS: rearfoot strike condition; FFS: forefoot strike condition. * Indicates a significant difference between bars.

4. Discussion

The present study investigated the effect of changing the swing-foot strike pattern (FFS vs. RFS) on ML APAs and active braking of the centre of mass fall, and on the related ML stability and force collisions. Specifically, it was hypothesised that the active braking of the centre of mass fall and the ML APAs associated with gait initiation over an obstacle to be cleared are both attenuated in an FFS compared to an RFS pattern, but with no alteration of ML stability or increase in collision forces. The results discussed below are in line with these expectations and suggest the existence of an interdependent relationship between balance control mechanisms for optimal stability control.

Do and colleagues first reported that, in young healthy adults, the centre of mass fall during the swing phase of gait initiation was actively braked in anticipation of the swing foot's collision with the ground [2,14,15,27,28]. This centre of mass brake, which was ascribed to an activation of the soleus of the stance leg (not recorded in the present study), aims to attenuate the vertical ground reaction forces at the time of foot contact (i.e., the collision forces). Therefore, this braking facilitates balance control following foot contact and attenuates the transmission of the collision forces throughout the entire musculoskeletal system by allowing a softer swing foot landing [6,14,15]. The results of the present study showed that the indicator of this centre of mass brake (the braking index) reached a significantly lower value under the FFS than under the RFS condition (the difference between the two conditions was Δ = 19.4%), thus revealing that the fall of the centre of mass was much less braked under the FFS condition. The much larger downward-oriented centre of mass velocity at foot contact under the FFS condition (Δ = 74.5%) can be considered a direct consequence of this reduced braking. Therefore, it could be surprising that the two indicators of the collision forces' perturbing effect (namely, peak collision forces and the slope of this peak) were both smaller under the FFS than under the RFS condition (Δ = 27.3% and Δ = 26.2%, respectively), revealing that the foot landing was softer under the first condition. In addition, the jerk value recorded at the C7 spine level was lower under the FFS than under the RFS condition. This finding suggests that the collision forces acting at the foot were absorbed to a greater extent by the swing leg under the FFS condition, reducing its transmission through the upper part of the musculoskeletal system. In turn, these results are in line with classical studies in the literature comparing collision forces during running [16,19–22,29,30], or during stepping down during ongoing gait [17]. According to van Dieën et al. [17], the FFS pattern allows the ankle plantar flexors of the swing leg to produce negative work that can absorb part of the collision forces. This negative work is absent in the RFS pattern which might explain the greater collision forces found under the RFS condition. As stressed in the introduction, it is noteworthy that centre of mass braking has not been investigated in any of these studies. Consequently, a possible modulation effect of this force braking on the collision forces cannot be discounted in these studies.

In fact, the result of the present study showing that the collision forces were smaller under the FFS than under the RFS condition, despite the centre of mass braking being attenuated, further stresses the efficiency of the FFS pattern in dampening collision forces. The present results may also suggest that the dampening effect specifically associated with the FFS alone was probably underestimated in the studies on foot strike pattern cited above.

The question arises as to why centre of mass braking is attenuated under the FFS condition compared to the RFS condition. To interpret this finding, it should be stressed that this braking has an energetic and attentional cost [14,15] and shares a dampening action on collision forces with the foot strike strategy. Therefore, when the FFS pattern is employed, the need to actively brake the centre of mass fall becomes less crucial than when the RFS pattern is used. The CNS would then attenuate this active brake in order to minimise the associated energetic and attentional cost, and would rely more heavily on the efficiency of the foot strike strategy to dampen the collision forces. This implies that the CNS takes into account the foot strike pattern and the related dampening effect when programming braking force generation. This statement is in line with the hypothesis of an interdependent relationship between these two mechanisms (active braking and foot strike pattern) for optimal stability control and better ground reaction force management. The results linking ML APAs to the foot strike pattern further strengthen this hypothesis (paragraph below).

During gait initiation or any locomotor task, imbalance occurs in the frontal plane due to the transition from a bipedal to a unipedal stance. Following swing foot-off, the centre of mass falls rapidly towards the swing leg side under gravity, and this lateral fall is braked by foot contact with the ground. This ML imbalance is greater when an obstacle is to be cleared, since the duration of the lateral fall of the centre of mass is increased compared to a control condition with no obstacle to be cleared [10,11,31–33]. The development of ML APAs is therefore of utmost importance to maintain stability as it acts to minimise this lateral fall by propelling the centre of mass towards the stance leg side and thus attenuating the ML gap between the centre of mass and the centre of pressure at the time of swing foot-off [7,8,10,11,34,35]. Previous modelling studies demonstrated the link between ML APAs and ML stability at foot contact [8,10,11,34].

In the present study, the results showed that the amplitude of these ML APAs, in terms of peak centre of pressure shift, was lower under the FFS than under the RFS condition (Δ = 9.4%). It also revealed that this lower amplitude was not accompanied by a ML APA duration lengthening strategy. Such a trade-off strategy between APA amplitude and APA duration has been repeatedly reported in the literature under various conditions of postural constraints, such as fear of falling [36], fatigue [37–39], leg dominance [40] and temporal pressure [41]. There was no such compensative trade-off strategy in the present study. Consequently, the ML centre of mass velocity at the time of swing foot-off, which represents the total amount of forces applied to the centre of mass during the APAs, reached a lower value under the FFS than under the RFS condition (Δ = 5.99%). For the above reasons, ML dynamic stability, as measured with the MOS [25], could be expected to be lower under the FFS than under the RFS condition. However, contrary to this expectation, the MOS reached much higher values under the FFS than under the RFS condition (Δ = 71.42%), that is, ML dynamic stability was significantly improved. Changes in two components of the MOS with the foot strike strategy can account for this result, as detailed below.

First, the ML centre of mass velocity at the time of foot contact reached a lower value under the FFS than under the RFS condition (Δ = 14%), that is, the velocity at which the centre of mass fell towards the swing leg side was attenuated. This finding could be ascribed to the lower duration of the swing phase of gait initiation under the FFS compared to the RFS condition (Δ = 6.70%), which implies that the period during which the centre of mass was falling laterally under gravity was reduced. In turn, this shorter duration of the swing phase under the FFS condition could be ascribed to the lower force braking under the FFS condition (Δ = 19.4%), which accelerated the fall of the centre of mass to a greater extent (as discussed above).

As a second factor contributing to enhanced stability, the size of the ML base of support was increased under the FFS compared to the RFS condition. This result could not be ascribed to a strategy of a more lateral placement of the swing foot [24,35] (since the distance between the two heels remained unchanged between the two foot strike conditions), but to foot anatomy. Under the FFS condition, the ML base of support size corresponded to the distance between the stance heel and the fifth metatarsal of the swing foot, while under the RFS condition, it corresponded to the distance between the two heels. It was therefore smaller than under the FFS condition. This difference reached up to 10 cm in some subjects.

The question arises as to why the amplitude of ML APAs is decreased when stability is increased with the change in the foot strike pattern.

Generating ML APAs is known to have an energetic cost, requiring a coordinated activation of hip abductors and ankle dorsiflexors [7,42]. Although not shown in the literature to date, it may also have an attentional cost, as has previously been documented for the control of static erect posture and level walking [43,44]. An optimal strategy for stability control would appear to be to reduce this cost by attenuating the ML APAs under the FFS condition while, at the same time, increasing stability with the mechanical and anatomical changes reported above—neither of which requires any additional energetic or attentional cost (even less for attenuated force braking). This optimal strategy implies that one (or both) of these factors associated with the foot strike pattern be accounted for in the programming of ML APAs. This statement further reinforces the hypothesis proposed above of an interdependent relationship between motor mechanisms for optimal stability control.

Interestingly, previous studies focusing on the effect of different sorts of constraints applied to the postural system during gait initiation (such as temporal pressure [41,45], progression velocity [24], the presence of an obstacle of different heights and distances to be cleared [10,11], the addition of a mass distributed (a)symmetrically over the body or the ankles [46–48], etc.) repeatedly showed that the CNS modulated the spatiotemporal features of ML APAs so as to maintain an equivalent MOS value. This motor invariance through these disparate postural constraints led the authors to suggest that the CNS would set a fixed MOS value before initiating gait and would then adjust ML APAs features to keep it unchanged (for review, see [2]). Thus, the present finding that the MOS was not invariant is not in line with these previous results and related hypothesis. Now, it may be recalled that the participants systematically used an RFS pattern in all of the previous studies. Therefore, the MOS was computed and compared across conditions at the moment the swing heel hit the ground. By contrast, the present study compared the MOS of two different postures of the landing foot. These foot postures were responsible for the difference in the ML size of the base of support, which was a major factor explaining the MOS differences between the two foot strike conditions. It is possible that, for a given foot strike pattern (RFS or FFS), a fixed MOS value is planned before initiating gait, and may be greater for the FFS than the RFS pattern due to foot anatomy.

5. Conclusions

In conclusion, these findings are in line with the hypothesis of an interdependent relationship between balance control mechanisms and foot strike pattern for optimal stability control. These findings are original since previous studies in the literature focused on either balance control mechanisms during gait initiation or changes in the collision forces with the foot strike pattern. Thus, the present study linked these two major lines of research to better highlight how balance is controlled in humans during a functional locomotor task. The question remains as to whether and how coordination between these balance control mechanisms is affected by aging as well as neurological and musculoskeletal disorders.

Author Contributions: R.A., P.F., C.T., T.C., A.D., E.Y.: made substantial contributions to the conception of the work, the acquisition, analysis, and interpretation of data; Drafted the work and revised it critically; made final approval of the version to be published; Agreed to be accountable for all aspects of the work in ensuring that questions related to the accuracy or integrity of any part of the work are appropriately investigated and resolved. All authors have read and agreed to the published version of the manuscript.

Funding: This research received no external funding.

Conflicts of Interest: The authors declare no conflict of interest.

References

1. Yiou, E.; Caderby, T.; Hussein, T. Adaptability of anticipatory postural adjustments associated with voluntary movement. *World J. Orthop.* **2012**, *3*, 75–86. [CrossRef] [PubMed]
2. Yiou, E.; Caderby, T.; Delafontaine, A.; Fourcade, P.; Honeine, J.-L. Balance control during gait initiation: State-of-the-art and research perspectives. *World J. Orthop.* **2017**, *8*, 815–828. [CrossRef] [PubMed]
3. Brenière, Y.; Cuong Do, M.; Bouisset, S. Are dynamic phenomena prior to stepping essential to walking? *J. Mot. Behav.* **1987**, *19*, 62–76. [CrossRef] [PubMed]
4. Breniere, Y.; Do, M.C. When and how does steady state gait movement induced from upright posture begin? *J. Biomech.* **1986**, *19*, 1035–1040. [CrossRef]
5. Lepers, R.; Brenière, Y. The role of anticipatory postural adjustments and gravity in gait initiation. *Exp. Brain Res.* **1995**, *107*, 118–124. [CrossRef]
6. Michel, V.; Chong, R.K.Y. The strategies to regulate and to modulate the propulsive forces during gait initiation in lower limb amputees. *Exp. Brain Res.* **2004**, *158*, 356–365. [CrossRef]
7. Honeine, J.-L.; Schieppati, M.; Crisafulli, O.; Do, M.-C. The Neuro-Mechanical Processes That Underlie Goal-Directed Medio-Lateral APA during Gait Initiation. *Front. Hum. Neurosci.* **2016**, *10*, 445. [CrossRef]
8. Lyon, I.N.; Day, B.L. Control of frontal plane body motion in human stepping. *Exp. Brain Res.* **1997**, *115*, 345–356. [CrossRef]

9. McIlroy, W.E.; Maki, B.E. The control of lateral stability during rapid stepping reactions evoked by antero-posterior perturbation: Does anticipatory control play a role? *Gait Posture* **1999**, *9*, 190–198. [CrossRef]

10. Yiou, E.; Fourcade, P.; Artico, R.; Caderby, T. Influence of temporal pressure constraint on the biomechanical organization of gait initiation made with or without an obstacle to clear. *Exp. Brain Res.* **2016**, *234*, 1363–1375. [CrossRef]

11. Yiou, E.; Artico, R.; Teyssedre, C.A.; Labaune, O.; Fourcade, P. Anticipatory Postural Control of Stability during Gait Initiation Over Obstacles of Different Height and Distance Made Under Reaction-Time and Self-Initiated Instructions. *Front. Hum. Neurosci.* **2016**, *10*, 449. [CrossRef] [PubMed]

12. Chong, R.K.Y.; Chastan, N.; Welter, M.-L.; Do, M.-C. Age-related changes in the center of mass velocity control during walking. *Neurosci. Lett.* **2009**, *458*, 23–27. [CrossRef] [PubMed]

13. Delafontaine, A.; Gagey, O.; Colnaghi, S.; Do, M.-C.; Honeine, J.-L. Rigid Ankle Foot Orthosis Deteriorates Mediolateral Balance Control and Vertical Braking during Gait Initiation. *Front. Hum. Neurosci.* **2017**, *11*, 214. [CrossRef] [PubMed]

14. Honeine, J.-L.; Schieppati, M.; Gagey, O.; Do, M.-C. By counteracting gravity, triceps surae sets both kinematics and kinetics of gait. *Physiol. Rep.* **2014**, *2*, e00229. [CrossRef] [PubMed]

15. Welter, M.-L.; Do, M.C.; Chastan, N.; Torny, F.; Bloch, F.; du Montcel, S.T.; Agid, Y. Control of vertical components of gait during initiation of walking in normal adults and patients with progressive supranuclear palsy. *Gait Posture* **2007**, *26*, 393–399. [CrossRef]

16. Lieberman, D.E.; Venkadesan, M.; Werbel, W.A.; Daoud, A.I.; D'Andrea, S.; Davis, I.S.; Mang'eni, R.O.; Pitsiladis, Y. Foot strike patterns and collision forces in habitually barefoot versus shod runners. *Nature* **2010**, *463*, 531–535. [CrossRef]

17. van Dieën, J.H.; Spanjaard, M.; Könemann, R.; Bron, L.; Pijnappels, M. Mechanics of toe and heel landing in stepping down in ongoing gait. *J. Biomech.* **2008**, *41*, 2417–2421. [CrossRef]

18. van Dieën, J.H.; Pijnappels, M. Effects of conflicting constraints and age on strategy choice in stepping down during gait. *Gait Posture* **2009**, *29*, 343–345. [CrossRef]

19. Almeida, M.O.; Davis, I.S.; Lopes, A.D. Biomechanical Differences of Foot-Strike Patterns During Running: A Systematic Review With Meta-analysis. *J. Orthop. Sports Phys. Ther.* **2015**, *45*, 738–755. [CrossRef]

20. Boyer, E.R.; Rooney, B.D.; Derrick, T.R. Rearfoot and midfoot or forefoot impacts in habitually shod runners. *Med. Sci. Sports Exerc.* **2014**, *46*, 1384–1391. [CrossRef]

21. Breine, B.; Malcolm, P.; Van Caekenberghe, I.; Fiers, P.; Frederick, E.C.; De Clercq, D. Initial foot contact and related kinematics affect impact loading rate in running. *J. Sports Sci.* **2017**, *35*, 1556–1564. [CrossRef] [PubMed]

22. Hatala, K.G.; Dingwall, H.L.; Wunderlich, R.E.; Richmond, B.G. Variation in foot strike patterns during running among habitually barefoot populations. *PLoS ONE* **2013**, *8*, e52548. [CrossRef] [PubMed]

23. Mickelborough, J.; van der Linden, M.L.; Richards, J.; Ennos, A.R. Validity and reliability of a kinematic protocol for determining foot contact events. *Gait Posture* **2000**, *11*, 32–37. [CrossRef]

24. Caderby, T.; Yiou, E.; Peyrot, N.; Begon, M.; Dalleau, G. Influence of gait speed on the control of mediolateral dynamic stability during gait initiation. *J. Biomech.* **2014**, *47*, 417–423. [CrossRef]

25. Hof, A.L.; Gazendam, M.G.J.; Sinke, W.E. The condition for dynamic stability. *J. Biomech.* **2005**, *38*, 1–8. [CrossRef]

26. Winter, D. Human balance and posture control during standing and walking. *Gait Posture* **1995**, *3*, 193–214. [CrossRef]

27. Chastan, N.; Westby, G.W.M.; du Montcel, S.T.; Do, M.C.; Chong, R.K.; Agid, Y.; Welter, M.L. Influence of sensory inputs and motor demands on the control of the centre of mass velocity during gait initiation in humans. *Neurosci. Lett.* **2010**, *469*, 400–404. [CrossRef]

28. Honeine, J.-L.; Schieppati, M.; Gagey, O.; Do, M.-C. The Functional Role of the Triceps Surae Muscle during Human Locomotion. *PLoS ONE* **2013**, *8*, e52943. [CrossRef]

29. Cunningham, C.B.; Schilling, N.; Anders, C.; Carrier, D.R. The influence of foot posture on the cost of transport in humans. *J. Exp. Biol.* **2010**, *213*, 790–797. [CrossRef]

30. Daoud, A.I.; Geissler, G.J.; Wang, F.; Saretsky, J.; Daoud, Y.A.; Lieberman, D.E. Foot strike and injury rates in endurance runners: a retrospective study. *Med. Sci. Sports Exerc.* **2012**, *44*, 1325–1334. [CrossRef]

31. Chou, L.-S.; Kaufman, K.R.; Hahn, M.E.; Brey, R.H. Medio-lateral motion of the center of mass during obstacle crossing distinguishes elderly individuals with imbalance. *Gait Posture* **2003**, *18*, 125–133. [CrossRef]

32. Chou, L.S.; Kaufman, K.R.; Brey, R.H.; Draganich, L.F. Motion of the whole body's center of mass when stepping over obstacles of different heights. *Gait Posture* **2001**, *13*, 17–26. [CrossRef]

33. Hahn, M.E.; Chou, L.-S. Age-related reduction in sagittal plane center of mass motion during obstacle crossing. *J. Biomech.* **2004**, *37*, 837–844. [CrossRef] [PubMed]

34. Lyon, I.N.; Day, B.L. Predictive control of body mass trajectory in a two-step sequence. *Exp. Brain Res.* **2005**, *161*, 193–200. [CrossRef] [PubMed]

35. Zettel, J.L.; McIlroy, W.E.; Maki, B.E. Environmental constraints on foot trajectory reveal the capacity for modulation of anticipatory postural adjustments during rapid triggered stepping reactions. *Exp. Brain Res.* **2002**, *146*, 38–47. [CrossRef] [PubMed]

36. Yiou, E.; Deroche, T.; Do, M.C.; Woodman, T. Influence of fear of falling on anticipatory postural control of medio-lateral stability during rapid leg flexion. *Eur. J. Appl. Physiol.* **2011**, *111*, 611–620. [CrossRef]

37. Mezaour, M.; Yiou, E.; Le Bozec, S. Effect of lower limb muscle fatigue on anticipatory postural adjustments associated with bilateral-forward reach in the unipedal dominant and non-dominant stance. *Eur. J. Appl. Physiol.* **2010**, *110*, 1187–1197. [CrossRef]

38. Vuillerme, N.; Nougier, V.; Teasdale, N. Effects of lower limbs muscular fatigue on anticipatory postural adjustments during arm motions in humans. *J. Sports Med. Phys. Fit.* **2002**, *42*, 289–294.

39. Yiou, E.; Ditcharles, S.; Le Bozec, S. Biomechanical reorganisation of stepping initiation during acute dorsiflexor fatigue. *J. Electromyogr. Kinesiol.* **2011**, *21*, 727–733. [CrossRef]

40. Mezaour, M.; Yiou, E.; Le Bozec, S. Does symmetrical upper limb task involve symmetrical postural adjustments? *Gait Posture* **2009**, *30*, 239–244. [CrossRef]

41. Yiou, E.; Hussein, T.; Larue, J. Influence of temporal pressure on anticipatory postural control of medio-lateral stability during rapid leg flexion. *Gait Posture* **2012**, *35*, 494–499. [CrossRef] [PubMed]

42. Nouillot, P.; Bouisset, S.; Do, M.C. Do fast voluntary movements necessitate anticipatory postural adjustments even if equilibrium is unstable? *Neurosci. Lett.* **1992**, *147*, 1–4. [CrossRef]

43. Fraizer, E.V.; Mitra, S. Methodological and interpretive issues in posture-cognition dual-tasking in upright stance. *Gait Posture* **2008**, *27*, 271–279. [CrossRef] [PubMed]

44. Woollacott, M.; Shumway-Cook, A. Attention and the control of posture and gait: A review of an emerging area of research. *Gait Posture* **2002**, *16*, 1–14. [CrossRef]

45. Hussein, T.; Yiou, E.; Larue, J. Age-related differences in motor coordination during simultaneous leg flexion and finger extension: Influence of temporal pressure. *PLoS ONE* **2013**, *8*, e83064. [CrossRef]

46. Caderby, T.; Yiou, E.; Peyrot, N.; de Viviés, X.; Bonazzi, B.; Dalleau, G. Effects of Changing Body Weight Distribution on ML Stability Control during Gait Initiation. *Front. Hum. Neurosci.* **2017**, *11*, 127. [CrossRef]

47. Caderby, T.; Dalleau, G.; Leroyer, P.; Bonazzi, B.; Chane-Teng, D.; Do, M.C. Does an additional load modify the Anticipatory Postural Adjustments in gait initiation ? *Gait Posture* **2013**, *37*, 144–146. [CrossRef]

48. Yiou, E.; Hussein, T.; LaRue, J. Influence of ankle loading on the relationship between temporal pressure and motor coordination during a whole-body paired task. *Exp. Brain Res.* **2014**, *232*, 3089–3099. [CrossRef]

 *applied sciences*

Article

The Switching of Trailing Limb Anticipatory Locomotor Adjustments is Uninfluenced by what the Leading Limb Does, but General Time Constraints Remain

Félix Fiset [1,2] and Bradford J. McFadyen [1,2,3,]*

[1] Centre for Interdisciplinary Research in Rehabilitation and Social Integration, CIUSSS-CN, IRDPQ, Québec City, QC G1M 2S8, Canada; felix.fiset.1@ulaval.ca

[2] Department of Rehabilitation, Faculty of Medicine, Université Laval, Québec City, QC G1V 0A6, Canada

[3] CIRRIS-CIUSSS-CN, 525 boulevard Hamel, Québec City, QC G1M2S8, Canada

* Correspondence: brad.mcfadyen@fmed.ulaval.ca; Tel.: 418-529-9141

Received: 24 February 2020; Accepted: 23 March 2020; Published: 26 March 2020

Featured Application: The fundamental knowledge from this work has potential to eventually inform clinical interventions for gait and mobility training.

Abstract: Research shows a blend of bilateral influence and independence between leading and trailing limbs during obstacle avoidance. Recent research also shows time constraints in switching leading limb strategies. The present study aimed to understand the ability to switch anticipatory locomotor adjustments (ALAs) in the trailing limb. Ten healthy young adults (24 ± 3 years) were immersed in a virtual environment requiring them to plan and step over an obstacle that, for the trailing limb, could change to a platform, requiring a switch in locomotor strategies to become a leading limb to step onto a new surface. Such perturbations were provoked at either late planning or early execution of the initial trailing limb obstacle avoidance. Sagittal plane trailing limb kinematics, joint kinetics and energetics were measured along with electromyographic activity of key lower limb muscles. Repeated measures ANOVAs compared dependent variables across conditions. To adjust to the new environment, knee flexor power around toe-off decreased ($p < 0.001$) and hip flexor power increased ($p < 0.001$) for late planning phase perturbations, while there was only an increase in mid-swing hip flexor power ($p < 0.05$) during perturbations at execution. Findings showed no influence of the leading limb function on the ability to switch trailing limb ALAs during late planning. However, the trailing limb was also constrained for modifying ALAs once execution began, but on-going limb control strategies could be exploited in a reactive mode.

Keywords: obstacle avoidance; anticipatory control; gait; locomotion; perturbation

---

1. Introduction

Navigating complex daily environments involves anticipatory control to safely avoid obstacles. There are robust and different lower limb strategies for stepping over an obstacle compared to stepping onto a platform. The anticipatory locomotor adjustment (ALA) to step over an obstacle involves a knee flexor generation strategy around toe-off (K5) [1,2] for both the leading and trailing limbs (respectively the first and second limbs to adapt) that puts energy in the lower limb to increase both knee and hip flexion. For the trailing limb, hip flexor generation power around toe-off (referred to as H3) for limb advancement can also be delayed until mid-swing (referred to as H3D) [3,4]. The higher the obstacle, the greater H3D [3]. For a level change accommodation, the leading limb instead increases the existing hip flexor generation at toe-off to help lift the limb to the new height [5].

In addition, there are differences in the modes of control between the leading and trailing limbs. Mohagheghi et al. [6] showed that the leading limb control is modifiable online by visual information during execution, whereas the trailing limb depends on visuomotor memory. The authors also concluded that there is independent control between the leading and the trailing limbs related to this difference in the use of visual information. The visual memory of the trailing limb for obstacle avoidance has been shown to be robust enough to be retained during cessation of the task up to two minutes after the leading limb has crossed in humans [7].

The hypothesis of independence between the two lower limbs was also reinforced by unexpected perturbations due to contact with the obstacle and the fact that only the limb that made contact was modified on subsequent obstacle crossings [8]. More recently, Howe et al. [9] found that lead limb toe trajectory was affected by the combined loss of the lower half of the visual field and ankle skin local anesthesia, whereas the trailing limb was only affected by loss of lower visual input. This emphasizes differences in motor control between limbs. Finally, such differences in adapted limb control have been highlighted during memory-guided obstacle crossing wherein the obstacle was removed and only its position replaced by a contrast tape [10]. The success rate of the trailing limb was four times lower than for the leading limb. The authors suggested that greater dependence on visual sampling during the approach phase for the trailing limb could explain this degraded performance.

However, Santos et al. [11] suggested that the hypothesis of independence between leading and trailing limbs could have been observed because of the predictability of the static environments used. By changing obstacle dimension one step before clearance, the authors showed that the trailing limb can use visual information from the leading limb's obstacle crossing. In studying foot–body balance geometries for adapted gait, Dugas et al. [12] perturbed obstacle placement both early and late in the approach to and beginning of obstacle avoidance. Secondary findings of differences between clearance and maximum foot heights for the leading and trailing limbs across perturbation conditions further supported limb independence in leading and trailing limb adaptations within non-predictable environments.

While limb independence seems well accepted, the above-noted studies using dynamic environments also point to the ability to modify obstacle avoidance once it has been planned. In an early example of this, Patla et al. [13] showed that relatively quick modifications of a planned obstacle avoidance can be made when a second obstacle is introduced unexpectedly. Yet, all of this work only supports the modification of an already planned strategy. Very recently, McFadyen et al. [14] used virtual reality (VR) to instantly change environmental demands, requiring one to switch between the respective knee and hip strategies described earlier either during the late planning stage at foot contact preceding the obstacle or at the beginning of step execution for the leading limb. Results showed that leading limb strategies could be switched within the late planning stage but appeared to be "locked-in" once execution began, at least to allow the planned strategy to be initiated uninterrupted, regardless of whether substituting a knee with a hip strategy or vice versa.

However, given the presumed independence between leading and trailing limb control for obstacle avoidance noted above and differences in the underlying motor control, it is not clear whether the same time-constrained control previously shown [14] exists for the trailing limb. Therefore, the purpose of the present work was to study the ability to switch from a trailing limb obstacle avoidance strategy to a level change accommodation strategy with the leading limb during either late planning or early execution when following obstacle avoidance by the contralateral leading limb. Knowing underlying locomotor control constraints could be relevant to clinicians for mobility training in pathologic populations.

2. Materials and Methods

2.1. Participants

Ten healthy young adults (24 ± 3 years; 5 males; height = 1.75 ± 0.79 m; mass = 67.5 ± 9.0 kg) with normal or corrected-to-normal vision (Snellen chart eye test) and without musculoskeletal or neurological problems affecting their walking (self-reported) were recruited and provided their written consent to participate to the experiment. The study was approved by the ethics committee of the Centre intégré universitaire de santé et de services sociaux de la Capitale-Nationale (#2018-433).

2.2. Experimental setup and protocol

Non-collinear triads of markers were placed on the lateral aspects of the feet, lower legs and thighs and at the level of the sacrum for the pelvis. Markers were tracked by nine cameras (Vicon Motion Systems, Inc/MX T-series, Denver, USA; 100 Hz). Joint centers, segment centers of mass and radii of gyration were estimated in relation to anatomical landmarks (heels, toes, 5th metatarsal heads, medial/lateral malleoli, medial/lateral femoral condyles, left/right iliac spines and left/right anterior superior iliac spines) digitized with respect to the segment marker triads during a calibration set-up. It is unknown how limb dominance influences anticipatory locomotor adjustments (ALAs); and to control for any possible variability, the dominant limb, as determined from the Waterloo Footedness Questionnaire [15] (testing dominance during both movement and stabilization), was designated as the trailing limb for initial obstacle crossing.

Surface electromyography (EMG) was recorded (Noraxon, Scottsdale, USA; 1000 Hz) with electrodes on the mid-bellies of the rectus femoris (RF), vastus lateralis (VL), semitendinosus (ST), biceps femoris (BF), tibialis anterior (TA), medial gastrocnemius (GM) and soleus (SOL) muscles of the trailing limb. Signals were amplified (gain = 1000) and band-pass filtered at collection (15–500 Hz) and then offline (20–400 Hz). Signal strength and crosstalk were verified by isometric testing of each muscle, except for the GM and SOL, for which participants were asked to rise onto their toes while standing in order to solicit these muscles.

Participants donned a security harness used to prevent a complete fall. A starting line was determined at a point to allow the participant to make 3 steps at a self-selected natural speed with the trailing foot landing on a force plate (Bertec Corporation, Columbus, USA, 1000 Hz) corresponding to the final position before clearing an obstacle (1.6 cm^2 wooden stick; Figure 1A). A second environment, involving the same starting position, introduced a 1.5 m deep platform (Figure 1B) positioned at 0.71 m (obstacle front to platform front) from where the obstacle was placed. Both surface changes were set to heights between 15% to 16% of participants' lower limb lengths. Participants were familiarized with both physical environments performing five trials of stepping over an obstacle and five trials to step onto the platform. Five additional trials of stepping over an obstacle with the leading limb only and then stepping onto the platform with both limbs (Figure 1C) were carried out to practice the experimental conditions requiring them to adjust for an obstacle with the leading limb first and a level change to the platform with the trailing limb immediately after. For all trials, participants were asked to walk at a natural pace.

After these trials in the real world, participants were fitted with a head mounted display (HMD, Oculus Rift V1, Menlo Park, USA) tracked by three non-colinear markers. A virtual reality system similar to the one used previously [14] produced a virtual environment (VE) resembling the laboratory with the above described surface changes (Figure 1A,B). This VE was modeled in Blender and programmed with Vizard (WorldViz, Inc, Santa Barbara, USA) to be synchronized to the participant's movements. The lower limbs of a first-person avatar were also aligned with those of the participant's, using a skeletal template from the Vicon data to allow participants to also see their lower limbs in real time (refresh rate of 60 Hz). This was important for more realistic control of lower limbs during ALAs.

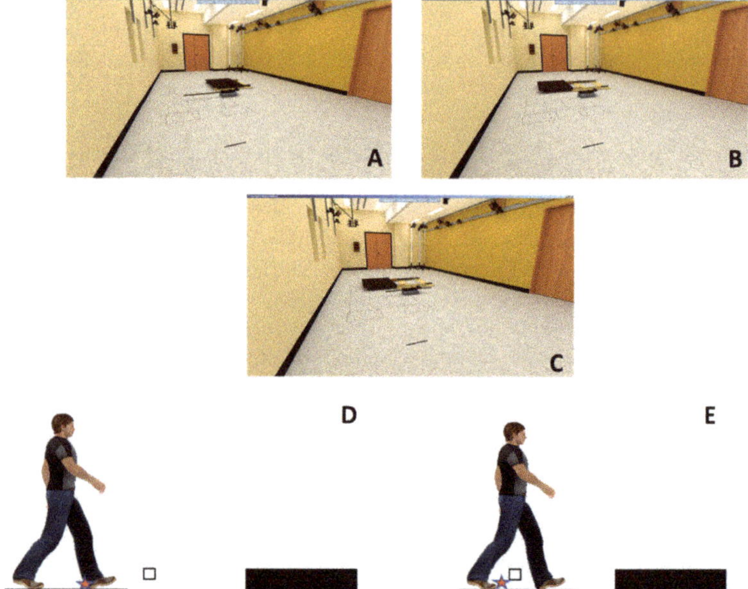

Figure 1. Virtual environment with the obstacle (**A**), platform (**B**) and the half obstacle and platform (**C**). The timing of instantaneous perturbations was triggered by a 20 N load at trailing limb heel contact (**D**; T1) or at 20 N of unloading at trailing limb toe-off (**E**; T2).

After participants were comfortably fitted in the HMD, both real obstacle and platform environments were also set in place and simultaneously presented in the VE, and participants were invited to step over the obstacle, step onto the platform and touch them both with a foot for as long as they needed to feel comfortable and to understand that the VE coincided spatially to the real environment. Participants then finally practiced both obstructed environments individually as the ALA tasks to be used in the protocol.

During data collection, participants were required to approach and step over the obstacle in the VE that could disappear and be instantaneously replaced with a platform appearing 71 cm further along the path requiring a new locomotor strategy by the trailing limb. Two experimental conditions, presented in blocks counterbalanced across participants, were related to the timing of this VE change during the trailing limb step preceding the obstacle at either foot contact (Figure 1D) or at the subsequent toe-off (Figure 1E). The environment changes were respectively triggered when the vertical ground reaction force under the trailing foot exceeded 20 N (T1 corresponding to late planning) and falling below 20 N (T2 corresponding to execution). Each block was comprised of a total of 15 trials, including 5 environment perturbations and 10 catch trials without environment perturbations. Participants were instructed to walk at a comfortable and natural rhythm and to adapt their gaits to the new environment if presented. Participants were assured that the physical environment always corresponded to the last environment that was presented in the HMD. To provide confidence to the participants, before collection of each experimental block, participants first observed the instantaneous change from obstacle to platform and then performed three practice trials for the environment, knowing it would change at the respective time. Between all data collection trials, the image in the HMD was blacked out and participants wore earphones that played pink noise at a volume determined to block all ambient noise within the laboratory to prevent them from anticipating the upcoming condition.

At the end of data collection, French versions of the simulator sickness questionnaire [16] were used to measure simulator sickness, and the Presence Questionnaire [17] and SUS questionnaire [18], both used to measure the degree of VE immersion, were administered.

2.3. Data analysis

Marker, force plate and full wave rectified EMG data were all passed through lowpass, zero lag, Butterworth filters with cut-off frequencies set respectively to 6, 50 and 100 Hz based on previous work [14,19]. Relative joint angles were calculated using Cardan rotation matrices prioritizing the sagittal plane. Maximum relative joint angles (MJRA) during the ALA swing phase, minimum foot clearance (MFC; the vertical distance between the heel and the obstacle or platform front edge) and mean COM velocity (COMv; as the mean velocity in sagittal plane of pelvis COM during the ALA stride) were also calculated. Newton–Euler inverse dynamics equations with a custom-made program were used to estimate lower limb kinetics and energetics respectively for net muscle moments of force (MMF) and muscle mechanical power (product of the MMF and relative joint angular velocities). These data were normalized to body mass, and muscle mechanical work of the trailing limb was calculated as the mathematical integral of targeted positive and negative power bursts corresponding to ankle plantar flexor generation (push-off; A2), knee extensor absorption at the end of stance (K3), the delayed knee extensor absorption around mid-swing (K3D), knee flexor generation at toe-off (K5) and hip flexor generation at both toe-off (H3) and mid-swing (H3D). Hip hiking work (HH) around toe-off was also calculated as the area under the positive power burst from the product of the vertical hip joint reaction force and vertical hip joint velocity with positive power corresponding to energy from the pelvis.

The areas under muscle bursts for BF and ST EMG activity that aligned with the K5 power burst for obstacle avoidance [1] were calculated. As in a previous study [14], we took electromechanical delays into consideration by choosing these muscle bursts to begin at the local minimum prior to the K5 power burst and continue for the duration found for the corresponding K5 power burst of that trial. This allowed EMG data to provide physiological confirmation of knee flexor strategies across conditions. To compare across participants, percentage changes in these EMG bursts from the mean unperturbed trials for each perturbation time within each environment block were also calculated. As there was no appropriate hip flexor muscle activity recorded (RF has not been shown to be a good measure of a hip flexor activity [20]), no corresponding EMG areas were presented for hip flexion.

For descriptive time series data, EMG, kinematic, kinetic and energetic (muscle mechanical power) data were normalized to 100% of stride, with toe-off fixed at 60% and ensemble averaged across trials for each condition.

2.4. Statistical analysis

Repeated measures ANOVAs for perturbation conditions were used (SPSS, IBM, New York, USA, v.24) to compare the three perturbation timings (unperturbed, early and late). Sphericity was verified for each variable with Mauchly's test of sphericity, and a Huynh–Feldt correction was applied when appropriate. When main effects were found, Bonferroni-corrected post-hoc comparisons were applied for further comparisons between conditions. A Wilcoxen signed-rank test was used to compare percentage changes in targeted knee flexor bursts between T1 and T2 perturbated conditions. Significance level was set to $p = 0.05$ (including after Bonferroni corrections) for all analyses. Effect sizes (partial eta squared, η^2) with 95 % confidence intervals (CI; {lower endpoint, upper endpoint}) are presented with ANOVA data.

3. Results

It appears that all participants tolerated the VE well given average low group scores of $0.31/3 \pm 0.19$ on the Simulator Sickness Questionnaire. In addition, average group scores of $5.51/7 \pm 0.77$ were found on the Presence Questionnaire, and they were $4.54/7 \pm 0.68$ on the SUS Questionnaire, showing that participants were relatively well immersed.

Two participants contacted the platform without tripping when it appeared at trail foot toe-off (T2). One of them contacted it twice and the other only once. These trials were kept for analysis because the participants did not lose their balance and patterns prior to platform contact did not show any outlying behavior.

Before presenting the dependent variables, the general pattern changes within the time series patterns for angular displacement, net muscle moment of force and net muscle power plots are described. Across the three joints, locomotor adjustments for the new VE were mostly seen around toe-off and during the swing phase (Figure 2). At the ankle, a greater dorsiflexion occurred at mid-swing when the obstacle changed to the platform at trailing foot toe-off (T2). The related kinetic and energetic changes were less evident in the plots because of the relatively small mass of a foot to control during swing phase compared to the whole body during the stance phase.

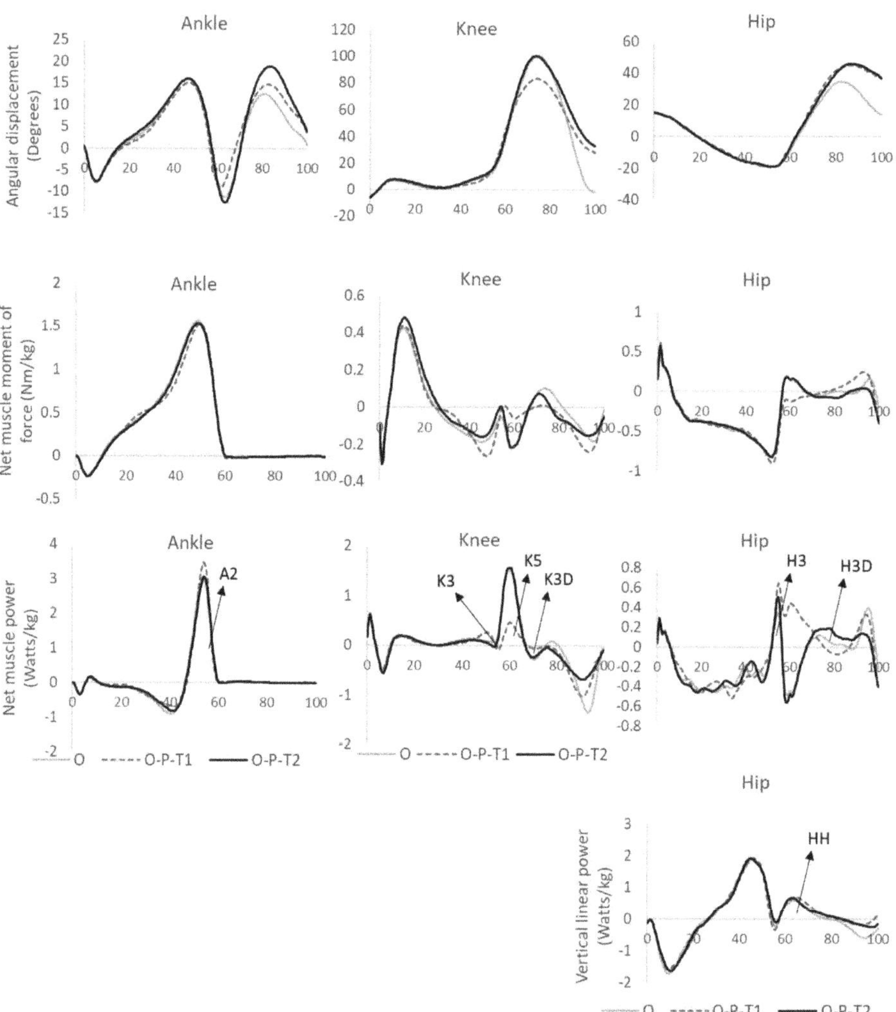

Figure 2. Average plots of angular displacement (top), net muscle moment of force (middle) and net muscle power (bottom) at the ankle (left), knee (middle) and hip (right) joints of the trailing limb during unperturbed conditions (gray line), and perturbations of obstacle to platform at T1 (O-P-T1; dashed line) and T2 (O-P-T2; black line). Gait stride was normalized to 100% with toe-off fixed at 60%.

At the knee joint, there was a decrease in knee flexion combined with a decreased knee flexor moment associated with decreased generation for the early change (T1) at toe-off. For the late perturbation (T2), kinetic and energetic curves were similar to those for the obstacle condition.

Finally, at the hip joint, hip flexion was higher when the obstacle changed to a platform, regardless of the timing of perturbations (T1 or T2). The kinetic and energetic plots showed a larger hip flexor moment combined with energy generation around toe-off (H3) when the platform appeared early (T1). For the obstacle-only condition and late perturbation (T2), there was some hip flexor energy generation at toe-off that switched to a hip extensor moment with energy absorption before returning, at mid-swing, to hip flexor power generation (H3D). There did not appear to be any change in linear hip hiking power (HH) across conditions.

To better understand how an obstacle avoidance strategy can change to level accommodation strategy, specific kinematic variables were calculated and presented in Figure 3. First, MJRA of dorsiflexion, knee and hip flexion during the obstacle crossing swing phase were all significantly different across conditions with a main effect of time ($p < 0.001$ for the three joints; ankle $\eta^2 = 0.6811$ CI{0.3308, 0.7906}; knee $\eta^2 = 0.8117$ CI{0.5647, 0.8761}; hip $\eta^2 = 0.8340$ CI{0.6107, 0.8907}). More precisely, at the ankle, dorsiflexion was greater for late perturbation than O ($p < 0.001$) and O-P-T1 ($p = 0.007$) conditions, while O-P-T1 was similar to obstacle condition ($p = 0.715$). At the knee, MJRA were only decreased for the early perturbation compared to the obstacle condition ($p < 0.001$) and to late perturbation ($p < 0.001$), but the late perturbation was not different from the obstacle condition ($p = 1.000$). At the hip, MJRA increased for both changes to Pl ($p < 0.001$ for T1 and T2), and no differences were observed between the two perturbation timings ($p = 1.000$).

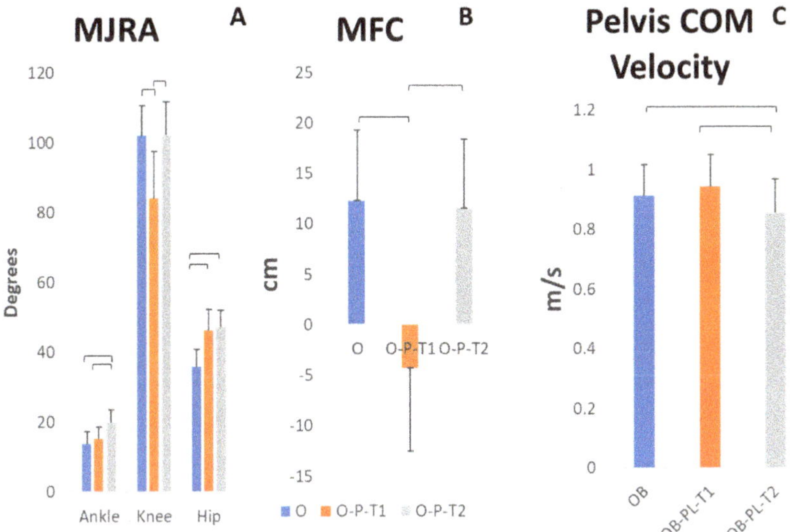

Figure 3. Average (standard deviation) of maximum joint relative angles (MJRA) during trailing limb swing phase (**A**). Minimal trailing limb foot clearance (MFC) of the obstacle—negative values correspond to the foot passing below obstacle level (**B**). Average pelvis center of mass (COM) velocity during trailing limb swing phase (**C**) across unperturbed (blue), early perturbation of obstacle to platform (O-P-T1, orange) and late perturbation (O-P-T2, gray) conditions.

Finally, MFC of the trailing foot over the obstacle had a main effect of time ($p < 0.001$; $\eta^2 = 0.7874$ CI{0.3939, 0.8743}) with decreased MFC for early VE change compared to obstacle only and to the T2 perturbation ($p < 0.001$ for both), and late perturbation was not different than obstacle only ($p = 1.000$). A main effect of time was also present for COMv ($p = 0.002$; $\eta^2 = 0.4987$ CI{0.1064, 0.6677}). COMv was significantly slower for O-P-T2 condition than for the obstacle condition ($p = 0.008$) and for O-P-T1 ($p = 0.025$), but these latter two conditions were not different from each other ($p = 0.673$). We can see in the plot of COMv (Figure 4) that participants appeared to slow down only during the swing phase.

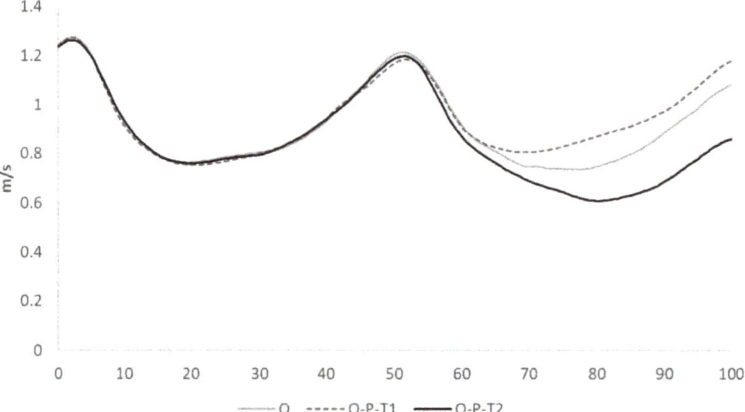

Figure 4. Average plots of pelvis COM velocity (COMv) across unperturbed (gray), early perturbation of obstacle to platform (O-P-T1, dashed line) and late perturbation (O-P-T2, black line) conditions. Gait stride was normalized to 100% with toe-off fixed at 60%.

For the accommodation strategies related to the muscle mechanical work for the power bursts around toe-off and during swing (Figure 5), ankle push-off (A2) remained unchanged across conditions ($p = 0.755$; $\eta^2 = 0.028$ CI{0, 0.1939}). Knee extensor absorption at the end of the stance phase (K3), however, showed a main effect of time ($p = 0.014$; $\eta^2 = 0.4357$ CI{0.0216, 0.6544}). Post-hoc analyses showed a tendency of K3 to be augmented for early change ($p = 0.053$), but not for late one ($p = 0.146$), and the O-P-T1 (obstacle-platform at T1) condition was not different than O-P-T2 conditions ($p = 0.141$). For knee flexor generation (K5) and the subsequent delayed knee extensor absorption (K3D), there were main effects of time for both power bursts ($p < 0.001$; $\eta^2 = 0.7894$ CI{0.3962, 0.8756}, $p = 0.001$; $\eta^2 = 0.6356$ CI{0.1944, 0.7786} respectively). Post-hoc analyses showed that K5 significantly decreased when the obstacle switched to a platform early (O-P-T1; $p < 0.001$), but not late (O-P-T2; $p = 1.000$). Significant differences were also observed between the two perturbation timings ($p < 0.001$). The delayed knee extensor absorption right after K5 (K3D) was also significantly decreased for the early change compared to the obstacle only ($p = 0.002$) and late perturbation conditions ($p = 0.019$). There was no difference between late perturbated trials and obstacle only ($p = 1.000$). Hip flexor generation around toe-off (H3) showed a general effect of time ($p < 0.001$; $\eta^2 = 0.7530$ CI{0.3192, 0.8555}) and was augmented in an early perturbation (O-P-T1; $p < 0.001$), but not for a late one (O-P-T2; $p = 0.408$). O-P-T1 and O-P-T2 conditions were also different from each other ($p = 0.003$). There was also a main effect of time for the hip flexor generation during mid-swing (H3D; $p = 0.005$; $\eta^2 = 0.4685$ CI{0.0637, 0.6578}). Post-hoc analyses showed that it was increased for the late perturbation only (O-P-T2) compared to the obstacle condition ($p = 0.037$) and to the O-P-T1 condition ($p = 0.034$) with no significant difference observed between O-P-T1 and obstacle conditions ($p = 0.379$). The positive hip linear power burst at toe-off showed a main effect of time ($p = 0.009$; $\eta^2 = 0.4830$ CI{0.0925, 0.6568}) with an increase of power burst for a late perturbation compared to the obstacle condition ($p < 0.001$). The obstacle condition was not different from O-P-T1 ($p = 0.519$). O-P-T1 was not different from O-P-T2 ($p = 0.242$).

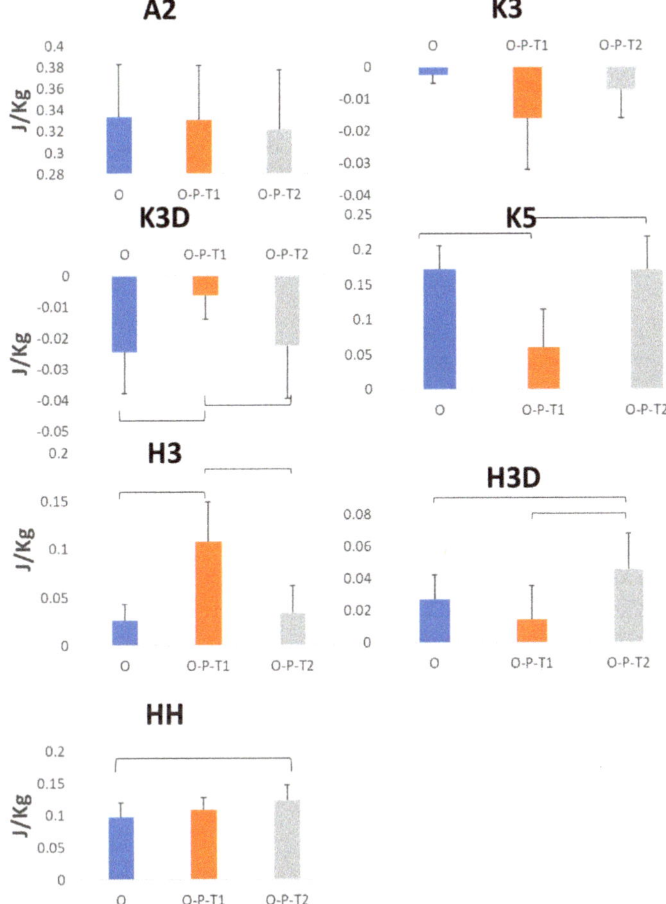

Figure 5. Average (standard deviation) of net muscle work corresponding to ankle push-off (A2); knee extensor absorption around (K3); after toe-off (K3D); after knee flexor generation (K5); hip flexor generation around toe-off (H3); hip flexor generation at mid-swing (H3D); and hip hiking around at toe-off (HH) across the unperturbed (O; blue), early perturbation of obstacle to platform (O-P-T1, orange) and late perturbation (O-P-T2, gray) trials.

The related muscle activity is presented in Figure 6 for a participant with patterns representing what was observed across all participants, specifically for the areas of interest for statistical testing. EMG plots of tibialis anterior showed increased activity in swing during late perturbation (T2). Semitendinosus (ST) and the biceps femoris (BF) activity showed higher activity at toe-off, corresponding to the K5 muscle power burst, for the obstacle and late perturbation (T2) conditions. All other muscles appeared to be qualitatively similar across conditions. To confirm the qualitative observations for greater ST and BF activity at toe-off corresponding to the K5 power burst, a percentage of change relative to obstacle condition was calculated for each perturbation condition (Figure 7). For both targeted bursts of ST and BF, muscle activity for the early perturbation condition was lower than for the late one ($p = 0.017$ for ST; $p = 0.007$ for BF).

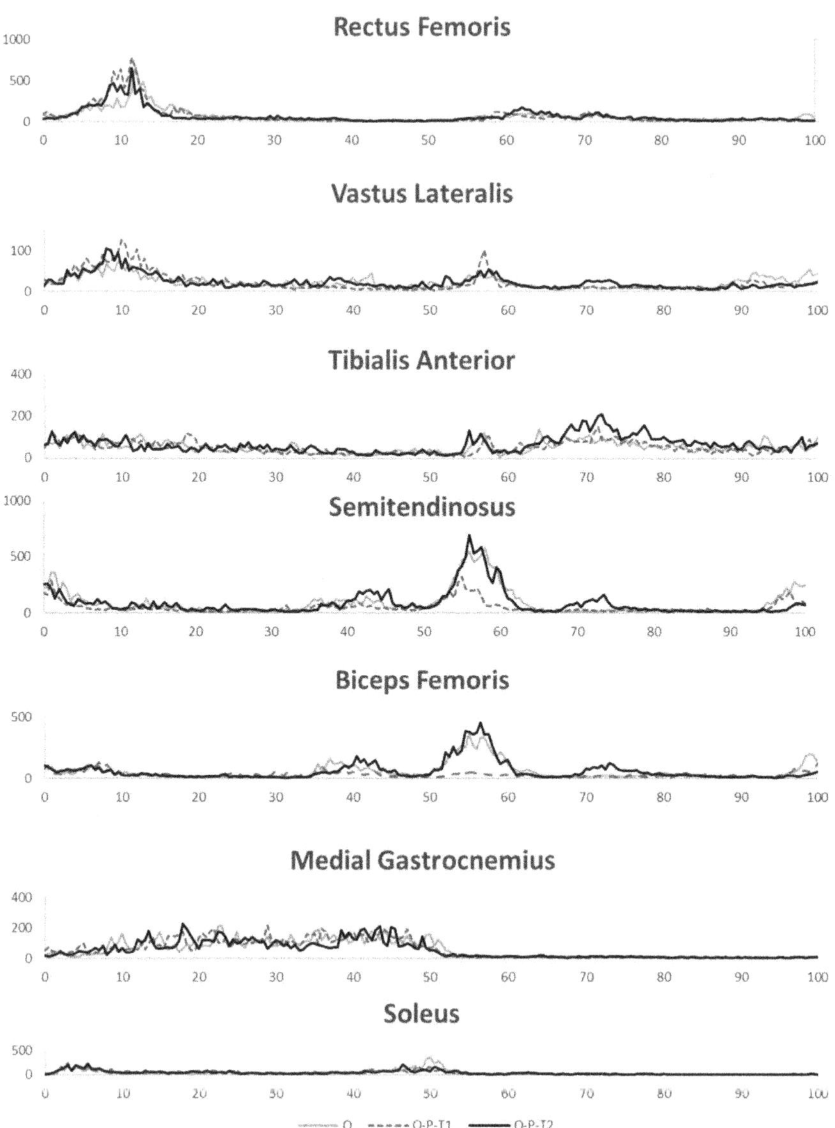

Figure 6. Surface EMG of seven lower limb muscles of one representative subject across conditions of unperturbed (O; gray), early perturbation of obstacle to platform (O-P-T1, dashed line) and late perturbation (O-P-T2, black line) conditions.

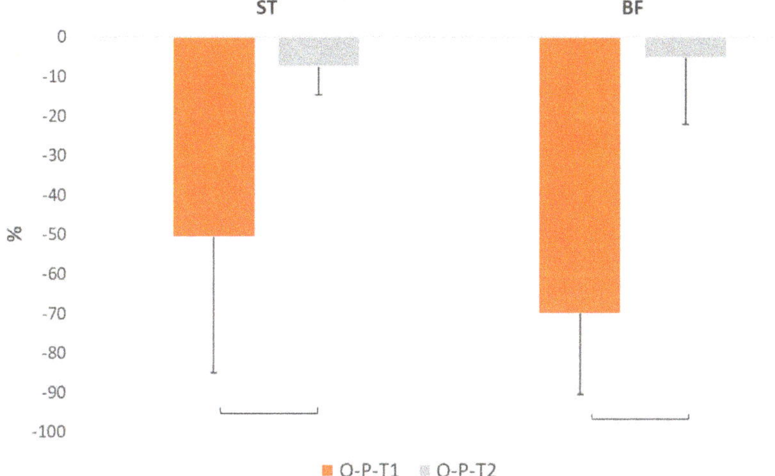

Figure 7. Percentage changes in the area under the curve of semitendinosus (ST) and biceps femoris (BF) EMG activity corresponding to the K5 power burst across the early perturbation of obstacle to platform (O-P-T1, orange) and late perturbation (O-P-T2, gray) conditions.

4. Discussion

ALAs are essential to maintaining equilibrium within the dynamic environments found daily. Earlier work has shown the capacity to add onto an already planned ALA online [13], while a more recent study has revealed that one can also switch from one ALA to another for the leading limb, but only before ALA execution begins [14]. Yet, there is ample research to suggest differences in sensorimotor control between leading and trailing limbs [6,7]. The present study involved a perturbation requiring participants to switch from an initially planned obstacle avoidance ALA in the trailing limb to a leading limb level change accommodation ALA. This strategy switch created a contrast between what the leading and trailing limb planned and executed. As previously observed by McFadyen et al. [14], it was also found here that a trailing limb ALA can be switched within the late planning stage, but not once execution begins. In addition, it was found that, differently than for the leading limb, the pre-existing hip flexor work at the transition to swing phase could be exploited after initial execution to accommodate the need to step up to a higher surface level.

The current observation of the ability to switch trailing limb ALAs during late planning (T1) at the same time as the leading limb executes obstacle avoidance, supports the fact that the leading limb obstacle avoidance function was of insignificant influence on the trailing limb successfully switching to a new ALA. Such an ALA switch must, therefore, be driven by visual information of the environmental change. This ability to use visual information for trailing limb control has already been suggested for a static obstacle avoidance situation when vision was manipulated during planning and approach [6]. In addition, Santos et al. [11] showed this in a more dynamic situation for changing obstacle heights at a trailing limb foot contact (similar timing to T1 in the present study). However, previous studies involved only modifying an already planned trailing limb strategy. In the present work, ALAs after switching were very different for both limbs, and the leading limb knee flexor strategy to step over the obstacle did not affect the capacity to switch ALAs within the trailing limb. This further supports independent control between limbs [6,8,12]. It appears that the previously suggested feedforward influence of leading limb function on trailing limb control [11] is thus context specific and likely only true when both limbs perform the same ALA (i.e., both stepping over an obstacle).

Despite this lack of influence of leading limb control at the early switch (T1) of ALAs in the trailing limb, it was observed that the initially planned obstacle avoidance strategy could not be canceled out at the beginning of its execution (T2). As noted above, this is probably not due to influence from the leading limb and is more likely for the same reason discussed by McFadyen et al. [14] for switching leading limb ALAs. Specifically, a locomotor strategy is not immediately modifiable once it is launched regardless of being a leading or trailing limb. This inability to switch as execution begins (T2) was clearly expressed in the lower limb kinematic, kinetic and EMG patterns with large effect sizes as shown by the partial eta squared values for most variables.

Yet, despite the fact that ALAs could not switch at the beginning of the execution, some modifications in locomotor patterns were seen at mid-swing. More precisely, for the T2 perturbation, the trailing limb increased the already existing mid-swing delayed hip flexor generation (H3D) seen for trailing limb obstacle avoidance. This was likely to compensate for the inability to increase the pre-planned hip flexor generation at toe-off (H3) for obstacle avoidance. This may be similar to what Patla et al. [13] proposed as an ability to exploit an on-going plan (in their case for knee flexion of the leading limb) and modulate it online at a later latency. That is, the delayed hip flexor generation at mid-swing would was already planned for trailing limb obstacle avoidance, but then further exploited to replace lost H3 strategy for a platform ALA. In a previous study [14], when the leading limb switched at execution, a hip extensor burst appeared during swing phase 300 ms after the environment perturbation. This timing is similar to the fine-tuning stage of gait adaptations [6] and was considered to be more a reactive response rather than a strategy substitution. The present study shows a similar reaction time for the trailing limb for the increased H3D in mid-swing. This reactive strategy would also be assisted by a slight increase in hip hiking power (HH) as well representing a frontal plane contribution. Yet, this latter change on its own was at the same time as when the initially planned obstacle ALA was first carried out. Thus, it would appear that HH is autonomous to the lower limb ALA and can be added on at early latency, even when the newly erroneous obstacle ALA must be initiated for the late perturbation at execution. The observed significant changes in H3D and HH power bursts are supported by good effect sizes with approximately half the variance explained by the conditions. However, with the larger confidence intervals, some caution is required.

Results at the ankle joint also showed greater mid-swing dorsiflexion and TA activity for the late perturbation condition (T2) but without changes immediately following toe-off. This again supports no immediate means to switch an ALA once executed, but an ability to adapt online at a later latency as a reactive strategy. Why the ankle creates more dorsiflexion than the other conditions even if it appears to finish the stride on the new surface level with the same relative joint angles between T1 and T2 perturbations is unclear. It could be a preventative overshoot to avoid tripping in this reaction to the new environment.

It is difficult from the present study to deduce the underlying neural control. With respect to the same observations for lead limb ALA switching, McFadyen et al. [14] offered some conjecture from the literature. They noted that perhaps motor plans involving posterior parietal cortex (PPC) [21] could lead to muscle synergies set in the motor cortex through pyramidal tract neurons [22]. These synergies may be initially locked-in upon execution, but can also be modified at a later point, perhaps through reticular spinal neuron pathways suggested to be involved in pattern modifications [23]. In the present study, wherein participants had to switch from a trailing limb obstacle crossing strategy to a leading limb level accommodation strategy, it is highly likely that similar neural control is involved. However, while PPC would likely be involved in trailing limb motor planning, it appears to also be important for the visual memory of the obstacle for this limb [24,25]. However, this visual memory would be useless in the current ALA switch. PPC cell discharge is related to bilateral function rather than any specific limb. Thus, it could perhaps reorganize motor planning such that cells related to trailing limb function would be inhibited while other cells related to the new leading limb function would need to newly discharge in relation to the new position of the limb relative to the platform that appeared. The premotor cortex (PMC) might also be involved in this transition from trailing to leading limb roles given recent literature suggesting some limb-dependent cells of PMC in cats that contribute the selection of the first limb to step over and the related interlimb coordination [26]. How such neural pathways are be related to the constraints in anticipatory and reactive control of trailing and leading limb function for ALAs discussed above will require more study.

Although more research is required to understand any clinical relevance of these present findings, given the differences in the underlying control of the leading and trailing limbs, it could be clinically relevant to train each side in leading and trailing functions to improve dynamic balance. As to the inability to switch ALAs once execution begins, but with online adaptation of hip flexors to assure final and safe accommodation, perhaps reactive training could be another point of focus.

The present study is not without limitations. Virtual reality is a great way to control environmental factors and produce protocols not possible in the physical environment. However, it is important participants see it as being as real as possible. The results showed acceptable presence levels, but as discussed in previous similar work [14], VR did also result in some behavior (higher foot clearance and maintaining some knee flexor generation power (K5) for a platform accommodation) that is different from what might be expected in an only physical environment. However, in this within-subject design, we are confident that we were able to show general control changes between the ALA switching conditions. The present study was also limited in number of participants and to healthy young adults only. Further work is needed to understand such adaptability constraints and their consequences for safe ALAs in other populations (e.g., older adults, populations with neurological impairments).

5. Conclusions

The present study involved switching from a planned trailing limb obstacle avoidance involving a knee flexor strategy to a leading limb surface level accommodation involving a hip flexor strategy. The findings showed that, despite presumed feedforward control involving the leading limb during the original plan, such leading limb influence appears context specific, and the trailing limb can benefit from ongoing visual information. However, there also appears to be a limb independent rule that anticipatory locomotor adjustments can be switched during the planning phase, but at execution, the planned pattern must be initiated. Yet, the control system appears able to exploit other aspects of the ongoing lower limb control strategies, such as the hip flexor power at mid-swing, to accommodate to the new locomotor demands of the environment. Underlying neural mechanisms for such flexibility and constraints in anticipatory locomotor adjustments are only speculative for now and require further study. More research is also needed to better understand the implications of such control for locomotor control changes with age and impairments.

Author Contributions: Conceptualization, methodology, data curation: B.J.M. and F.F.; formal analysis, writing—original draft preparation: F.F.; writing—review and editing, supervision, project administration, funding acquisition: B.J.M. All authors have read and agreed to the published version of the manuscript.

Funding: Research was funded Natural Sciences and Engineering Research Council of Canada (RGPIN/191782-2017).

Acknowledgments: The authors thank N. Robitaille (programming), S. Forest (infrastructure construction), G. St-Vincent and F. Dumont (motion capture) for valuable technical assistance and Dr. Jean Leblond for statistical consulting. We also thank Dr. Philippe Jackson for the use of the HMD.

Conflicts of Interest: The authors declare no conflict of interest.

References

1. McFadyen, B.; Winter, D. Anticipatory locomotor adjustments during obstructed human walking. *Neurosci. Res. Commun.* **1991**, *9*, 37–44.
2. Patla, A.E.; Prentice, S.D. The role of active forces and intersegmental dynamics in the control of limb trajectory over obstacles during locomotion in humans. *Exp. Brain Res.* **1995**, *106*, 499–504. [CrossRef]
3. Niang, A.E.; McFadyen, B.J. Adaptations in bilateral mechanical power patterns during obstacle avoidance reveal distinct control strategies for limb elevation versus limb progression. *Motor Control* **2004**, *8*, 160–173. [CrossRef]

4. Rietdyk, S. Anticipatory locomotor adjustments of the trail limb during surface accommodation. *Gait Posture* **2006**, *23*, 268–272. [CrossRef]
5. McFadyen, B.J.; Carnahan, H. Anticipatory locomotor adjustments for accommodating versus avoiding level changes in humans. *Exp. Brain Res.* **1997**, *114*, 500–506. [CrossRef]
6. Mohagheghi, A.A.; Moraes, R.; Patla, A.E. The effects of distant and on-line visual information on the control of approach phase and step over an obstacle during locomotion. *Exp. Brain Res.* **2004**, *155*, 459–468. [CrossRef]
7. Lajoie, K.; Bloomfield, L.W.; Nelson, F.J.; Suh, J.J.; Marigold, D.S. The contribution of vision, proprioception, and efference copy in storing a neural representation for guiding trail leg trajectory over an obstacle. *J. Neurophysiol.* **2012**, *107*, 2283–2293. [CrossRef] [PubMed]
8. Rhea, C.K.; Rietdyk, S. Influence of an unexpected perturbation on adaptive gait behavior. *Gait Posture* **2011**, *34*, 439–441. [CrossRef] [PubMed]
9. Howe, E.E.; Toth, A.J.; Bent, L.R. Online visual cues can compensate for deficits in cutaneous feedback from the dorsal ankle joint for the trailing limb but not the leading limb during obstacle crossing. *Exp. Brain Res.* **2018**, *236*, 2887–2898. [CrossRef] [PubMed]
10. Heijnen, M.J.; Romine, N.L.; Stumpf, D.M.; Rietdyk, S. Memory-guided obstacle crossing: More failures were observed for the trail limb versus lead limb. *Exp. Brain Res.* **2014**, *232*, 2131–2142. [CrossRef]
11. Santos, L.C.; Moraes, R.; Patla, A.E. Visual feedforward control in human locomotion during avoidance of obstacles that change size. *Motor Control* **2010**, *14*, 424–439. [CrossRef] [PubMed]
12. Dugas, L.P.; Bouyer, L.J.; McFadyen, B.J. Body-foot geometries as revealed by perturbed obstacle position with different time constraints. *Exp. Brain Res.* **2018**, *236*, 711–720. [CrossRef] [PubMed]
13. Patla, A.; Beuter, A.; Prentice, S. A two stage correction of the limb trajectory to avoid obstacles during stepping. *Neurosci. Res. Commun.* **1991**, *8*, 153–159.
14. McFadyen, B.J.; Fiset, F.; Charette, C. Substituting anticipatory locomotor adjustments online is time constrained. *Exp. Brain Res.* **2018**, *236*, 1985–1996. [CrossRef]
15. Elias, L.J.; Bryden, M.P.; Bulman-Fleming, M.B. Footedness is a better predictor than is handedness of emotional lateralization. *Neuropsychologia* **1998**, *36*, 37–43. [CrossRef]
16. Kennedy, R.; Lane, N.; Berbaum, K.S. Simulator sickness questionnaire: An enhanced method for quantifying simulator sickness. *Int. J. Aviat. Psychol.* **1993**, *3*, 203–220. [CrossRef]
17. Witmer, B.G.; Singer, M.J. Measuring presence in virtual environments: A presence questionnaire. *Presence* **1998**, *7*, 225–240. [CrossRef]
18. Slater, M.; Steed, A. A Virtual Presence Counter. *Presence* **2000**, *9*, 413–434. [CrossRef]
19. Winter, D.A.; Sidwall, H.G.; Hobson, D.A. Measurement and reduction of noise in kinematics of locomotion. *J. Biomech.* **1974**, *7*, 157–159. [CrossRef]
20. Neptune, R.R.; Zajac, F.E.; Kautz, S.A. Muscle force redistributes segmental power for body progression during walking. *Gait Posture* **2004**, *19*, 194–205. [CrossRef]
21. Andujar, J.E.; Lajoie, K.; Drew, T. A contribution of area 5 of the posterior parietal cortex to the planning of visually guided locomotion: Limb-specific and limb-independent effects. *J. Neurophysiol.* **2010**, *103*, 986–1006. [CrossRef] [PubMed]
22. Yakovenko, S.; Krouchev, N.; Drew, T. Sequential activation of motor cortical neurons contributes to intralimb coordination during reaching in the cat by modulating muscle synergies. *J. Neurophysiol.* **2011**, *105*, 388–409. [CrossRef] [PubMed]
23. Drew, T.; Prentice, S.; Schepens, B. Cortical and brainstem control of locomotion. *Prog. Brain Res.* **2004**, *143*, 251–261. [CrossRef]
24. McVea, D.A.; Taylor, A.J.; Pearson, K.G. Long-lasting working memories of obstacles established by foreleg stepping in walking cats require area 5 of the posterior parietal cortex. *J. Neurosci.* **2009**, *29*, 9396–9404. [CrossRef] [PubMed]

Appl. Sci. **2020**, *10*, 2256

25. Wong, C.; Pearson, K.G.; Lomber, S.G. Contributions of Parietal Cortex to the Working Memory of an Obstacle Acquired Visually or Tactilely in the Locomoting Cat. *Cereb. Cortex* **2018**, *28*, 3143–3158. [CrossRef] [PubMed]
26. Nakajima, T.; Fortier-Lebel, N.; Drew, T. Premotor Cortex Provides a Substrate for the Temporal Transformation of Information During the Planning of Gait Modifications. *Cereb. Cortex* **2019**. [CrossRef]

 *applied sciences*

Article

Age-Related Differences in Muscle Synergy Organization during Step Ascent at Different Heights and Directions

Remco J. Baggen [1,2], Jaap H. van Dieën [2], Evelien Van Roie [1], Sabine M. Verschueren [3], Georgios Giarmatzis [4], Christophe Delecluse [1] and Nadia Dominici [2,*]

[1] Department of Movement Sciences, Physical Activity, Sports and Health Research Group, KU Leuven, 3001 Leuven, Belgium

[2] Department of Human Movement Sciences, Faculty of Behavioural and Movement Sciences, Vrije Universiteit Amsterdam, Institute for Brain and Behavior Amsterdam & Amsterdam Movement Sciences, 1081 BT Amsterdam, The Netherlands

[3] Department of Rehabilitation Sciences, Research Group for Musculoskeletal Rehabilitation, KU Leuven, 3001 Leuven, Belgium

[4] Department of Electrical and Computer Engineering, Visualization & Virtual Reality Group, University of Patras, 26504 Rio Achaia, Greece

[*] Correspondence: n.dominici@vu.nl; Tel.: +31-20-59-88-591

Received: 30 January 2020; Accepted: 5 March 2020; Published: 14 March 2020

Abstract: The aim of this study was to explore the underlying age-related differences in dynamic motor control during different step ascent conditions using muscle synergy analysis. Eleven older women (67.0 y ± 2.5) and ten young women (22.5 y ± 1.6) performed stepping in forward and lateral directions at step heights of 10, 20 and 30 cm. Surface electromyography was obtained from 10 lower limb and torso muscles. Non-negative matrix factorization was used to identify sets of (n) synergies across age groups and stepping conditions. In addition, variance accounted for (VAF) by the detected number of synergies was compared to assess complexity of motor control. Finally, correlation coefficients of muscle weightings and between-subject variability of the temporal activation patterns were calculated and compared between age groups and stepping conditions. Four synergies accounted for >85% VAF across age groups and stepping conditions. Age and step height showed a significant negative correlation with VAF during forward stepping but not lateral stepping, with lower VAF indicating higher synergy complexity. Muscle weightings showed higher similarity across step heights in older compared to young women. Neuromuscular control of young and community-dwelling older women could not be differentiated based on the number of synergies extracted. Additional analyses of synergy structure and complexity revealed subtle age- and step-height-related differences, indicating that older women rely on more complex neuromuscular control strategies.

Keywords: EMG; muscle synergies; forward stepping; lateral stepping; aging; neural control

---

1. Introduction

Aging is associated with gradual changes in neuromuscular control [1,2]. Eventually, these changes can have a major impact on fall risk, mobility and independence in older adults [3–5], which may be exacerbated in post-menopausal women due to accelerated muscle wasting [5]. One major challenge with assessing changes in neuromuscular control in healthy community-dwelling older adults is that, due to the gradual nature of age-related neuromuscular deterioration, pre-clinical changes in neuromuscular control of everyday tasks may go undetected [6]. In healthy older adults, changes in neuromuscular control can be revealed by increasing task challenge. For example, by increasing gait speed, cadence, step length, and step height [7–9].

Step ascent, a functional task in daily life, provides a challenge that can easily be modified by increasing step height and has been demonstrated to be an effective training stimulus to improve muscle volume and functional ability in older women [10,11]. However, the effects of aging and task challenge on the neuromuscular control strategies behind step ascent have not yet been thoroughly explored. In a previous study conducted with older women, we found that peak activation of several major lower limb muscles occurred during the ascent phase of stepping and that there is a positive dose-response relationship between step height and peak muscle activation [10]. However, the increase in step height was also accompanied by increased between-subject variance of peak activation magnitudes [10]. This could be attributable to a tendency of older adults to modulate their motor strategies (e.g., shifting joint moment generation from the knee to the ankle), in order to operate within the limits of their physical capacity when ascending steps [12,13]. Additionally, it is unknown if other observed age-related changes in neuromuscular control strategies, such as increased antagonistic co-contraction of quadriceps and hamstrings to help maintain postural control during dynamic tasks [14–17], may interact with increased task challenge.

One way to assess neuromuscular control is through muscle synergy analysis. Previous studies have found that low-dimensional sets of motor modules, also known as muscle synergies, can be used to reconstruct muscle activation patterns during various motor tasks [18–22]. These synergies are composed of groups of muscles that are assumed to be activated by a single neural command [23]. It is thought that the central nervous system employs this modular organization to reduce the large number of degrees of freedom inherent to the redundancy of the human musculoskeletal system [24], and to allow for flexible but accurate response selection during motor tasks [25]. However, some researchers have argued that modular recruitment of muscles might represent predetermined control strategies and could merely be an effect of task constraints or optimized performance criteria, rather than reflecting neural control strategies employed by the central nervous system [23,26]. Regardless of the mechanisms underlying modular organization of muscle activation, extracting muscle synergies from electromyographic (EMG) signals can provide important insights about neuromuscular control strategies used to perform functional tasks [27]. In older adults with a history of falls, declines in neuromuscular control are reflected in a decreased number of extracted synergies from walking tasks that challenge dynamic balance [6]. A decreased number of synergies is indicative of decreased complexity of motor control or a decreased motor repertoire. These underlying changes in synergy complexity can be quantified using variance account for (VAF) by a given set of extracted synergies and defining the number of synergies required to adequately reconstruct the original EMG signals (indicated by a-priori or a-posteriori set thresholds for VAF) [28]. For example, high VAF by a limited number of synergies represents decreased complexity of motor control, which is often associated with neuromuscular pathologies such as cerebral palsy and stroke [6,27,29] and characterized by increased levels of co-activation between individual muscles [30]. However, healthy aging is associated with a more gradual process of physical decline and thus changes in complexity of motor control may not manifest in a decreased number of synergies [8]. Consequently, age-related changes in motor control may be better reflected by comparisons of VAF for a fixed number of synergies [29,31,32] or by assessing changes in spatio-temporal organization of muscle synergies, such as altered module composition and shifts in activation patterns [2,6]. Differences in spatio-temporal organization within a stable number of synergies can arguably be considered more subtle than differences in the total number of synergies as they tend to reflect compensatory or alternative motor strategies in order to overcome increased or altered task challenges, whereas a decreased number of synergies is usually used as an indication of neural impairments. Although it is currently unknown how aging and task intensity affect muscle synergy organization during stair ascent, analyses of underlying differences could provide a basis to improve detection of pre-clinical age-related deterioration in neuromuscular control and more effectively target fall prevention programs at individuals most at risk.

Therefore, the purpose of this study was to explore muscle synergy recruitment during step ascent in forward and lateral directions and with incremental step heights in young and older adults.

Specifically, we aimed to assess the effects of age-related changes, task intensity, and their interaction on complexity and organization of motor control by comparing the number of extracted synergies, variance accounted for (VAF) by a fixed number of synergies, and spatio-temporal characteristics of the extracted synergies across conditions.

2. Materials and Methods

2.1. Participants

Eleven older women (67.0 y ± 2.5, 161.3 cm ± 4.9, 64.4 kg ± 6.8) and ten young women (22.5 y ± 1.6, 168.9 cm ± 1.7, 64.2 kg ± 7.9) were recruited for this study. Potential participants were excluded if they suffered from neurological or motor disorders, impaired balance control, or if they had been involved in a structured training program in the last 6 months prior to participation in the study. This study was approved by the Human Ethics Committee of KU Leuven in accordance with the Declaration of Helsinki and registered with the Clinical Trial Center UZ Leuven (S56405). All participants signed informed consent prior to participation in the study.

2.2. Experimental Protocol

Participants performed a series of stepping tasks consisting of stepping onto a wooden block in forward direction (Fstep) and in lateral direction (Lstep). Task intensity was determined by the height of the block (10, 20 and 30 cm) [10]. A previous study by Singh and Latash revealed that muscle fatigue can cause higher variability of muscle activation patterns and composition of synergy components [33]. In contrast with more common tasks used for muscle synergy analyses (e.g., gait and perturbation trials), step ascent with incremental heights presents a relatively high physical challenge for older adults [10], thus increasing the risk of fatigue with a high number of repetitions. Pilot testing showed that, despite regular breaks, execution of more than three repetitions per trial, combined with the number of trials, was already quite fatiguing for the older women in this study. Therefore, each trial consisted of three repetitions to avoid confounding effects of fatigue on synergy composition. Every repetition was performed with the dominant leg first, followed by the trailing leg, and ending in double support on top of the block. Left-right dominance was determined during familiarization by noting with which foot participants preferred to take the first step. As a control question, participants were asked with which foot they would prefer to kick a ball [34]. Only two participants (both young women) showed left-side dominance. Step ascent was assessed in both forward and lateral directions, as these are functional tasks that require simultaneous coordination of the hip, knee and ankle musculature and can be challenging for older adults [12,13]. Stepping up in forward direction shows close functional resemblance to stair-climbing [34], which is an activity of daily life associated with high fall risk in older adults [35]. Stepping up in lateral direction is a less common task for older adults and can provide an additional challenge to the hip abductors, which play an important role in medio-lateral balance control [36]. The speed of task execution was controlled by a metronome at 1 second for ascent, 1 second stance, and 1 second descent to avoid differences in muscle activation due to explosive movements.

2.3. Kinematics

Kinematics were recorded at a sampling rate of 100 samples/s by means of 3D motion capture cameras (Vicon®, Oxford Metrics, Oxford, UK). Infrared reflective markers (diameter 14 mm) were placed on both heels and the sacrum. Only data from the ascent phase were used for analysis. Based on the kinematic data, the start of the ascent phase was defined at 200 ms prior to initial vertical displacement of the leading heel marker beyond 2× the standard deviation obtained during normal stance. The end of the ascent phase was defined at 500 ms after maximum knee extension, defined as the maximum relative distance between the heel and sacrum. Two sub-phases (foot clearance and pull-up) of step ascent were defined using the dominant heel marker trajectories. The shift from foot clearance to pull-up phase was defined at 100 ms after vertical displacement of the leading heel marker dropped below 2× the standard deviation obtained during normal stance. Kinematic data from the trailing heel were included to detect and eliminate trials with undesirable events, such as the toe getting caught on the edge of the step. No such events were detected.

2.4. Electromyography

Muscular activation was collected unilaterally from ten lower limb and trunk muscles on the dominant side using surface electromyography (EMG) (Aurion®, ZeroWire, Milan, Italy) sampled at 1000 samples/s. Activation was recorded from the following 10 muscles: the tibialis anterior (TA), the lateral head of the gastrocnemius (GL), soleus (SOL), vastus lateralis (VL), rectus femoris (RF), biceps femoris (BF), semitendinosus (ST), gluteus maximus (GMAX), gluteus medius (GMED), and the erector spinae (ERS), in accordance with SENIAM guidelines [37]. The skin was shaved and thoroughly rubbed with an alcohol swab to ensure optimal conductivity. Bi-polar Ag/Ag-Cl electrodes (Ambu® BlueSensor P, Ballerup, Denmark) were then placed on the belly of the muscles with an inter-electrode distance of 25 mm. Sampling of kinematic and EMG data was synchronized.

2.5. Synergy Extraction and Data Analyses

All EMG and kinematic data were processed using custom MATLAB scripts (MATLAB R2014b, MathWorks®, Natick, MA, USA). The EMG signals were high-pass filtered with a 1st order Butterworth filter with a cut-off at 20 Hz, full-wave rectified and smoothed with a 0.1-s moving average window [10,38]. EMG signals from forward and lateral stepping were normalized to the respective maximum activation obtained over all trials performed in the congruent direction so that activation could not exceed 100% [39,40]. The EMG signals were time-synchronized with the kinematically defined start and end points and subsequently normalized over time to define 0%–100% of the step cycle. Finally, because EMG data were only collected during step ascent and therefore represented intervals, rather than continuous activation patterns (as would be the case during gait trials), signals were averaged over the three repetitions performed in each condition. The choice to average signals rather than concatenating them was made in order to obtain the best reconstruction quality for our relatively short intervals, at the risk of losing information on step-to-step variability [40].

Muscle synergies were extracted from the individual average EMG data matrix using non-negative matrix factorization (NNMF). NNMF calculates muscle synergies (W) and their relative temporal activation patterns (C), resulting in muscle activations being represented as W × C + e. W represents the relative muscle co-activation, defined as the relative weight of each muscle per synergy, and is constructed as an $m \times n$ matrix where m is the total number of muscles and n is the selected number of synergies. C represents the temporal activation patterns and is constructed as an $n \times t$ matrix where t represents the number of data points over normalized time (100 per individual trial) and e is the residual error matrix [22,41]. The algorithm was repeated 1000 times for each subject to avoid local minima. The appropriate number of synergies was defined using two criteria. First, using an iterative process where the number of synergies varied between 1 and 10, the minimum number of synergies was selected based on the number required to reach ≥ 85% of group-averaged variance accounted for (VAF). As an additional local criterion, synergies had to account for ≥75% VAF for each individual muscle [42]. This double criterion approach was selected in order to adequately reproduce relevant features of the synergy compositions. VAF was defined as the uncentered Pearson correlation coefficient between W × C and the EMG amplitude time series. To compare spatio-temporal characteristics, individual synergies obtained from different subjects were pooled and matched based on the correlation of their structure (muscle weightings in each synergy of W) using a custom cluster analysis algorithm [43]. If a synergy showed equal correlation to more clusters, that synergy remained in the pool it was initially assigned to. Each synergy of W and C was subsequently averaged over all participants in that age group. For comparisons between age groups, the group-averaged synergies were also matched based on their structure using cluster analysis. Finally, we computed time-averaged standard deviations of the synergy activation patterns, with a fixed number of synergies, to assess if age-related differences in group-averaged VAF could be affected by between-subject variability of temporal activation patterns.

2.6. Statistical Analyses

Statistical analyses were performed with SPSS (IBM® SPSS v23 Statistics for Windows, Armonk, NY, USA). For both step directions, two-way repeated measures ANOVA (age × synergy number) for VAF was used to assess the interaction effect of age and number (n) synergies on VAF [39]. The number of synergies needed to adequately reconstruct the EMG signals was then determined using the iterative process described in the previous paragraph. Subsequently, two-way repeated measures ANOVA (age × step height) was used to assess the main and interaction effects of age and step height on synergy complexity, which was defined as VAF obtained with n synergies [29] fixed to four. Data were tested for normality with a Kolmogorov–Smirnov test. Sphericity was checked using Mauchly's test for sphericity. If a significant main effect was found, post-hoc tests comparing differences between age groups were performed using independent samples t-tests, while related-samples t-tests were used to compare differences per step height and synergy number. Alpha was set to 0.05 for all statistical tests.

Similarity of muscle synergies (based on muscle weightings, W) was quantified based on Pearson's correlation coefficients where $r > 0.7$ represented significant similarity and $r > 0.45$ represented marginal similarity [20,44]. Correlated synergies within age groups between step heights, and between age groups for each step height, were considered to be shared synergies, while non-correlated synergies were considered task-specific or age-related synergies [44]. Differences in muscle contributions to each synergy (W) between age groups were checked using Mann–Whitney U tests.

3. Results

Time-normalized kinematic data from the heel and pelvic markers showed high similarity ($r > 0.9$) in averaged vertical displacement over time between young and older women for all step heights. An example of the averaged vertical displacement patterns and standard deviations at 30 cm step height is provided in Figure 1.

Two-way ANOVA (age × synergy number) of VAF revealed significant main effects of synergy number for all step directions and heights ($p < 0.001$), but no interaction effects with age ($p \geq 0.05$). A significant main effect of age ($p = 0.028$) was detected only for lateral stepping at 30 cm. For the group-averaged VAF, four muscle synergies were required to achieve a threshold level of 85% VAF for reconstructed signals across both age groups, step directions and step heights (Figure 2). This indicates that age, step direction and step height did not affect the number of synergies needed to reconstruct the EMG data. Consequently, the following results were obtained assuming $n = 4$ synergies. For forward stepping, four synergies accounted for 90.5%, 89.8% and 91.8% of variance in young women and 88.5%, 87.3% and 87.4% in older women for step heights of 10, 20 and 30 cm respectively. In lateral stepping, VAF by four synergies was 90.3%, 90.0% and 91.7% in young women and 88.2%, 88.0% and 87.4% in older women for step heights of 10, 20 and 30 cm respectively. Two-way ANOVA (age × step height) on VAF obtained from $n = 4$ synergies revealed a significant main effect of step height ($p = 0.002$) and age ($p = 0.026$) in forward direction, but not in lateral direction ($p = 0.187$ and $p = 0.138$ respectively). No significant age × step height interaction effect was found for either step direction ($p > 0.05$). For forward stepping, related-samples t-tests revealed a significant difference between step heights of 10 cm versus 20 and 30 cm ($p = 0.009$ and 0.014 respectively), but not between 20 and 30 cm ($p > 0.05$). Independent samples t-tests revealed a significant difference between age groups for each step height ($p = 0.005$, $p = 0.041$ and $p = 0.019$ for 10, 20 and 30 cm respectively).

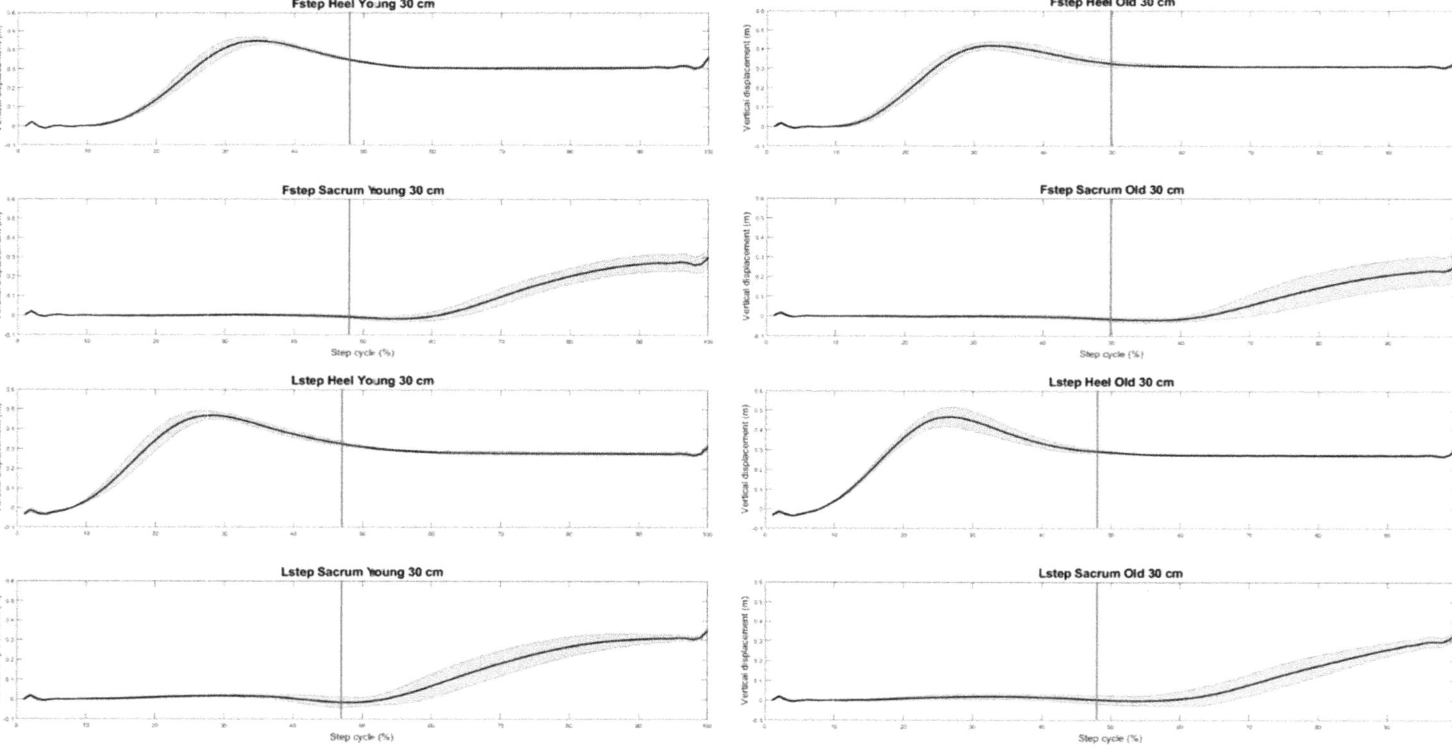

Figure 1. Averaged vertical displacement of the heel and sacrum in young (left column) and older women (right column) at 30 cm step height in forward direction (Fstep) and lateral direction (Lstep). Vertical lines indicate average point of transition from foot clearance to pull-up phase.

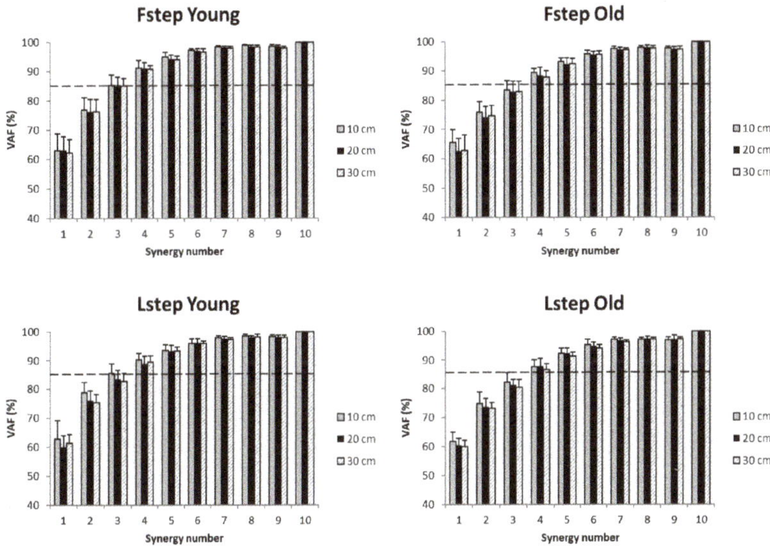

Figure 2. Averaged variance accounted for (VAF) by number of extracted synergies at step heights of 10, 20 and 30 cm in forward (Fstep) and lateral (Lstep) stepping directions for young and older women. The appropriate number of synergies was defined as the least number of synergies required to reach >85% VAF (indicated by dashed lines) and a reconstruction quality of ≥75% VAF for each individual muscle.

Comparisons between muscle weightings (Table 1, Figures 3 and 4) showed that synergy 2 and 4 had high inter-step height similarity for both age groups and step directions. Synergy 4 also appeared to be highly similar between age groups. In synergy 2, a lower similarity between age groups was found, probably due to a difference in quadriceps/hamstring co-activation, as characterized by a significantly decreased contribution of the quadriceps and increased contribution of the hamstrings for most stepping conditions in older women. The composition of synergy 3 appeared to be the most variable. In contrast with forward stepping, which showed a robust synergy organization within and between age groups, lateral stepping resulted in lower correlations between step heights for the young group when compared to the older group.

Table 1. Similarity index (Pearson's *r*) of synergy weightings, across step heights for each age group, and across age groups for each step height. Results are displayed separately for forward stepping (Fstep) and lateral stepping (Lstep). Grey = marginal similarity (*r* > 0.45) and black = significant similarity (*r* > 0.7).

	Young Women			Older Women			Young vs. Older Women		
	10 cm vs. 20 cm	*20 cm vs. 30 cm*	*10 cm vs. 30 cm*	*10 cm vs. 20 cm*	*20 cm vs. 30 cm*	*10 cm vs. 30 cm*	*10 cm*	*20 cm*	*30 cm*
Fstep Correlation coefficients (r)									
Synergy 1	0.64	0.56	0.87	0.84	0.76	0.88	0.61	0.02	0.69
Synergy 2	0.95	0.96	0.87	0.88	0.84	0.83	0.59	0.76	0.73
Synergy 3	0.44	0.70	0.89	0.74	0.43	0.85	0.88	0.71	0.85
Synergy 4	0.96	0.95	0.89	0.91	0.78	0.86	0.89	0.85	0.74
Lstep Correlation coefficients (r)									
Synergy 1	0.94	0.48	0.38	0.56	0.95	0.54	0.48	0.67	0.03
Synergy 2	0.61	0.90	0.40	0.83	0.91	0.95	0.19	0.51	0.79
Synergy 3	0.90	0.43	0.15	0.88	0.94	0.71	0.90	0.82	0.29
Synergy 4	0.70	0.66	0.95	0.76	0.64	0.80	0.90	0.75	0.76

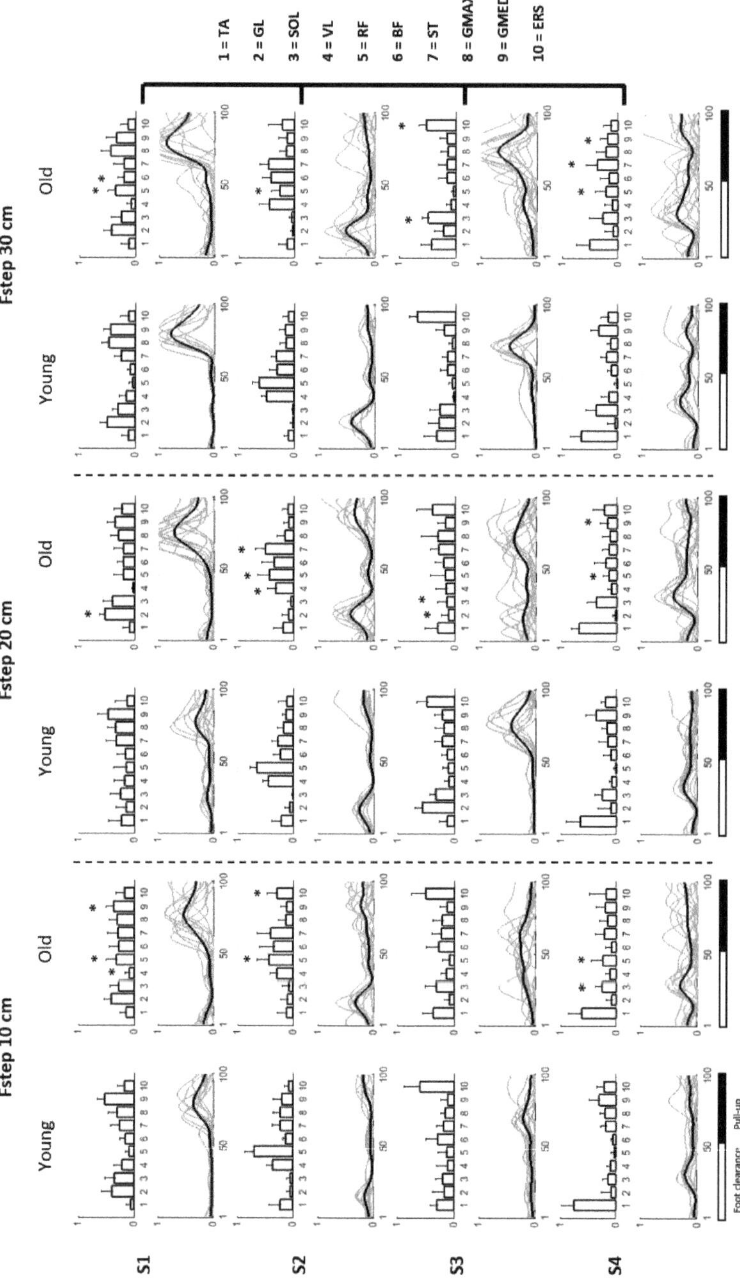

Figure 3. Muscle weightings and temporal activation patterns for four synergies (S1–S4) extracted during forward stepping (Fstep). Each set of two columns represents muscle synergies extracted from step heights of 10, 20 or 30 cm per age group with young women represented in the left column and older women in the right column. * indicates a significant difference between contributions of individual muscles between age groups. Time-averaged foot clearance (white bars) and pull-up (black bars) phases of ascent are depicted at the bottom. TA = tibialis anterior, GL = gastrocnemius lateralis, SOL = soleus, VL = vastus lateralis, RF = rectus femoris, BF = biceps femoris, ST = semitendinosus, GMAX = gluteus maximus, GMED = gluteus medius, ERS = erector Spinae.

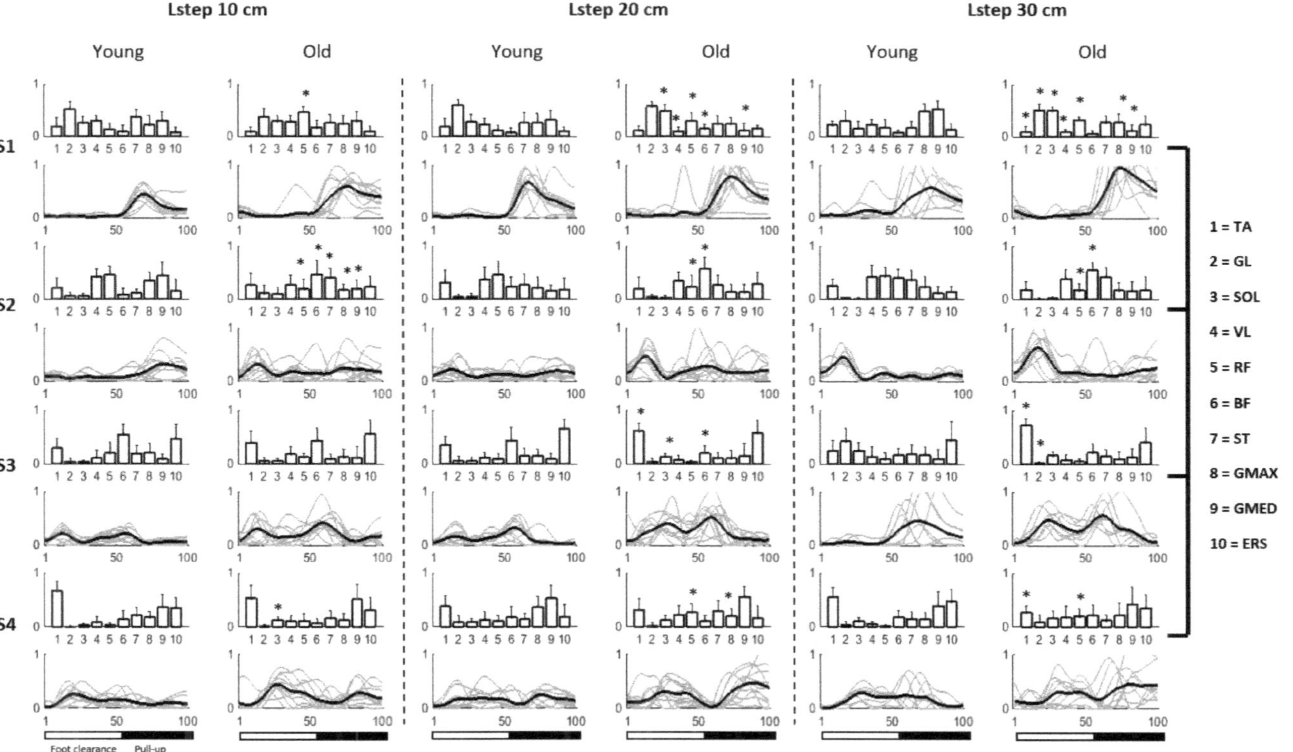

Figure 4. Muscle weightings and temporal activation patterns for four synergies (**S1–S4**) extracted during lateral stepping (Lstep). Each set of two columns represents muscle synergies extracted from step heights of 10, 20 or 30 cm per age group with young women represented in the left column and older women in the right column. * indicates a significant difference between contributions of individual muscles between age groups. Time-averaged foot clearance (white bars) and pull-up (black bars) phases of ascent are depicted at the bottom. TA = tibialis anterior, GL = gastrocnemius lateralis, SOL = soleus, VL = vastus lateralis, RF = rectus femoris, BF = biceps femoris, ST = semitendinosus, GMAX = gluteus maximus, GMED = gluteus medius, ERS = erector Spinae.

Analyses of the temporal activation patterns (Figures 3 and 4) and the time-averaged standard deviation of these temporal activation patterns (Figure 5) showed that the between-subject variability of activation timing for most step heights and directions was significantly higher in the older cohort.

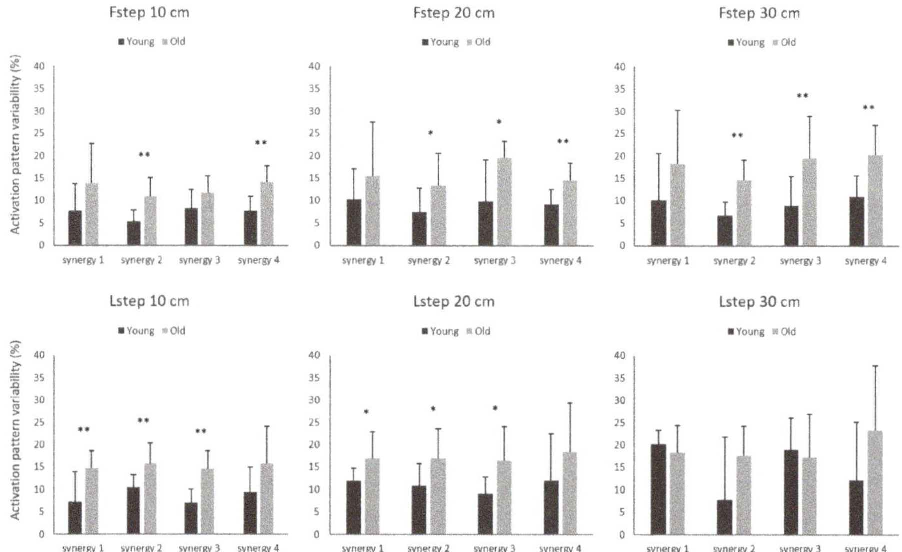

Figure 5. Between-subject variability (time-averaged standard deviation) of temporal activation patterns across time-normalized trials, expressed as a percent of 100 (in arbitrary units, *n* synergies = 4). Data are displayed separately for forward (Fstep) and lateral (Lstep) step directions with step heights of 10, 20 and 30 cm. * indicates a significant difference at $p = 0.05$, ** indicates a significant difference at $p = 0.01$.

4. Discussion

The purpose of this study was to investigate the effects of age and task challenge on neuromuscular control and resultant complexity of neuromuscular activation patterns of women during step ascent by examining muscle synergy organization during stepping tasks with incremental step heights in both forward and lateral directions.

Our results show that complexity of motor control is quite robust across step heights and age groups for stepping in forward and lateral directions. We found no differences in the number of synergies between age groups and step heights for either age group. Further analyses of VAF by four synergies revealed main effects, but no interaction effect, of age and step height on complexity of motor control during forward stepping, indicating subtle differences that could not be detected by analyses of the number of extracted synergies. These analyses revealed that older women actually exhibited more complex motor control strategies, indicated by lower VAF [29], for each step height. Additionally, forward step heights of 20 and 30 cm yielded a significantly lower VAF at $n = 4$ synergies compared to 10 cm. Comparisons of muscle synergy organization revealed that muscle contributions to individual synergies (e.g., muscle weightings) during forward stepping were quite robust between step heights for both age groups. Finally, age-related differences in synergy organization were characterized by a notably lower similarity index between step heights for lateral stepping in young women and increased between-subject variability of the temporal activation patterns in older women.

The extraction of four synergies from step ascent is in agreement with previous studies of locomotion in healthy adults that included a maximum of ten lower limb muscles [24,40]. An important aspect of this study is the inclusion of EMG signals obtained from 10 lower limb muscles, including the gluteus medius, gluteus maximus and the erector spinae. Our results were in line with a previous study by Oliveira et al., who also found four synergies were sufficient to reconstruct the EMG signals of ten lower limb muscles. However, they also proposed that the addition of EMG measurements from hip extensors and abductors during gait would likely increase dimensionality [40]. Our data show that this is not necessarily true for step ascent. For example, during forward stepping, gluteus medius activation coincided with activation of the triple extensors (gluteus maximus, plantar flexors and rectus femoris). During lateral stepping, inclusion of the gluteus medius and erector spinae did not increase dimensionality in the form of an additional synergy as their activation coincided mainly with tibialis anterior activation during lift-off of the trailing foot and trunk stabilization prior to the double support phase after ascent. In line with previous studies including healthy older adults, our results indicate that age did not affect the number of synergies required to reconstruct the muscle activation, which implies that the complexity of motor control (or motor repertoire) of our healthy older cohort was not reduced [6,8,17,45]. However, we were surprised to find that the VAF at a fixed number of synergies, which can be used as an alternative way to assess complexity of motor control [29], was decreased in older women and with step height, indicating increased rather than decreased complexity. We propose that this may be due to the increased challenge posed by step ascent for older adults, forcing some to adopt different motor strategies to compensate for decreased physical capacity [15,46]. As a consequence, the organization or timing of motor modules may be altered, inevitably leading to higher between-subject variability in older adults compared young adults. The assumption that shifts in muscle synergy organization are attributable to (relative) increases in task challenge is in line with previous findings by Routson et al., showing that changes in speed, cadence, step length, and step height during gait can lead to altered temporal activation patterns [7]. Additionally, other studies have explored the interaction between age and task challenge during gait and found that walking at a higher than preferred cadence revealed small differences in spatio-temporal characteristics of neuromuscular control in older but not in young adults [8,9], although this proposed interaction effect was not confirmed by our results.

More detailed analyses of synergy organization in our older cohort revealed a trend towards increased contribution of the hamstrings and decreased contribution of the quadriceps in synergy 2, indicating an increase in quadriceps/hamstring co-activation during the initial foot clearance phase for both step directions compared to young women. These results are in line with findings from previous studies that have shown elevated muscle co-activation in older adults to increase joint stiffness and enhance stability during activities of daily life, such as stair climbing and single step descent [17,47–49], and directly affect muscle synergy organization [8,17]. Additionally, comparisons of synergy organization across step heights revealed that, in young women, increasing the step height from 20 to 30 cm was associated with an increased contribution of the gluteus medius and maximus to synergy 1 and of the calf muscles to synergy 3 during foot placement and the initial pull-up phase of ascent in lateral direction, while the composition of the remaining synergies remained similar. These changes indicate that some synergies reflect basic motor patterns which are activated during a variety of tasks, whereas other synergies can be flexibly recruited to match task-specific demands [20,44], such as the increased challenge to medio-lateral stability imposed by increased step heights. This is reflected by kinematic analyses of motor strategies for stair negotiation, showing both common and variable patterns, with the highest variability often seen at the hip joint [50]. The age-related differences in neuromotor strategies found in this study were reflected by subtle differences in kinematic profiles of the heel and sacrum. For example, visual inspection (Figure 1) revealed that, for older women, average vertical displacement of the sacrum was more linear in both directions with distinctly higher between-subject variability during the pull-up phase of forward stepping and higher between-subject variability of heel peak height during the foot clearance phase. However, these kinematic profiles were primarily included to define start and end points of each step cycle. As such, they provide limited information about how muscle synergy organization affects strategies such as total lower limb extension patterns and joint kinematics and kinetics and should be a focus of future studies [50].

Additional analyses were included to detect possible age-related differences of temporal activation patterns. Our results did reveal higher between-subject variability of temporal activation patterns, indicating increased heterogeneity of motor control strategies within the older cohort. This may reflect a relative increase in functional demand imposed by step heights of 20–30 cm in this age group, necessitating individual modulations of synergy timing to compensate for decreased physical capacity [7,12]. Higher between-subject variability of temporal activation patterns associated with increased task challenge in young women indicates that synergy organization is likely also associated with differences in motor skill levels [27]. This is illustrated by notable changes in between-subject variability of activation patterns in synergies 1 and 3 between step heights of 20 and 30 cm in lateral direction. Finally, although not the primary focus of this study, the differences in synergy organization between forward and lateral stepping are in line with findings from previous studies, showing that EMG recruitment patterns are task-specific for forward and lateral stepping [34,51].

Some limitations of this study have to be recognized. We chose to include only women in this study. As such, additional research is required to assess if these findings also apply to older males. EMG was only collected from the dominant leg. For this reason, differences in motor strategies involving additional push-off force of the trailing leg could not be analyzed [25]. Additionally, due to the technical limitations of surface EMG recording, no data were collected from the deeper thigh muscles such as the hip adductors. Future studies involving step ascent should consider including the hip adductors in order to increase dimensionality and provide information regarding the effects of antagonistic co-contraction of the hip ab- and adductors. Finally, a possible limitation lies with averaging EMG signals of individual participants over three repetitions of the same trial, rather than concatenating them prior to running the factorization algorithm. This may have led to a decrease of detail in the data [40].

5. Conclusions

Neuromuscular control of young and community-dwelling older women in stepping up in forward and lateral direction could not be differentiated based on the number of synergies. However, additional analyses of synergy complexity, such as VAF by the given number of synergies, and synergy structure revealed several age- and step-height related differences. These findings show that the ability to modulate synergy composition is well preserved in healthy older women and that they respond to more challenging physical tasks by adapting basic muscle recruitment patterns. This results in more complex motor control patterns despite evidence of increased antagonistic co-activation, likely indicating increased involvement of mechanisms for balance control.

Author Contributions: R.J.B., E.V.R., S.M.V., C.D., J.H.v.D. and N.D.: Conceptualization and experimental design; R.J.B. and G.G.: Participant recruitment and data collection; R.J.B., J.H.v.D. and N.D.: Data analysis and interpretation; R.J.B., E.V.R., S.M.V., J.H.v.D., C.D. and N.D.: Manuscript drafting and/or reviewing; R.J.B., J.H.v.D., E.V.R., S.M.V., G.G., C.D. and N.D.: Final approval of submitted manuscript. All authors have read and agreed to the published version of the manuscript.

Funding: This study was funded by the European Commission through MOVE-AGE, an Erasmus Mundus Joint Doctorate programme (#2014-0691) fellowship awarded to R.J.B., the Fund for Scientific Research Flanders (FWO-Vlaanderen, #G0521-05) awarded to S.M.V., and the ERC-Starting Grant "Learn2Walk" (#715945) and the NWO VIDI-Grant "FirSTeps" (#016.156.346) both awarded to N.D.

Acknowledgments: The authors would like to thank Tine De Reys for her assistance with data collection and processing and Hans Essers for his advice on the cluster analyses.

Conflicts of Interest: The authors declare no conflict of interest.

References

1. Erim, Z.; Beg, M.F.; Burke, D.T.; de Luca, C.J. Effects of aging on motor-unit control properties. *J. Neurophysiol.* **1999**, *82*, 2081–2091. [CrossRef]
2. Vernooij, C.A.; Rao, G.; Berton, E.; Retornaz, F.; Temprado, J.J. The effect of aging on muscular dynamics underlying movement patterns changes. *Front. Aging Neurosci.* **2016**, *8*, 1–12. [CrossRef]

3. Pijnappels, M.; van der Burg, P.J.C.E.; Reeves, N.D.; van Dieën, J.H. Identification of elderly fallers by muscle strength measures. *Eur. J. Appl. Physiol.* **2008**, *102*, 585–592. [CrossRef]

4. Hakkinen, K.; Kraemer, W.J.; Kallinen, M.; Linnamo, V.; Pastinen, U.-M.; Newton, R.U. Bilateral and unilateral neuromuscular function and muscle cross-sectional area in middle-aged and elderly men and women. *J. Gerontol. Ser. A Biol. Sci. Med. Sci.* **1996**, *51*, 21–29. [CrossRef]

5. Montero, N.; Serra, J.A. Role of sarcopenia in elderly. *Eur. J. Phys. Rehabil. Med.* **2013**, *49*, 131–143.

6. Allen, J.L.; Franz, J.R. The motor repertoire of older adult fallers may constrain their response to balance perturbations. *J. Neurophysiol.* **2018**, *120*, 2368–2378. [CrossRef] [PubMed]

7. Routson, R.L.; Kautz, S.A.; Neptune, R.R. Modular organization across changing task demands in healthy and poststroke gait. *Physiol. Rep.* **2014**, *2*, 1–14. [CrossRef] [PubMed]

8. Monaco, V.; Ghionzoli, A.; Micera, S. Age-Related modifications of muscle synergies and spinal cord activity during locomotion. *J. Neurophysiol.* **2010**, *104*, 2092–2102. [CrossRef] [PubMed]

9. Artoni, F.; Monaco, V.; Micera, S. Selecting the best number of synergies in gait: Preliminary results on young and elderly people. *IEEE Int. Conf. Rehabil. Robot.* **2013**, 1–4. [CrossRef]

10. Baggen, R.J.; Van Roie, E.; van Dieën, J.H.; Verschueren, S.M.; Delecluse, C. Weight bearing exercise can elicit similar peak muscle activation as medium–high intensity resistance exercise in elderly women. *Eur. J. Appl. Physiol.* **2018**, *118*, 531–541. [CrossRef]

11. Baggen, R.J.; Van Roie, E.; Verschueren, S.M.; Van Driessche, S.; Coudyzer, W.; van Dieën, J.H.; Delecluse, C. Bench stepping with incremental heights improves muscle volume, strength and functional performance in older women. *Exp. Gerontol.* **2019**, *120*, 6–14. [CrossRef] [PubMed]

12. Reeves, N.D.; Spanjaard, M.; Mohagheghi, A.A.; Baltzopoulos, V.; Maganaris, C.N. Older adults employ alternative strategies to operate within their maximum capabilities when ascending stairs. *J. Electromyogr. Kinesiol.* **2009**, *19*, e57–e68. [CrossRef] [PubMed]

13. Hortobágyi, T.; Mizelle, C.; Beam, S.; DeVita, P. Old adults perform activities of daily living near their maximal capabilities. *J. Gerontol. A Biol. Sci. Med. Sci.* **2003**, *58*, M453–M460. [CrossRef]

14. Di Nardo, F.; Mengarelli, A.; Maranesi, E.; Burattini, L.; Fioretti, S. Assessment of the ankle muscle co-contraction during normal gait: A surface electromyography study. *J. Electromyogr. Kinesiol.* **2015**, *25*, 347–354. [CrossRef] [PubMed]

15. Mair, J.L.; Laudani, L.; Vannozzi, G.; De Vito, G.; Boreham, C.; Macaluso, A. Neuromechanics of repeated stepping with external loading in young and older women. *Eur. J. Appl. Physiol.* **2014**, *114*, 983–994. [CrossRef] [PubMed]

16. Hortobágyi, T.; Solnik, S.; Gruber, A.; Rider, P.; Steinweg, K.; Helseth, J.; DeVita, P. Interaction between age and gait velocity in the amplitude and timing of antagonist muscle coactivation. *Gait Posture* **2009**, *29*, 558–564. [CrossRef] [PubMed]

17. Wang, Y.; Watanabe, K.; Asaka, T. Aging effect on muscle synergies in stepping forth during a forward perturbation. *Eur. J. Appl. Physiol.* **2017**, *117*, 201–211. [CrossRef]

18. Oliveira, A.S.; Silva, P.B.; Lund, M.E.; Gizzi, L.; Farina, D.; Kersting, U.G. Effects of perturbations to balance on neuromechanics of fast changes in direction during locomotion. *PLoS ONE* **2013**, *8*, e59029. [CrossRef]

19. Torres-Oviedo, G.; Macpherson, J.M.; Ting, L.H. Muscle synergy organization is robust across a variety of postural perturbations. *J. Neurophysiol.* **2006**, *96*, 1530–1546. [CrossRef]

20. Torres-Oviedo, G.; Ting, L.H. Subject-Specific muscle synergies in human balance control are consistent across different biomechanical contexts. *J. Neurophysiol.* **2010**, *103*, 3084–3098. [CrossRef]

21. Cheung, V.C.K.; D'Avella, A.; Tresch, M.C.; Bizzi, E. Central and sensory contributions to the activation and organization of muscle synergies during natural motor behaviors. *J. Neurosci.* **2005**, *25*, 6419–6434. [CrossRef] [PubMed]

22. Ivanenko, Y.P.; Cappellini, G.; Dominici, N.; Poppele, R.E.; Lacquaniti, F. Coordination of locomotion with voluntary movements in humans. *J. Neurosci.* **2005**, *25*, 7238–7253. [CrossRef] [PubMed]

23. De Groote, F.; Jonkers, I.; Duysens, J. Task constraints and minimization of muscle effort result in a small number of muscle synergies during gait. *Front. Comput. Neurosci.* **2014**, *8*, 1–11. [CrossRef] [PubMed]

24. Saito, A.; Tomita, A.; Ando, R.; Watanabe, K.; Akima, H. Similarity of muscle synergies extracted from the lower limb including the deep muscles between level and uphill treadmill walking. *Gait Posture* **2018**, *59*, 134–139. [CrossRef] [PubMed]

25. MacLellan, M.J. Modular organization of muscle activity patterns in the leading and trailing limbs during obstacle clearance in healthy adults. *Exp. Brain Res.* **2017**, *235*, 2011–2026. [CrossRef] [PubMed]

26. Hart, C.B.; Giszter, S.F. A neural basis for motor primitives in the spinal cord. *J. Neurosci.* **2010**, *30*, 1322–1326. [CrossRef] [PubMed]

27. Safavynia, S.A.; Torres-Oviedo, G.; Ting, L.H. Muscle synergies: Implications for clinical evaluation and rehabilitation of movement. *Top. Spinal Cord Inj. Rehabil.* **2011**, *17*, 16–24. [CrossRef]

28. Torres-Oviedo, G. Muscle synergies characterizing human postural responses. *J. Neurophysiol.* **2007**, 2144–2156. [CrossRef]

29. Steele, K.M.; Rozumalski, A.; Schwartz, M.H. Muscle synergies and complexity of neuromuscular control during gait in cerebral palsy. *Dev. Med. Child Neurol.* **2015**, *57*, 1176–1182. [CrossRef]

30. Torricelli, D.; Barroso, F.; Coscia, M.; Alessandro, C.; Lunardini, F.; Esteban, E.B.; d'Avella, A. Muscle synergies in clinical practice: Theoretical and practical implications. *Emerg. Ther. Neurorehabil. II* **2016**, *10*, 251–272. [CrossRef]

31. Goudriaan, M.; Shuman, B.R.; Steele, K.M.; Van den Hauwe, M.; Goemans, N.; Molenaers, G.; Desloovere, K. Non-Neural muscle weakness has limited influence on complexity of motor control during gait. *Front. Hum. Neurosci.* **2018**, *12*, 1–11. [CrossRef] [PubMed]

32. Yu, Y.; Chen, X.; Cao, S.; Wu, D.; Zhang, X.; Chen, X. Gait synergetic neuromuscular control in children with cerebral palsy at different gross motor function classification system levels. *J. Neurophysiol.* **2019**, *121*, 1680–1691. [CrossRef] [PubMed]

33. Singh, T.; Latash, M.L. Effects of muscle fatigue on multi-muscle synergies. *Exp. Brain Res.* **2011**, *214*, 335–350. [CrossRef] [PubMed]

34. Wang, M.-Y.; Flanagan, S.; Song, J.-E.; Greendale, G.A.; Salem, G.J. Lower-Extremity biomechanics during forward and lateral stepping activities in older adults. *Clin. Biomech.* **2003**, *18*, 214–221. [CrossRef]

35. Startzell, J.K.; Owens, D.A.; Mulfinger, L.M.; Cavanagh, P.R. Stair negotiation in older people: A review. *J. Am. Geriatr. Soc.* **2000**, *48*, 567–580. [CrossRef] [PubMed]

36. Arvin, M.; Van Dieën, J.H.; Faber, G.S.; Pijnappels, M.; Hoozemans, M.J.M.; Verschueren, S.M.P. Hip abductor neuromuscular capacity: A limiting factor in mediolateral balance control in older adults? *Clin. Biomech.* **2016**, *37*, 27–33. [CrossRef]

37. Hermens, H.J.; Freriks, B.; Disselhorst-Klug, C.; Rau, G. Development of recommendations for SEMG sensors and sensor placement procedures. *J. Electromyogr. Kinesiol.* **2000**, *10*, 361–374. [CrossRef]

38. De Luca, C.J.; Donald Gilmore, L.; Kuznetsov, M.; Roy, S.H. Filtering the surface EMG signal: Movement artifact and baseline noise contamination. *J. Biomech.* **2010**, *43*, 1573–1579. [CrossRef]

39. Banks, C.L.; Pai, M.M.; McGuirk, T.E.; Fregly, B.J.; Patten, C. Methodological choices in muscle synergy analysis impact differentiation of physiological characteristics following stroke. *Front. Comput. Neurosci.* **2017**, *11*, 1–12. [CrossRef]

40. Oliveira, A.S.; Gizzi, L.; Farina, D.; Kersting, U.G. Motor modules of human locomotion: Influence of EMG averaging, concatenation, and number of step cycles. *Front. Hum. Neurosci.* **2014**, *8*, 1–9. [CrossRef]

41. Ting, L.H.; Chvatal, S.A. Decomposing muscle activity in motor tasks: Methods and interpretation. *Mot. Control Theor. Exp. Appl.* **2011**, 102–138. [CrossRef]

42. Chvatal, S.A.; Ting, L.H. Voluntary and reactive recruitment of locomotor muscle synergies during perturbed walking. *J. Neurosci.* **2012**, *32*, 12237–12250. [CrossRef] [PubMed]

43. Essers, J.M.N.; Peters, A.; Meijer, K.; Peters, K.; Murgia, A. Superficial shoulder muscle synergy analysis in facioscapulohumeral dystrophy during humeral elevation tasks. *IEEE Trans. Neural Syst. Rehabil. Eng.* **2019**, *99*, 1–10. [CrossRef] [PubMed]

44. Nazifi, M.M.; Yoon, H.U.; Beschorner, K.; Hur, P. Shared and task-specific muscle synergies during normal walking and slipping. *Front. Hum. Neurosci.* **2017**, *11*, 1–14. [CrossRef] [PubMed]

45. Ting, L.H.; Chiel, H.J.; Trumbower, R.D.; Allen, J.L.; McKay, J.L.; Hackney, M.E.; Kesar, T.M. Neuromechanical principles underlying movement modularity and their implications for rehabilitation. *Neuron* **2015**, *86*, 38–54. [CrossRef]

46. Bice, M.R.; Hanson, N.; Eldridge, J.; Reneau, P.; Powell, D.W. Neuromuscular adaptations in elderly adults are task-specific during stepping and obstacle clearance tasks. *Int. J. Exerc. Sci.* **2011**, *4*, 77–85.

47. Hortobágyi, T.; Devita, P. Muscle pre-and coactivity during downward stepping are associated with leg stiffness in aging. *J. Electromyogr. Kinesiol.* **2000**, *10*, 117–126. [CrossRef]

48. Macaluso, A.; Nimmo, M.A.; Foster, J.E.; Cockburn, M.; McMillan, N.C.; De Vito, G. Contractile muscle volume and agonist-antagonist coactivation account for differences in torque between young and older women. *Muscle Nerve* **2002**, *25*, 858–863. [CrossRef]
49. Larsen, A.H.; Puggaard, L.; Hämäläinen, U.; Aagaard, P. Comparison of ground reaction forces and antagonist muscle coactivation during stair walking with ageing. *J. Electromyogr. Kinesiol.* **2008**, *18*, 568–580. [CrossRef]
50. McFadyen, B.; Winter, D. An integrated biomechanical analysis of normal stair ascent and descent. *J. Biomech.* **1988**, *21*, 733–744. [CrossRef]
51. Cook, T.M.; Zimmermann, C.L.; Lux, K.M.; Neubrand, C.M.; Nicholson, T.D. EMG comparison of lateral step-up and stepping machine exercise. *J. Orthop. Sport. Phys. Ther.* **1992**, *16*, 108–113. [CrossRef] [PubMed]

Article

Adaptation in Gait to Body-Weight Unloading

Rakshatha Kabbaligere [1,2,*] and Charles S. Layne [1,2,3]

[1] Department of Health and Human Performance, University of Houston, Houston, TX 77004, USA; clayne2@central.uh.edu
[2] Center for Neuromotor and Biomechanics Research, University of Houston, Houston, TX 77004, USA
[3] Center for Neuro-Engineering and Cognitive Science, University of Houston, Houston, TX 77004, USA
* Correspondence: rakshatha.b@gmail.com

Received: 30 September 2019; Accepted: 20 October 2019; Published: 23 October 2019

Abstract: Modifications in load-related sensory input during unloaded walking can lead to recalibration of the body schema and result in aftereffects. The main objective of this study was to identify the adaptive changes in gait and body-weight perception produced by unloaded walking. Gait performance during treadmill walking was assessed in 12 young participants before and after 30 min of unloaded walking (38% body weight) by measuring lower limb kinematics, temporal gait measures, and electromyography (EMG). A customized weight-perception scale was used to assess perception of body weight. Participants perceived their body weight to be significantly heavier than normal after unloading while walking. Angular displacement about ankle and knee was significantly reduced immediately after unloaded walking, while temporal gait parameters remained unchanged. The EMG activity in some muscles was significantly reduced after unloading. These findings indicate that walking at reduced body weight results in alterations in segmental kinematics, neuromuscular activity, and perception of body weight, which are the aftereffects of motor adaptation to altered load-related afferent information produced by unloading. Understanding the adaptive responses of gait to unloading and the time course of the aftereffects will be useful for practitioners who use body-weight unloading for rehabilitation.

Keywords: motor adaptation; body-weight unloading; gait adaptation; treadmill walking; spaceflight; lower-body positive pressure

---

1. Introduction

The sensorimotor system is comprised of sensory systems, motor systems, and the central integration processes that help in producing and controlling movements. The vestibular, proprioceptive, and visual systems provide sensory information from the external environment as well as that related to body position and movement. The sensory inputs from these systems are integrated in the central nervous system (CNS) following which appropriate motor commands are generated and sent through the descending pathways to the various body segments. Any changes in the sensory information either due to changes in the environmental condition or body condition will affect movement control. The process that enables us to modify and maintain accurate movements as sensory condition changes is called motor adaptation [1].

Several studies have investigated motor adaptation using a variety of experimental paradigms. One of the ways is by studying the changes in the movement characteristics produced by adding and subsequently removing sensory (visual, proprioceptive, acoustic, and vestibular) distortions or perturbations while performing motor tasks [2–5]. One of the common locomotor adaptation paradigms involves studying locomotor changes by altering the normal interlimb relationship during walking by changing the speeds of treadmill belts relative to each other (commonly called split-belt treadmill walking [6–8]). Adaptive changes in step length and double support time are observed during split-belt walk. Similarly, in another paradigm, the effect of increased trunk rotation during walking was studied by having the subjects walk along the circumference of a rotating disc for 2 h [9]. Subsequently, after the adaptation session, the subjects were found to produce curved walking trajectories. Likewise, adaptive changes in heading direction in response to modifications of the direction of optical flow was also observed after exposing subjects to a visual scene that gave the perception of walking along the perimeter of a room [10].

Body-weight unloading (BWU) using various types of body-weight support systems is used to study locomotor adaptation in response to reduced load input [11–14]. Load-related sensory information is essential for regulating the timing, phasing, and magnitude of neural activities that generate locomotor patterns during stepping [15–17]. They also help in maintaining balance control during locomotion and gait termination [18–20]. Lower-body positive pressure (LBPP) is an emerging technology that is used to provide body-weight support [21]. It consists of an air chamber that covers the lower part of the body. When inflated with positive pressure, the lower part of the body is lifted upwards from the hips and the body weight is reduced. LBPP is regarded as one of the superior methods of unweighting when compared to upper-body harness [21]. Upper-body harness partially supports the body weight and results in the formation of pressure points. LBPP on the other hand results in uniform distribution of air pressure around the entire surface of the body while still maintaining normal muscle activation and is thus considered superior [22,23]. Body weight as large as 80% can be unloaded in increments as small as 1% using this system. An antigravity treadmill is a special type of treadmill that is equipped with an LBPP air chamber and provides an opportunity to study locomotor adaptation to BWU. Several studies using either the antigravity treadmill or the vertical harness system have investigated the immediate effects of BWU on metabolic energy expenditure and locomotor performance [12,13,24–26]. A linear decrease in stride frequency, vertical contact forces, stance time, peak hip and knee flexion, and extensor and flexor muscles' activity burst during the stance phase with BWU has been reported [24,26–28].

Besides having acute effects on gait performance, exposure to BWU for a short period of time in the order of a few minutes and subsequent reloading can alter movement characteristics [29,30]. Further, prolonged exposure to BWU for a longer period of time in the order of a few months can also produce significant changes in movement characteristics that last for a significant period of time as seen in crewmembers returning from space [20,31–34]. Due to the fact that adaptive changes produced during spaceflight is coupled with changes in structural and functional characteristics of the neuromuscular system, we cannot compare it with short-term changes that we observe on the ground that are solely produced by reduced body load. However, studying the adaptive changes to short-term BWU can improve our understanding of the mechanism adopted by the sensorimotor system during adaptation. Also, with the increase in use of BWU for treadmill training in patients with neurological impairments, studying the adaptive changes to short-term BWU may provide valuable information to practitioners who administer gait or balance rehabilitation using it [21,35,36].

The main objective of this study was to investigate changes in locomotion produced by 30 min of walking at a reduced body weight, which roughly relates to that of Martian gravity (38% of normal body weight). Locomotion was assessed by measuring temporal gait parameters, changes in joint angles, and neuromuscular activation of the lower limbs before, during, and after the adaptation session. We also assessed the adaptive changes in subjective body-weight perception as a function of movement context (walking, standing, and sitting). Body-weight perception was assessed using a customized weight scale. It was hypothesized that the effect of unloaded walking on weight perception would vary as a function of movement context.

2. Materials and Methods

2.1. Participants

Twelve young and healthy participants (23.7 ± 3.3 years; 6 males and 6 females) participated in the study. The experimental protocol was approved by the institutional review board at the University of Houston and was conducted in accordance with the Declaration of Helsinki. The participants were screened using a modified physical activity readiness questionnaire (PAR-Q) [37]. This questionnaire was used to assess if the participants were physically fit and for any conditions that might affect gait and balance. The participants were excluded if their response was "yes" to any of the questions in the modified PAR-Q.

All the participants attended three data collection sessions, which were conducted on three separate days. Their gait performance was assessed during one of the three sessions before, during, and after the adaptation sessions. On the other two days, participants performed two different tests before and after the adaptation sessions while being seated and while standing (Figure 1). These tests were part of another study. Weight perception was assessed on all the three days before, during, and after the two adaptation sessions. During the adaptation sessions, the participants walked continuously for 30 min while being loaded (100% body-weight/control condition) or unloaded (38% body-weight/unloading condition). Post-adaptation sessions consisted of either continuing to walk, sitting, or standing without any movements. The order of the three sessions as well as the order of the two adaptation sessions (control and unloading condition) within each session were randomized across the participants (Figure 1). An overview of the experimental design is depicted in the flow chart below.

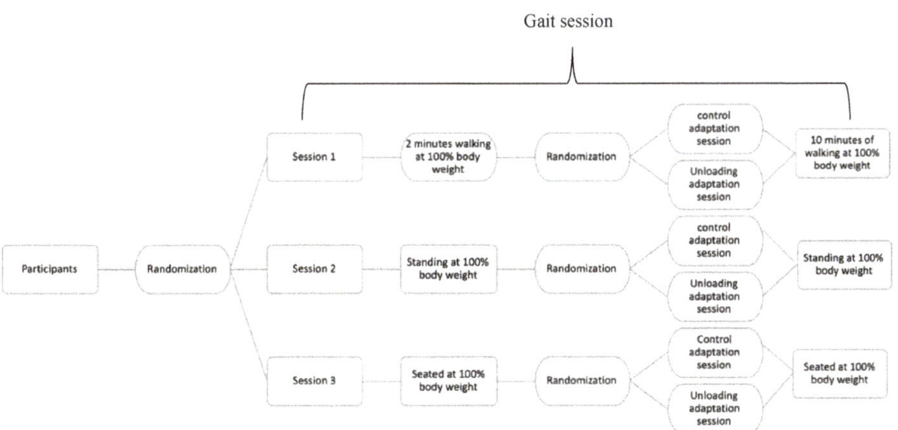

Figure 1. Flow chart describing the experimental design of the study.

2.2. Body-Weight Unloading

Participants were unloaded using an LBPP antigravity treadmill to unload the body weight. The participants donned a pair of neoprene shorts that were used to zip them into the antigravity treadmill. Once inside the chamber, the inside pressure was calibrated and set to a level that corresponded either to 100% body weight or 38% body weight.

2.3. Gait Assessment

After a familiarization period, baseline gait performance (preadaptation) was measured for a duration of two minutes prior to each adaptation session as the participants walked inside the chamber at normal (100%) body weight at a speed of 1 m/s (Figure 2). Following the baseline assessment, they performed the adaptation session for a period of 30 min. In the case of the control adaptation session, they continued to walk at 100% body weight while, during the unloading adaptation session, they were unloaded to 38% body weight. After the adaptation session, the participants performed the post-adaptation session during which their gait was once again assessed at normal body weight (Figure 1). After this, they were given a mandatory break of at least 10 min before starting the other session.

Participants' gait performance was assessed by measuring kinematics, electromyography (EMG), and temporal parameters before, during, and after the two adaptation sessions. Lower limb kinematics along the sagittal plane were measured by using goniometers (Biometrics©) attached unilaterally on the right ankle and knee joint. Joint kinematics were restricted to ankle and knee joints since the chamber frame obscures the pelvis and hip making recording of the hip motion prohibitive. Kinematic data were collected at a sampling rate of 1000 Hz. EMG was recorded using bipolar surface electrodes (Delsys© Trigno TM wireless EMG system) at a rate of 2000 Hz. EMG was recorded from four muscles viz tibialis anterior (TA), medial head of gastrocnemius (GA), rectus femoris (RF), and biceps femoris (BF) muscles of the right leg. A pair of foot switches (Biometrics©) were attached to the sole of the right foot, one under the great toe and the other under the heel. The data from the foot switches were recorded at a rate of 1000 Hz and were used to determine foot-fall events in the gait cycle.

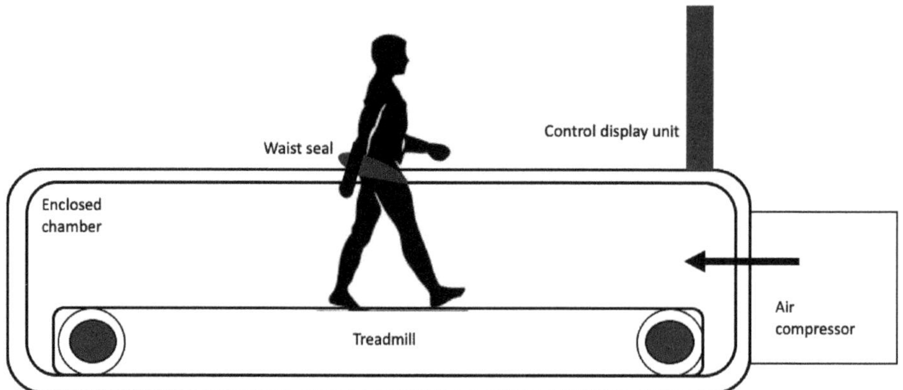

Figure 2. Figure illustrating the setup of lower-body positive pressure (LBPP) or antigravity treadmill.

2.4. Weight Perception

Weight perception was assessed during all three sessions, both during and after the adaptation sessions. We used a 7-point scale, with 1 being very light and 7 being very heavy. The participants were instructed to mentally assign the weight they perceived prior to the adaptation session as 4 on the 7-point scale that translated to "normal" weight. They were instructed to report their perception of weight relative to this during the adaptation session and at three different time intervals after the adaptation session, which were immediately after (T_{+0}), 5 min (T_{+5}) after, and 10 min (T_{+10}) after the session.

2.5. Data Analysis

Data analysis for temporal gait measures, kinematics, and EMG was primarily focused on 7 epochs for each of the control and unloading conditions (Figure 3). These were:

1. Mean of the first 30 strides after one minute of walking before the unloaded or control adaptation session (T_0)
2. Mean of 30 strides just prior to the completion of the control or unloading adaptation session starting from 29 min (T_{during})

3. Mean of the first 30 strides immediately after 30 min of the adaptation session (control and unloading), i.e., immediately post adaptation (T_1)
4. Mean of 30 strides immediately after the first 100 strides of the post-adaptation session (101–130 strides) (T_2)
5. Mean of the first 30 strides immediately after 3 min of the post-adaptation session (T_3)
6. Mean of the first 30 strides immediately after 6 min of the post-adaptation session (T_4)
7. Mean of the first 30 strides immediately after 9 min of the post-adaptation session (T_5).

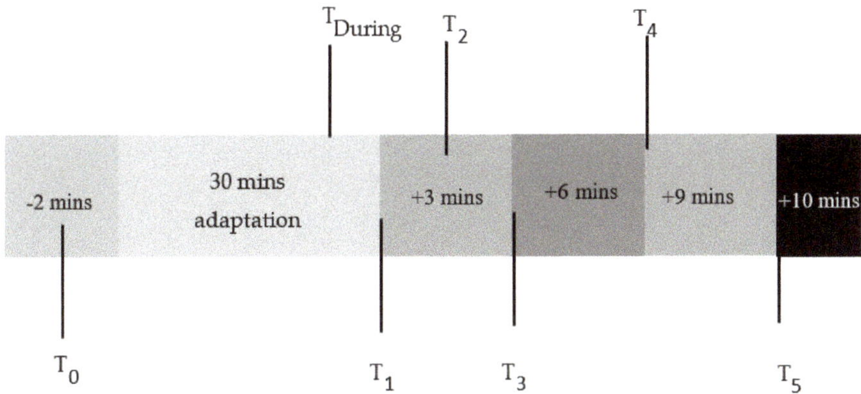

Figure 3. Flow chart indicating the epochs of interest for data analysis.

2.6. Temporal Gait Measures

Temporal gait measures were calculated using the gyroscope data obtained from the Trigno EMG sensor attached to the anterior part of the lower right leg (TA muscle). We used a gait feature extraction algorithm that was used previously in an earlier study [38] to identify three main gait-events: (1) initial contact (IC), (2) mid swing (MS), and (3) terminal contact (TC). These gait events were then used to compute stride time, stance time, and swing time. Stride time was defined as the duration between successive IC points. Stance time was defined as the time between IC and TC of each gait cycle. Swing time was defined as the time between TC of one gait cycle to the IC of the next gait cycle. The temporal gait parameters were then averaged within each epoch of interest.

2.7. Kinematic Measures

Joint angular data recorded from the goniometers were low-pass filtered with a cutoff frequency of 10 Hz [39]. Following this, they were divided into individual strides based on heel-strike events obtained from the foot switches and reduced to 100 data points. Then, the joint-angle waveforms were averaged across 30 strides within each epoch of interest. To facilitate comparison across subjects, the joint-angle waveforms of each of the epochs were normalized to the peak joint angle of the T_0 waveform by dividing all the points of the waveform by that peak angle value. Range of motion (ROM) of the joints was then computed as an outcome measure at each epoch as the difference between the maximum and minimum angular displacements of the normalized waveforms.

2.8. EMG Measures

The EMG signals were first band-pass filtered using a second-order Butterworth filter using cutoff frequencies of 10 Hz and 450 Hz, followed by full wave rectification [29]. The rectified signals were then low-pass filtered at a cutoff frequency of 25 Hz to obtain the linear envelope [40]. Following this, the signals were separated into individual strides based on heel-strike events and reduced to 100 data points per stride. Then, the EMG waveforms were averaged across 30 strides within each epoch of interest [41]. Waveforms of all the epochs were normalized to the mean EMG activity of the T_0 waveform by dividing all the points of the waveform by that mean value [42]. After this, within each epoch, the EMG activity across 7 different gait phases was calculated individually. This was calculated by taking the average of the points that represent the respective gait phases. For example, the average of points 1 through 12 would represent the mean EMG activity of the loading phase. A breakdown of the percentage of the time represented by each of the 7 gait phases within a gait cycle is described below [43].

1. Loading response (1–12%)
2. Mid-stance (12–31%)
3. Terminal stance (31–50%)
4. Pre-swing (50–62%)
5. Initial swing (62–75%)
6. Mid-swing (75–87%)
7. Terminal swing (87–100%)

2.9. Statistical Analyses

All statistical analyses were performed using SPSS V.20 (IBM Corp, Somers, NY, USA). Two-way repeated measures analysis of variance (RANOVA) was performed on each of the temporal (stride time, stance time, and swing time), kinematic (range of motion of knee and ankle joint), and EMG (mean EMG activity) outcome measures separately. The EMG activity of each of the 7 phases of the gait cycle was analyzed separately. The two factors in these analyses were "condition" (control vs. unloading) and epochs (T_0, T_{during}, and T_1–T_5). Simple planned contrasts were used to assess changes in the outcome measures across epochs relative to T_0 whenever there was a main effect of epoch. Additionally, a simple effects analysis was performed on each of the conditions whenever there was a significant interaction effect. Before performing the ANOVA analysis, data were analyzed to assess whether all the required assumptions were met. Whenever the assumption of sphericity was not fulfilled, the degrees-of-freedom was adjusted using Huyn–Feldt correction.

A nonparametric test (Friedman's ANOVA) was conducted to test for changes in body-weight perception separately for each of the three sessions. This was followed by a Wilcox signed ranked test to compare differences between different time intervals.

3. Results

3.1. Body-Weight Perception

Perception of body weight remained unchanged during and after the control adaptation session regardless of the movement context. During the unloading condition, there was a significant decrease in perceived body weight during unloaded walking across all the three sessions ($Z < -2.71$, $p < 0.007$). There was a significant increase in perceived body weight after unloading at T_{+0} ($Z = -3.06$, $p < 0.05$) and T_{+5} ($Z = -2.81$, $p < 0.05$) during the walking session. Although there was a trend towards increased body-weight perception while seated and standing after unloading, the changes in the scores were not significant (Figure 4).

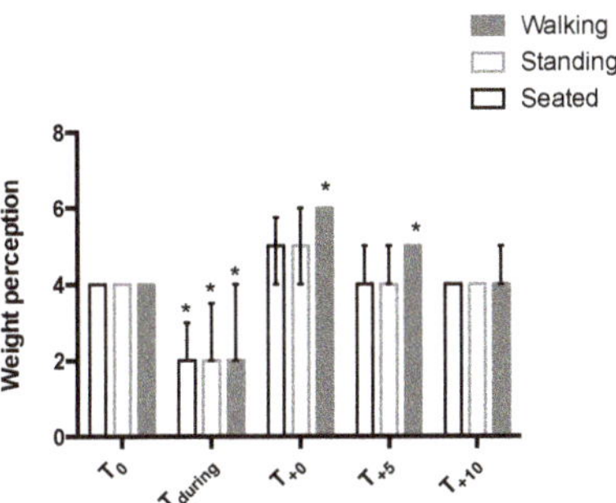

Figure 4. Group median along with the 25th and 75th percentile scores of weight perception during sitting, standing, and walking across different time intervals for the unloading condition. * Significant difference ($p < 0.05$) relative to T_0 of the corresponding condition.

3.2. Temporal Gait Measures

There was a significant main effect of time ($F_{(3.12, 37.49)} = 4.56$, $p = 0.007$) on swing time and a significant interaction effect between condition and time ($F_{(2.75, 33.00)} = 4.04$, $p = 0.017$). Results from the simple effects analysis indicated that the swing time was significantly increased at T_{during} (0.54 ± 0.01) relative to T_0 (0.52 ± 0.01) during the unloading condition and that it returned to baseline immediately after the adaptation session. However, swing time remained unchanged across time during the control condition. Stride time and stance time remained unchanged across time for both the control and unloading conditions. The average stance and swing times across different epochs for the two conditions are summarized in Figure 5.

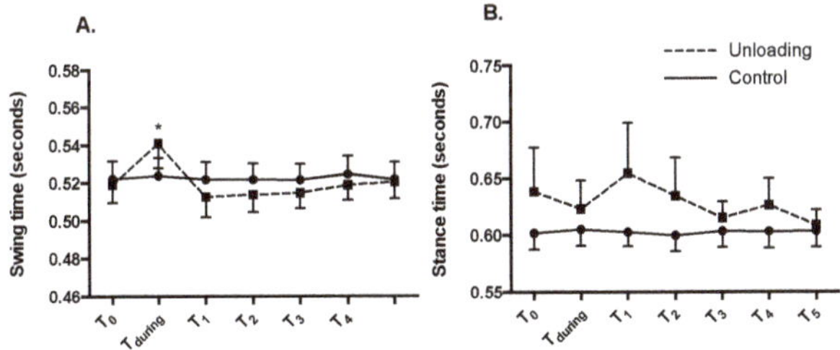

Figure 5. Group means ($\pm$SEM) of (**A**) swing time and (**B**) stance time across different epochs for the control and unloading conditions. * Significant difference ($p < 0.05$) relative to T_0 of the corresponding condition.

3.3. Kinematics

Overall, the ROMs of ankle and knee joints during the unloading condition were significantly less when compared to the control condition (main effect of condition, ankle: $F_{(1, 10)} = 15.29$, $p = 0.003$; knee: $F_{(1, 10)} = 8.47$, $p = 0.016$). There were no changes in ankle and knee ROMs across time in the control condition. For the unloading condition, the ROM for both the joints was significantly reduced at T_{during} (ankle: 2.44 ± 0.20; knee: 1.29 ± 0.04) and T_1 (ankle: 2.93 ± 0.17; knee: 1.46 ± 0.05) relative to T_0 (ankle: 3.31 ± 0.22; knee: 1.58 ± 0.03). Knee ROM was also reduced at T_2 (1.47 ± 0.05). However, it was not different from T_0, starting from T_3 through T_5 (Figure 6).

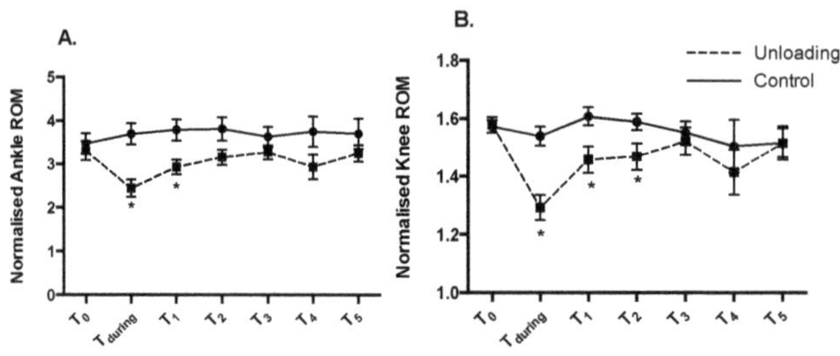

Figure 6. Group means (±SEM) of normalized range of motion of (**A**) ankle and (**B**) knee joint across different epochs during control and unloading conditions. * Significant difference ($p < 0.05$) relative to T_0 of the corresponding condition.

3.4. EMG

3.4.1. Rectus Femoris Activity

EMG activity during all seven gait phases remained unchanged after walking in the control condition. For the unloading condition, there was a significant effect of time on EMG activity during loading phase (F (3.07, 30.70) = 7.62, $p = 0.001$), mid-stance (F (3.24, 32.36) = 5.051, $p = 0.005$), pre-swing (F (3.15, 31.53) = 2.94, $p = 0.046$), mid-swing (F (2.94, 29.41) = 4.557, $p = 0.01$), and terminal swing (F (5.38, 53.85) = 8.90, $p < 0.0001$). There was reduced activity during unloaded walking (1.08 ± 0.11) relative to T_0 (1.45 ± 0.15) in the loading phase. Immediately after unloading (T_1), the activity increased (1.80 ± 0.18), following which the activity returned to baseline at T_2 (1.65 ± 1.7) and remained unchanged through the subsequent epochs (T_2–T_5). EMG activity during the mid-stance phase was significantly reduced during unloaded walking (0.66 ± 0.07) relative to T_0 (0.96 ± 0.10) and returned to baseline at T_1 (0.97 ± 0.11). However, it was subsequently reduced at epochs T_2 (0.83 ± 0.08) and T_3 (0.86 ± 0.09), following which it returned to baseline at T_4 (0.84 ± 0.1) and remained unchanged at T_5 (0.83 ± 0.09). During pre-swing, there was no significant change in activity during unloading and immediately after unloading at T_1, even though there was a major trend towards reduced activity. However, the activity was significantly reduced starting from T_2 (0.8 ± 0.07) through subsequent epochs (T_3: 0.81 ± 0.08, T_4: 0.80 ± 0.07, and T_5: 0.80 ± 0.06) relative to T_0 (1.08 ± 0.11). The EMG activity during mid-swing was significantly reduced during unloaded walking (0.61 ± 0.06) relative to T_0 (0.80 ± 0.05) and returned to baseline immediately after unloading at T_1 and was not different from that of T_0. Subsequently, it was reduced from T_2 through T_5 (T_2: 0.66 ± 0.05, T_3: 0.69 ± 0.06, T_4: 0.66 ± 0.07, and T_5: 0.68 ± 0.07). Similar to the mid-swing phase, EMG activity was also reduced in the terminal swing phase, during unloaded walking (0.85 ± 0.09) relative to T_0 (1.25 ± 0.10). However, it was not different from that of T_0 after unloading from T_1 through T_5. These changes are summarized in Figure 7.

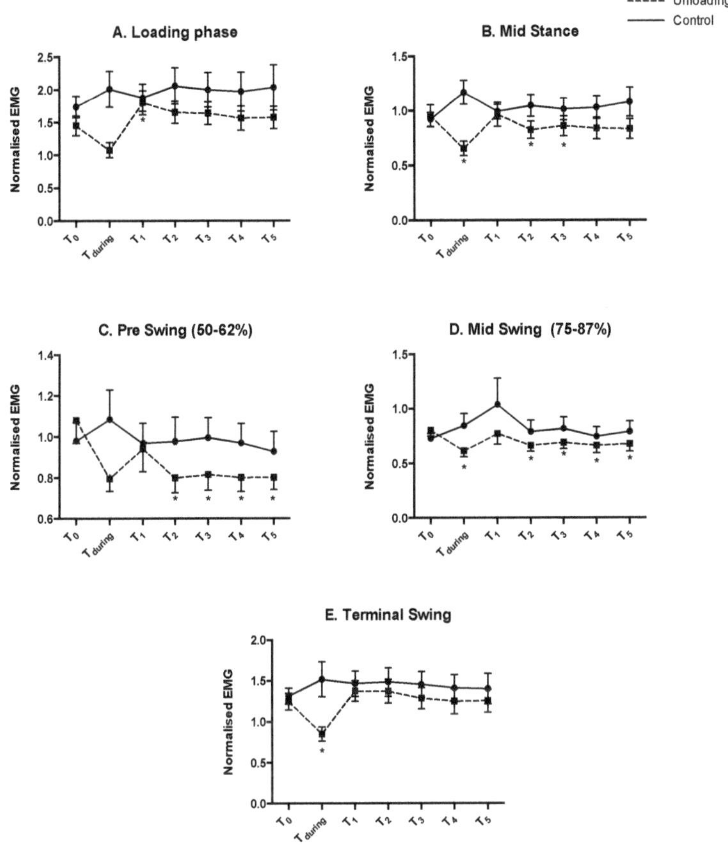

Figure 7. Group means (±SEM) of EMG activity of the rectus femoris (RF) muscle across different epochs during the (**A**) loading, (**B**) mid-stance, (**C**) pre-swing, (**D**) mid-swing, and (**E**) terminal swing phases for control and unloading conditions. * Significant difference ($p < 0.05$) relative to T_0 of the corresponding condition.

3.4.2. Biceps Femoris Activity

There were no changes in BF activity after the adaptation session during any of the gait phases in the control condition. However, the BF activity changed in the unloading condition during the terminal stance and mid-swing phases as a function of time. During the terminal stance phase, EMG activity remained unchanged during and immediately after unloaded walking at T_1 and T_2. However, it was reduced relative to T_0 (0.58 ± 0.07) starting from T_3 (T_3: 0.43 ± 0.07; T_4: 0.43 ± 0.07) through T_5 (0.41 ± 0.05). BF activity during the mid-swing phase was reduced during unloaded walking (0.83 ± 0.13) relative to T_0 (1.37 ± 0.17) and was not different from that of T_0 from T_1 through T_5 (Figure 8).

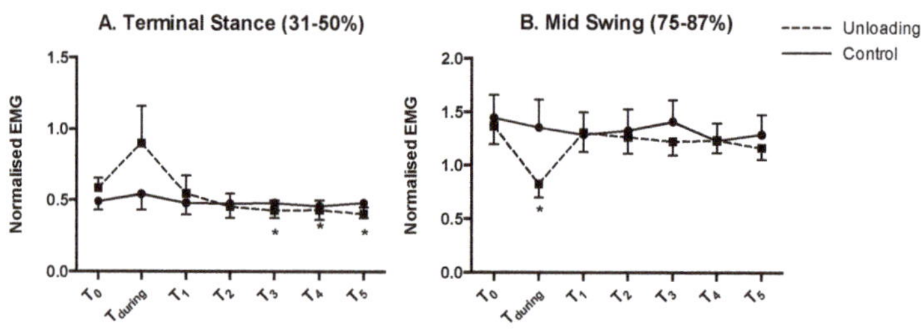

Figure 8. Group means (±SEM) of biceps femoris (BF) muscle activity during the (**A**) terminal stance and (**B**) mid-swing phases of the gait cycle during control and unloading conditions across different epochs. * Significant difference ($p < 0.05$) relative to T_0 of the corresponding condition.

3.4.3. Gastrocnemius Activity

EMG activity of GA remained unchanged relative to baseline during each of the seven phases in the control condition. There was a significant effect of time for the unloading condition during the mid-stance (F (2.14, 21.43) = 8.46, $p = 0.002$) and terminal stance phases (F (5.33,53.28) = 32.902, $p < 0.001$). The EMG activity during both phases was significantly reduced during unloaded walking (mid-stance: 0.77 ± 0.15, terminal stance: 0.81 ± 0.13). However, it was not different relative to T_0 (mid-stance: 1.58 ± 0.22, terminal stance: 1.93 ± 0.20) from T_1 through T_5 (Figure 9).

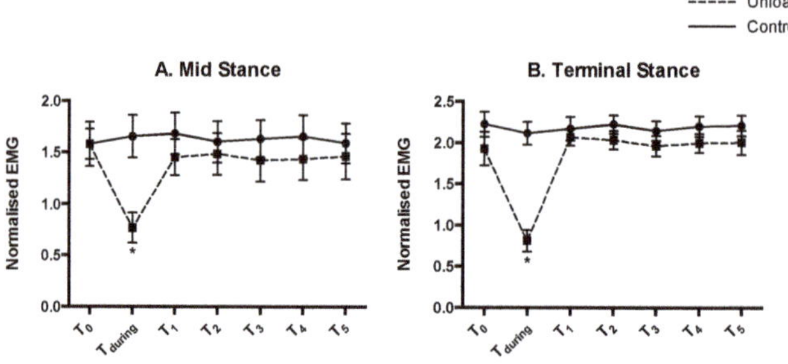

Figure 9. Group means (±SEM) of the medial head of gastrocnemius (GA) muscle activity during (**A**) mid-stance and (**B**) terminal stance during control and unloading conditions across different epochs. * Significant difference ($p < 0.05$) relative to T_0 of the corresponding condition.

3.4.4. Tibialis Anterior Activity

As in the case with other muscles, TA activity remained unchanged after the adaptation session in the control condition. There were changes in the muscle activity in the unloading condition only during the mid-swing phase. Specifically, the activity significantly increased immediately after unloaded walking relative to T_0 (1.00 ± 0.07) from T_1 (T_1: 1.19 ± 0.07, T_2: 1.20 ± 0.08, T_3: 1.18 ± 01, and T_4: 1.19 ± 0.1) through T_5 (1.24 ± 0.1) (Figure 10). Table 1 provides a summary of changes in EMG activity during and after unloaded walking across different epochs and gait phases.

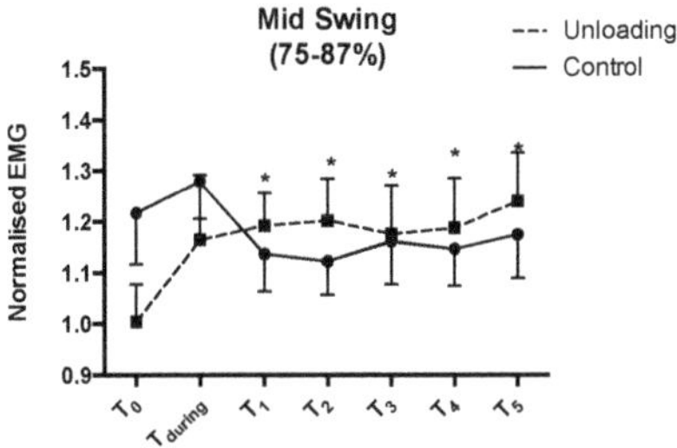

Figure 10. Group means (±SEM) of tibialis anterior (TA) muscle activity during the mid-swing phase of the gait cycle during control and unloading conditions across different epochs. * Significant difference ($p < 0.05$) relative to T_0 of the corresponding condition.

Table 1. A summary of changes in EMG activity of all four muscles across different epochs and gait phases: The upward arrow indicates a significant ($p < 0.05$) increase, and the downward arrow indicates a significant ($p < 0.05$) decrease in activity relative to T_0, while "-" indicates no significant changes relative to T_0.

RF	Loading Phase	Mid-Stance	Terminal Stance	Pre-Swing	Initial Swing	Mid-Swing	Terminal Swing
T_{during}	-	⇓	-	-	-	⇓	⇓
T_1	⇑	-	-	-	-	-	-
T_2	-	⇓	-	⇓	-	⇓	-
T_3	-	⇓	-	⇓	-	⇓	-
T_4	-	-	-	⇓	-	⇓	-
T_5	-	-	-	⇓	-	⇓	-
BF	Loading Phase	Mid-Stance	Terminal Stance	Pre-Swing	Initial Swing	Mid-Swing	Terminal Swing
T_{during}	-	-	-	-	-	⇓	-
T_1	-	-	-	-	-	-	-
T_2	-	-	-	-	-	-	-
T_3	-	-	⇓	-	-	-	-
T_4	-	-	⇓	-	-	-	-
T_5	-	-	⇓	-	-	-	-
GA	Loading Phase	Mid-Stance	Terminal Stance	Pre-Swing	Initial Swing	Mid-Swing	Terminal Swing
T_{during}	-	⇓	⇓	-	-	-	-
T_1	-	-	-	-	-	-	-
T_2	-	-	-	-	-	-	-
T_3	-	-	-	-	-	-	-
T_4	-	-	-	-	-	-	-
T_5	-	-	-	-	-	-	-
TA	Loading Phase	Mid-Stance	Terminal Stance	Pre-Swing	Initial Swing	Mid-Swing	Terminal Swing
T_{during}	-	-	-	-	-	-	-
T_1	-	-	-	-	-	⇑	-
T_2	-	-	-	-	-	⇑	-
T_3	-	-	-	-	-	⇑	-
T_4	-	-	-	-	-	⇑	-
T_5	-	-	-	-	-	⇑	-

4. Discussion

The main objective of this study was to investigate the changes in temporal, kinematics, and neuromuscular activity during locomotion produced by 30 min of walking at 38% bodyweight and the second was to test the adaptive changes in body-weight perception in response to 30 min of walking at reduced body weight as a function of movement context (walking, standing, and sitting). The results indicate that 30 min of unloaded walking modifies body-weight perception as a function of movement context. Further, it also modifies gait performance characterized by alterations in the movement of ankle and knee joints and neuromuscular activation patterns, some of which persist up to 10 min after the adaptation period. These results are individually discussed in detail in this section.

4.1. Body-Weight Perception

As expected, participants felt lighter during unloaded walking in all three test sessions. Their perception of body weight after unloaded walking varied as a function of the movement context. It significantly increased after unloading only during the gait session, while it returned to baseline immediately after unloading during the sessions when the participants were seated or stood statically. The increase in perceived body weight observed during the gait session aligns with the finding of increased rate of perceived exertion found after 3 min of unloaded running in a similar study [29]. The increase in perceived body weight only during the gait session could be due to the presence of active movements during walking as opposed to during sitting and standing.

In the literature related to sensory psychology, perception of active touch or touching has been shown to be different from passive touch or the act of being touched [44,45]. Active touch involves a combination of active movement and the sensory feedback that results from touching. Thus, the resulting stimulus during active touch contains two components: exterospecific and propriospecific information. Passive touch, on the other hand, involves only the sensory feedback of the stimulus that is applied on the skin. For example, when we use our hand to touch an object, the movement about all the joints present in the hand and the cutaneous inputs arising from contact are continuously integrated while shaping the percept of the object. However, when being touched by an external stimulus, only the sensory inputs from the cutaneous receptors in the skin and its underlying tissue are available. In other words, active touching involves both objective and subjective sensory information, thus making it comparatively a more enriching experience.

Extending the above theory to the context of the current study, the contexts of sitting and standing can be associated to a "passive experience" of the sensory environment. Conversely, walking can be associated with an "active experience" where the participants had the opportunity to "actively explore" the new sensory environment. Thus, it is possible that the perception of body weight during the active experience was modified as a result of changes in the relationship between the sensory and motor elements that were used in forming the percept of body weight. On the other hand, since only the sensory element related to body-load perception was used in forming the percept of body weight during passive experiences such as during sitting and standing, there was no change in perception of body weight. In summary, the findings of the current study support the notion that different movement contexts can have different sensory consequences associated with them.

4.2. Temporal Gait Measures

The results indicate that there were no significant changes in any of the temporal gait measures during and after unloaded walking, except for swing time, which was significantly increased during unloaded walking. The increase in swing time during unloaded walking corroborates the findings from previous studies [11,18,24,46,47]. This observation of increased swing time with body-weight unloading is in line with the predictions of the ballistic pendulum model of walking, which states that the motion of the swing leg is like that of a pendulum of which the oscillation period is inversely related to gravity. Additionally, the setup of LBPP by itself might also have contributed to this increase in swing time. Specifically, since the lower body is supported in this type of setup, there is less of a threat to postural instability, due to which one can afford to spend a longer time in the swing phase.

The lack of significant changes in temporal parameters following adaptation to unloading could be due to the constraints imposed on the walking performance by the speed and the dynamics of the treadmill and particularly due to the lack of stride to stride variability inherently associated with treadmill walking [48]. Thus, potentially testing the aftereffects of adaptation to unloading during over-ground walking as opposed to treadmill walking could provide additional insights into the adaptive process.

Although not statistically significant, there was a trend towards increased stance time accompanied by a decrease in swing time immediately after unloaded walking (at T_1) similar to that reported in another similar study [30]. These changes in stance and swing time after unloading have been proposed as a control strategy adopted by the motor control system to maintain prolonged foot contact with the ground and, hence, to maintain stable balance by increasing the stance time while also decreasing the time spent during the unstable swing phase to prevent the risk of falls following adaptation to unloaded walking. However, since the lower body was securely attached to the antigravity chamber in the current study, there was no risk of falling. Thus, similar trends in the adaptive changes of stance and swing times between the two studies indicate that these changes might have been a direct result of modulation of somatosensory information resulting from changes in body-load sensing mechanisms and not that due to postural instability.

4.3. Kinematics

The range of ankle and knee joint angular displacement was significantly reduced during unloaded walking as in other locomotor studies that were focused on assessing immediate online changes in kinematics in response to body-weight unloading [24,46,47,49]. There was reduced angular motion of ankle and knee joints in the form of aftereffects following the adaptation session (post-unloading) for up to three minutes. These modifications in kinematics similar to those observed during the adaptation phase were found to slowly decay and return to baseline levels by the end of the testing session. There were no changes in ankle or knee angular movement during or after the control adaptation session. This confirms that the kinematic changes observed after unloading are not caused by muscular fatigue.

Ruttley [40] found a significant increase in total ankle and knee angular movement from heel strike to peak knee motion during post-adaptation to unloaded walking. The disparity with the current results could be related to the differences in the nature of the unloading system. Unloading systems using a vertical harness produce inertial forces as a result of the systems mass. Thus, excessive joint excursions are produced to overcome these inertial forces while walking, which could have resulted in an aftereffect manifested as increases in knee and ankle joint excursion after unloaded walking in Reference [40]. Since unloaded walking inside the antigravity treadmill is not influenced by such mechanical constraints, the modifications observed in lower-limb movements in the current study might represent the pure aftereffects of adaptation to unloaded walking. Although the body-weight unloading techniques and the outcome measures used to quantify joint motion in the two studies were different, the fact that there was a distinct recovery curve in both studies indicates that there was an adaptive change in kinematics that was produced by unloaded walking.

4.4. EMG

4.4.1. During Adaptation

In line with the findings from other locomotor studies related to body-weight unloading, we found a significant reduction in the EMG activity of the extensor muscle GA during unloaded walking. This change was observed specifically during the mid-stance and terminal stance phases, which (during loaded walking) are the periods of peak muscle activity. GA enables controlled plantar flexion of the ankle joint in order to shift the center of gravity towards the front and allows lifting of the heel from the ground. The need for forward propulsion during push-off is expected to be reduced with reduced body load. A significant decrease in GA activity combined with reduced ankle-joint motion observed during unloaded walking in the current study supports this argument.

Additionally, we found a decrease in RF activity during mid-stance and later swing phases during unloaded walking. RF does not have a dominant role during mid-stance; however, it is known to play a predominant role during the transitory phases from stance to swing phases and vice versa. During the transition from swing to stance phases, it begins to prepare in a feedforward manner for the large ground reaction forces that the limbs will encounter upon heel strike during the next stance phase. Since the magnitude of the ground reaction forces decreases with unloading, the muscular effort required to counteract these forces at reduced body load is also less. This might be the reason why there was a reduced activity in RF during the later phases of the swing phase during unloaded walking. During the transition from the stance to swing phases, i.e., during the pre-swing phase, it acts as a hip flexor and helps in lifting and propelling the limb into swing. One would expect that the muscular effort required to propel the limb is reduced during unloaded walking. In line with this hypothesis, there was a trend towards reduced RF activity during the pre-swing phase as well, which approached significance ($p = 0.065$).

4.4.2. Post-Adaptation

Neuromuscular changes after unloading were most evident in RF and TA muscles. Additionally, these changes occurred close to two discrete events in the gait cycle which are characterized by large amounts of energy transfers, namely the heel strike and toe off. These observations are similar to those reported in spaceflight-related locomotor studies pointing towards some potentially common adaptive mechanisms [20,33,42]. As in the case with kinematics, there were no significant changes in neuromuscular activity of any of the muscles either during or after the control adaptation session. This further reiterates that the observed changes in neuromuscular activities after unloading are not caused due to fatigue but rather due to adaptation to unloaded walking. Although statistically insignificant, some visually evident changes in activity of some muscles were observed during certain phases of the gait cycle in the control condition. We speculate that these changes are mere random fluctuations as there are no specific methodological or physiological reasons as to why such changes might occur.

Around heel strike, we saw an increase in RF activity during the loading phase immediately after unloaded walking, combined with a reduction in knee and ankle angular excursion. This could be in response to the large ground reaction forces that the body encountered during the loading phase relative to that encountered during unloaded walking. Layne et al. [20] also found an increase in RF activity during the stance phase of the gait cycle after long-duration spaceflight. It has been suggested that modifications in RF activity combined with increased kinematic variability after spaceflight are attempts to attenuate the energy generated by the ground reaction forces around heel strike that are transmitted to the head [42,48,50,51]. Apart from heel-strike-specific modulations, we also observed a small reduction in RF and BF activity during the mid-stance and terminal stance phases of the gait cycle during some of the epochs. Although RF and BF have limited functional roles during these phases, these changes reflect an overall reinterpretation of sensory inputs as a result of adaptation to unloading.

Around toe off, there was a significant reduction in RF activity during the pre-swing phase after the first 100 strides (T_2) post-unloading adaptation session. As mentioned earlier, the RF activity during the pre-swing phase is responsible for lifting the leg so that it can be propelled forward during the swing phase. During unloading, as expected, we saw a reduction in activity during this phase as the muscular effort required to propel the leg is less. Extending this logic, we would expect the muscular effort post-unloading to increase due to the increase in body-weight load. As expected, the RF activity increased at T_1 relative to T_{during}. This increase in RF activity was however transient, as it was significantly reduced from T_2 up to 10 min. A reduction in RF activity was also combined with reduced angular excursion of ankle and knee joints during some epochs.

With regards to TA, we observed an increase in activity during the mid-swing phase. This must have been an attempt by the motor control system to compensate for reduced knee flexion by increasing its activity to increase the dorsiflexion about the ankle joint for adequate clearance of the toe. Since we did not observe any foot-scraping events in any of the subjects, it seems that the ankle dorsiflexion during the mid-swing phase was sufficient for toe clearance.

The above patterns of reduced RF activity during the pre-swing phase and increased TA activity during the mid-swing phase of the gait cycle were also observed after long-duration spaceflight during treadmill walking [20]. The fact that a short exposure of 30 min of body-weight unloading on Earth resulted in similar patterns of neuromuscular changes as those observed after long-duration spaceflight is interesting. This finding suggests that there might be some common adaptive mechanisms regardless of the duration of exposure to unloading. In general, modifications in neuromuscular activations after spaceflight have been suggested to be driven by alterations in the neural drive to the motor neurons. There is compelling evidence in the spaceflight literature that shows alterations in postural muscle activity caused by exposure to weightlessness [52–55]. Muscle disuse as well as muscle loss are also known to cause changes in neural drive after spaceflight. Furthermore, modifications in proprioceptive functioning has also been suggested to contribute to changes in neuromuscular activation.

4.5. Clinical Implications

Gait-training programs with BWU are designed to provide support to the patient's body weight to allow for the reestablishment of damaged sensorimotor pathways or emergence of new ones to restore normal walking patterns [22]. Gait alterations observed during and after unloaded walking in the current study suggests that lowering body weight to as low as 38% during gait training is not recommendable if we want to reduce the risk of altering normal gait characteristics. Practitioners should be vigilant about choosing the right combination of body-weight level and walking or running speed for training purposes. Particularly, they have to choose a level that does not alter normal kinematic patterns and neuromuscular activities to a large extent.

5. Conclusions

Alterations in kinematics and neuromuscular activities observed during unloaded walking are a result of the adaptation of the neuromuscular system to the reduction in ground reaction forces, shear forces, foot sole pressure, and joint loads associated with unloading. In particular, these changes are caused due to somatosensory-mediated central changes in the body schema produced by new relationships between sensory and motor elements that are characteristic to an unloaded environment. The continued alterations in kinematics and neuromuscular activities observed after unloading are aftereffects of the adapted state of the body schema. The recovery of kinematics and neuromuscular activity over the course of the post-adaptation phase are indicative of the recalibration process that the body schema undergoes in order to restore the original sensory motor relationships. Similarities in the pattern of changes in neuromuscular activation amplitudes between spaceflight and the current study indicate that there might be some common adaptive mechanisms that are mediated by load-related somatosensory changes. Additionally, these alterations in kinematics and neuromuscular characteristics caution practitioners to choose the optimal level of body-weight unloading for gait training.

6. Future Direction

Two important methodological limitations of this study are worth noting. Firstly, the choice of the level of body-weight unloading was limited to 38% body weight. This prevented us from capturing the adaptive effects across different levels of unloading. Future studies should aim to capture the dose-response relationship between level of unloading and the magnitude of adaptive changes in movement characteristics. This will allow us to determine the optimal level of unloading that can be useful for improving gait in patients while producing the least amount of alterations in the movement and neuromuscular activation patterns. Secondly, assessing gait performance during treadmill walking as opposed to during over-ground walking must have limited us from capturing true adaptive effects of unloading with regards to temporal gait measures.

Additionally, it will be worth exploring the effects of passive unloading during standing inside the antigravity chamber for extended periods of time. Comparing the effects of passive unloading and unloaded walking (active loading) as in the current study will help isolate the effects of inactivity from that of unloading.

Author Contributions: R.K. and C.S.L. participated in the conceptualization, methodology, validation, formal analysis, and writing –review and editing. R.K. participated in the software, investigation, data curation, writing-original draft preparation and visualization. C.S.L. participated in resources, supervision and project administration.

Funding: This research received no external funding.

Acknowledgments: The authors thank Mai Lee and David Young for their assistance in running experiments, Beom-Chan Lee for providing us access to the EMG recording system, and Ajitkumar Mulavara for providing valuable inputs during the conception of the study.

Conflicts of Interest: The authors declare no conflict of interest.

References

1. Shadmehr, R.; Smith, M.A.; Krakauer, J.W. Error correction, sensory prediction, and adaptation in motor control. *Annu. Rev. Neurosci.* **2010**, *33*, 89–108. [CrossRef] [PubMed]
2. Mazzoni, P.; Krakauer, J.W. An implicit plan overrides an explicit strategy during visuomotor adaptation. *J. Neurosci.* **2006**, *26*, 3642–3645. [CrossRef] [PubMed]
3. Adams, H.; Narasimham, G.; Rieser, J.; Creem-Regehr, S.; Stefanucci, J.; Bodenheimer, B. Locomotive Recalibration and Prism Adaptation of Children and Teens in Immersive Virtual Environments. *IEEE Trans. Vis. Comput. Graph.* **2018**, *24*, 1408–1417. [CrossRef] [PubMed]
4. Martin, T.A.; Keating, J.G.; Goodkin, H.P.; Bastian, A.J.; Thach, W.T. Throwing while looking through prisms. II. Specificity and storage of multiple gaze-throw calibrations. *Brain* **1996**, *119 Pt 4*, 1199–1211. [CrossRef]
5. Temple, D.R.; De Dios, Y.E.; Layne, C.S.; Bloomberg, J.J.; Mulavara, A.P. Efficacy of Stochastic Vestibular Stimulation to Improve Locomotor Performance During Adaptation to Visuomotor and Somatosensory Distortion. *Front. Physiol.* **2018**, *9*. [CrossRef]
6. Reisman, D.S.; Bastian, A.J.; Morton, S.M. Neurophysiologic and rehabilitation insights from the split-belt and other locomotor adaptation paradigms. *Phys. Ther.* **2010**, *90*, 187–195. [CrossRef]
7. Torres-Oviedo, G.; Vasudevan, E.; Malone, L.; Bastian, A.J. Locomotor adaptation. *Prog. Brain Res.* **2011**, *191*, 65–74.
8. Helm, E.E.; Reisman, D.S. The Split-Belt Walking Paradigm: Exploring Motor Learning and Spatiotemporal Asymmetry Poststroke. *Phys. Med. Rehabil. Clin. N. Am.* **2015**, *26*, 703–713. [CrossRef]
9. Gordon, C.R.; Fletcher, W.A.; Melvill Jones, G.; Block, E.W. Adaptive plasticity in the control of locomotor trajectory. *Exp. Brain Res.* **1995**, *102*, 540–545. [CrossRef]
10. Mulavara, A.P.; Richards, J.T.; Ruttley, T.; Marshburn, A.; Nomura, Y.; Bloomberg, J.J. Exposure to a rotating virtual environment during treadmill locomotion causes adaptation in heading direction. *Exp. Brain Res.* **2005**, *166*, 210–219. [CrossRef]
11. Davis, B.L.; Cavanagh, P.R.; Sommer, H.J.; Wu, G. Ground reaction forces during locomotion in simulated microgravity. *Aviat. Space Environ. Med.* **1996**, *67*, 235–242. [PubMed]
12. Ivanenko, Y.P.; Grasso, R.; Macellari, V.; Lacquaniti, F. Control of foot trajectory in human locomotion: Role of ground contact forces in simulated reduced gravity. *J. Neurophysiol.* **2002**, *87*, 3070–3089. [CrossRef] [PubMed]
13. Grabowski, A.M. Metabolic and biomechanical effects of velocity and weight support using a lower-body positive pressure device during walking. *Arch. Phys. Med. Rehabil.* **2010**, *91*, 951–957. [CrossRef] [PubMed]
14. Kurz, M.J.; Deffeyes, J.E.; Arpin, D.J.; Karst, G.M.; Stuberg, W.A. Influence of lower body pressure support on the walking patterns of healthy children and adults. *J. Appl. Biomech.* **2012**, *28*, 530–541. [CrossRef] [PubMed]
15. Takakusaki, K. Neurophysiology of gait: From the spinal cord to the frontal lobe. *Mov. Disord.* **2013**, *28*, 1483–1491. [CrossRef]
16. Nielsen, J.B.; Sinkjaer, T. Afferent feedback in the control of human gait. *J. Electromyogr. Kinesiol.* **2002**, *12*, 213–217. [CrossRef]
17. Pearson, K.G. Generating the walking gait: Role of sensory feedback. *Prog. Brain Res.* **2004**, *143*, 123–129.
18. Dietz, V.; Müller, R.; Colombo, G. Locomotor activity in spinal man: Significance of afferent input from joint and load receptors. *Brain* **2002**, *125*, 2626–2634. [CrossRef]
19. Harkema, S.J.; Hurley, S.L.; Patel, U.K.; Requejo, P.S.; Dobkin, B.H.; Edgerton, V.R. Human lumbosacral spinal cord interprets loading during stepping. *J. Neurophysiol.* **1997**, *77*, 797–811. [CrossRef]

20. Layne, C.S.; Lange, G.W.; Pruett, C.J.; McDonald, P.V.; Merkle, L.A.; Mulavara, A.P.; Smith, S.L.; Kozlovskaya, I.B.; Bloomberg, J.J. Adaptation of neuromuscular activation patterns during treadmill walking after long-duration space flight. *Acta Astronaut.* **1998**, *43*, 107–119. [CrossRef]

21. Takacs, J.; Anderson, J.E.; Leiter, J.R.; MacDonald, P.B.; Peeler, J.D. Lower body positive pressure: An emerging technology in the battle against knee osteoarthritis? *Clin. Interv. Aging* **2013**, *8*, 983–991. [PubMed]

22. Kurz, M.J.; Corr, B.; Stuberg, W.; Volkman, K.G.; Smith, N. Evaluation of lower body positive pressure supported treadmill training for children with cerebral palsy. *Pediatr. Phys. Ther.* **2011**, *23*, 232–239. [CrossRef] [PubMed]

23. Ruckstuhl, H.; Kho, J.; Weed, M.; Wilkinson, M.W.; Hargens, A.R. Comparing two devices of suspended treadmill walking by varying body unloading and Froude number. *Gait Posture* **2009**, *30*, 446–451. [CrossRef] [PubMed]

24. Finch, L.; Barbeau, H.; Arsenault, B. Influence of body weight support on normal human gait: Development of a gait retraining strategy. *Phys. Ther.* **1991**, *71*, 842–855, discussion 855–856. [CrossRef] [PubMed]

25. Sylos-Labini, F.; Lacquaniti, F.; Ivanenko, Y.P. Human locomotion under reduced gravity conditions: Biomechanical and neurophysiological considerations. *Biomed. Res. Int.* **2014**, *2014*, 547242. [CrossRef] [PubMed]

26. Newman, D.J.; Alexander, H.L.; Webbon, B.W. Energetics and mechanics for partial gravity locomotion. *Aviat. Space Environ. Med.* **1994**, *65*, 815–823.

27. Fischer, A.G.; Wolf, A. Assessment of the effects of body weight unloading on overground gait biomechanical parameters. *Clin. Biomech. (Bristol, Avon)* **2015**, *30*, 454–461. [CrossRef]

28. Apte, S.; Plooij, M.; Vallery, H. Influence of body weight unloading on human gait characteristics: A systematic review. *J. NeuroEng. Rehabil.* **2018**, *15*, 53. [CrossRef]

29. Sainton, P.; Nicol, C.; Cabri, J.; Barthelemy-Montfort, J.; Berton, E.; Chavet, P. Influence of short-term unweighing and reloading on running kinetics and muscle activity. *Eur. J. Appl. Physiol.* **2015**, *115*, 1135–1145. [CrossRef]

30. Ruttley, T.M. *The Role of Load-Regulating Mechanisms in Gaze Stabilization during Locomotion*; University of Texas Medical Branch: Galveston, TX, USA, 2007.

31. Reschke, M.F.; Bloomberg, J.J.; Harm, D.L.; Paloski, W.H.; Layne, C.; McDonald, V. Posture, locomotion, spatial orientation, and motion sickness as a function of space flight. *Brain Res. Rev.* **1998**, *28*, 102–117. [CrossRef]

32. Mulavara, A.P.; Peters, B.T.; Miller, C.A.; Kofman, I.S.; Reschke, M.F.; Taylor, L.C.; Lawrence, E.L.; Wood, S.J.; Laurie, S.S.; Lee, S.M.C.; et al. Physiological and Functional Alterations after Spaceflight and Bed Rest. *Med. Sci. Sports Exerc.* **2018**. [CrossRef] [PubMed]

33. Layne, C.; Mulavara, A.P.; McDonald, P.V.; Pruett, C.J.; Kozlovskaya, I.B.; Bloomberg, J. Alterations in human neuromuscular activation during overground locomotion after long-duration spaceflight. *J. Gravit. Physiol.* **2004**, 1–16.

34. Layne, C.S.; Spooner, B.S. Microgravity effects on "postural" muscle activity patterns. *Adv. Space Res.* **1994**, *14*, 381–384. [CrossRef]

35. Berthelsen, M.P.; Husu, E.; Christensen, S.B.; Prahm, K.P.; Vissing, J.; Jensen, B.R. Anti-gravity training improves walking capacity and postural balance in patients with muscular dystrophy. *Neuromuscul. Disord.* **2014**, *24*, 492–498. [CrossRef] [PubMed]

36. Birgani, P.M.; Ashtiyani, M.; Rasooli, A.; Shahrokhnia, M.; Shahrokhi, A.; Mirbagheri, M.M. Can an anti-gravity treadmill improve stability of children with cerebral palsy? In Proceedings of the 2016 38th Annual International Conference of the IEEE Engineering in Medicine and Biology Society (EMBC), Orlando, FL, USA, 16–20 August 2016; pp. 5465–5468.

37. Thomas, S.; Reading, J.; Shephard, R.J. Revision of the Physical Activity Readiness Questionnaire (PAR-Q). *Can. J. Sport Sci.* **1992**, *17*, 338–345. [PubMed]

38. Fraccaro, P.; Coyle, L.; Doyle, J.; O'Sullivan, D. Real-world Gyroscope-based Gait Event Detection and Gait Feature Extraction. In Proceedings of the eTELEMED, The Sixth International Conference on eHealth, Telemedicine, and Social Medicine, Barcelona, Spain, 24–27 March 2014; pp. 247–252.

39. Winter, D.A. Human balance and posture control during standing and walking. *Gait Posture* **1995**, *3*, 193–214. [CrossRef]

40. den Otter, A.R.; Geurts, A.C.H.; Mulder, T.; Duysens, J. Speed related changes in muscle activity from normal to very slow walking speeds. *Gait Posture* **2004**, *19*, 270–278. [CrossRef]

41. Shiavi, R.; Frigo, C.; Pedotti, A. Electromyographic signals during gait: Criteria for envelope filtering and number of strides. *Med. Biol. Eng. Comput.* **1998**, *36*, 171–178. [CrossRef]

42. Layne, C.S.; McDonald, P.V.; Bloomberg, J.J. Neuromuscular activation patterns during treadmill walking after space flight. *Exp. Brain Res.* **1997**, *113*, 104–116. [CrossRef]

43. Kharb, A.; Saini, V.; Jain, Y.K.; Dhiman, S. A review of gait cycle and its parameters. *IJCEM Int. J. Computat. Eng. Manag.* **2011**, *13*, 78–83.

44. Gibson, J.J. *The Senses Considered as Perceptual Systems*; Houghton Mifflin: Oxford, UK, 1966.

45. Gibson, J.J. Observations on active touch. *Psychol. Rev.* **1962**, *69*, 477–491. [CrossRef] [PubMed]

46. Awai, L.; Franz, M.; Easthope, C.S.; Vallery, H.; Curt, A.; Bolliger, M. Preserved gait kinematics during controlled body unloading. *J. Neuroeng. Rehabil.* **2017**, *14*, 25. [CrossRef] [PubMed]

47. Donelan, J.M.; Kram, R. The effect of reduced gravity on the kinematics of human walking: A test of the dynamic similarity hypothesis for locomotion. *J. Exp. Biol.* **1997**, *200*, 3193–3201. [PubMed]

48. McDonald, P.V.; Basdogan, C.; Bloomberg, J.J.; Layne, C.S. Lower limb kinematics during treadmill walking after space flight: Implications for gaze stabilization. *Exp. Brain Res.* **1996**, *112*, 325–334. [CrossRef]

49. Lewek, M.D. The influence of body weight support on ankle mechanics during treadmill walking. *J. Biomech.* **2011**, *44*, 128–133. [CrossRef]

50. Mulavara, A.P.; Ruttley, T.; Cohen, H.S.; Peters, B.T.; Miller, C.; Brady, R.; Merkle, L.; Bloomberg, J.J. Vestibular-somatosensory convergence in head movement control during locomotion after long-duration space flight. *J. Vestib. Res.* **2012**, *22*, 153–166.

51. Bloomberg, J.J.; Peters, B.T.; Smith, S.L.; Huebner, W.P.; Reschke, M.F. Locomotor head-trunk coordination strategies following space flight. *J. Vestib. Res.* **1997**, *7*, 161–177. [CrossRef]

52. Clément, G.; André-Deshays, C. Motor activity and visually induced postural reactions during two-g and zero-g phases of parabolic flight. *Neurosci. Lett.* **1987**, *79*, 113–116. [CrossRef]

53. Clément, G.; Gurfinkel, V.S.; Lestienne, F.; Lipshits, M.I.; Popov, K.E. Changes of posture during transient perturbations in microgravity. *Aviat. Space Environ. Med.* **1985**, *56*, 666–671.

54. Lestienne, F.G.; Gurfinkel, V.S. Postural control in weightlessness: A dual process underlying adaptation to an unusual environment. *Trends Neurosci.* **1988**, *11*, 359–363. [CrossRef]

55. Pöyhönen, T.; Avela, J. Effect of head-out water immersion on neuromuscular function of the plantarflexor muscles. *Aviat. Space Environ. Med.* **2002**, *73*, 1215–1218. [PubMed]

Article

Postural Control in Children with Cerebellar Ataxia

Veronica Farinelli [1], **Chiara Palmisano** [2,3], **Silvia Maria Marchese** [1,*],
Camilla Mirella Maria Strano [1], **Stefano D'Arrigo** [4], **Chiara Pantaleoni** [4], **Anna Ardissone** [5],
Nardo Nardocci [5], **Roberto Esposti** [1] and **Paolo Cavallari** [1]

[1] Human Physiology Section of the Department of Pathophysiology and Transplantation, Università degli
   Studi di Milano, 20133 Milano, Italy; veronica.farinelli@unimi.it (V.F.); camistra94@gmail.com (C.M.M.S.);
   roberto.esposti@unimi.it (R.E.); paolo.cavallari@unimi.it (P.C.)
[2] Movement Biomechanics and Motor Control Lab, Department of Electronic Information and Bioengineering,
   Politecnico di Milano, 20133 Milano, Italy; chiara.palmisano@polimi.it
[3] Department of Neurology, University Hospital and Julius-Maximilian University Würzburg,
   97080 Würzburg, Germany
[4] Developmental Neurology Department, Fondazione IRCCS Istituto Neurologico C. Besta, 20133 Besta, Italy;
   Stefano.Darrigo@istituto-besta.it (S.D.); chiara.pantaleoni@istituto-besta.it (C.P.)
[5] UOC Neuropsichiatria Infantile, Fondazione IRCCS Istituto Neurologico C. Besta, 20133 Besta, Italy;
   Anna.Ardissone@istituto-besta.it (A.A.); nardo.nardocci@istituto-besta.it (N.N.)
* Correspondence: silvia.marchese@unimi.it

Received: 5 December 2019; Accepted: 20 February 2020; Published: 28 February 2020

Abstract: Controlling posture, i.e., governing the ensemble of involuntary muscular activities that manage body equilibrium, represents a demanding function in which the cerebellum plays a key role. Postural activities are particularly important during gait initiation when passing from quiet standing to locomotion. Indeed, several studies used such motor task for evaluating pathological conditions, including cerebellar disorders. The linkage between cerebellum maturation and the development of postural control has received less attention. Therefore, we evaluated postural control during quiet standing and gait initiation in children affected by a slow progressive generalized cerebellar atrophy (SlowP) or non-progressive vermian hypoplasia (Joubert syndrome, NonP), compared to that of healthy children (H). Despite the similar clinical evaluation of motor impairments in NonP and SlowP, only SlowP showed a less stable quiet standing and a shorter and slower first step than H. Moreover, a descriptive analysis of lower limb and back muscle activities suggested a more severe timing disruption in SlowP. Such differences might stem from the extent of cerebellar damage. However, literature reports that during childhood, neural plasticity of intact brain areas could compensate for cerebellar agenesis. We thus proposed that the difference might stem from disease progression, which contrasts the consolidation of compensatory strategies.

Keywords: children; gait initiation; postural control; generalized cerebellar atrophy; cerebellar vermis hypoplasia; progressive ataxia; compensatory strategies

---

1. Introduction

Postural adjustments are involuntary muscular activities that accompany the voluntary movement. These activities spread over adjacent muscles and thus create "chains" that reach the available support points (in many cases, the ground). Such chains allow fine-tuning the body equilibrium, in order to adapt it to the mechanical needs of the ensuing movement. For example, when flexing both arms at the shoulder, postural actions develop in a dorsal muscle chain, including Erector Spinae (ES), Biceps Femoris (BF), and Soleus (SOL), to counteract the reaction force due to arm movement [1]. Instead, when rising on tiptoes, involuntary bursts of activity develop in Tibialis Anterior (TA) muscles, so as to induce a forward fall of the Centre of Mass (COM); in this way, COM reaches the forefoot, and the voluntary contraction of Soleus muscles (SOL) rises the body [2]. Otherwise, simply recruiting SOL would produce a backward fall. Whenever the mechanical needs of action may be estimated beforehand, like when programming a voluntary movement, appropriate postural actions are usually produced in advance of the movement itself, witnessing that such Anticipatory Postural Adjustments (APAs) are programmed in a feed-forward way [3–6].

APAs are particularly evident in gait initiation, in which they maintain the body's dynamic balance and create the propulsive forces to move the COM forwards. In healthy adults performing gait initiation [7–12], the Center of Pressure (CoP), i.e., the barycenter of the ground reaction forces, first moves backward and towards the future swing foot. The onset of such a CoP shift is usually called APA onset, while its time period is called the *imbalance phase*. Indeed, the ensuing horizontal gap between CoP and COM (where the gravity force vector is applied) produces an "imbalance" torque that pushes COM forwards and towards the future stance foot. Then, CoP moves laterally towards the stance foot, continuing to promote the forward acceleration of COM while braking its lateral fall. At the same time, this CoP shift withdraws body weight from the swing foot, hence the name *unloading phase*. Finally, as COM proceeds, CoP travels forwards along the stance foot, from toe-off to heel-strike of the swing foot (*first swing*) [13]. Considering the muscular actions that drive gait initiation, before the beginning of the imbalance phase, an inhibition occurs in the background EMG (electromyographic) activity of both SOL muscles, which are tonically active during quiet stance. SOL inhibitions are shortly followed by the recruitment of TA muscles, which are silent during quiet stance and activate close to the APA onset. In particular, in the stance leg, the SOL inhibition precedes TA excitation by about 100 ms [7]. A drop in the background activity is also observed in other dorsal muscles, like BF and ES, while bursts of activity occur in ventral muscles, like Rectus Femoris (RF) [14].

Less literature is available on APAs during gait initiation in children. A systematic survey by Ledebt et al. (1998) [15] showed that APAs start developing at 2–3 years of age but complete their maturation well after the age of 8, a result in line with the observations carried out on toddlers up to 5 years old children [16] and in 4–6 years old children [17]. Another interesting study was published by Isaias et al. (2014) [18], who analyzed SOL and TA in 10 ± 3 years-old children and reported inhibition-excitation patterns like in adults, but with a lower time interval between SOL inhibition and TA excitation.

Several studies showed altered gait initiation in those neurological diseases characterized by poor motor control, as Rett syndrome [18], Parkinson's disease [19], and cerebellar pathologies [20,21]. In particular, Timmann and Horak [21] reported that adults with cerebellar deficits showed a decreased force production and a significant reduction of the length and peak velocity of the first step, accompanied by impairments in the predictive adaptation of APAs to the mechanical needs of gait initiation. Despite these authors also found that the temporal parameters of APAs were overall preserved in patients with cerebellar disease, several other works [22–26] addressed the role of cerebellum in postural control and provided evidence that such structure is involved not only in modulating rate and force of muscle activities but also in determining their relative timing. Indeed, cerebellar deficits often lead to dysfunctional co-contractions [22,24,26]. In this regard, it is worth recalling the involvement of the cerebellum in building up the temporal pattern of APAs, in particular, its ability to create and store internal models of body mechanics. This is proved by the contribution of the cerebellum in modulating sensory-motor interactions and integrating feed-forward and feed-back modes [27].

The cerebellum is also known to play an important role in many developmental disorders [28]. Nevertheless, very little attention has been given to the linkage between the development of postural control and the maturation of such neural structure. Aiming to elucidate this topic, we explored quiet stance and gait initiation in children affected by Pediatric Cerebellar Ataxia (PCA), used as a model of cerebellar dysfunction vs. a healthy control group of comparable age.

PCAs are a heterogeneous group of cerebellar developmental disorders characterized by dysfunctional motor coordination and very early cerebellar symptoms. The first clinical signs are marked hypotonia, wobbling gait, dysmetria, dysarthria, and a significant developmental delay. Most children show also marked speech impairment and cognitive deficits. In some cases, the cerebellum degenerates with time, but so slowly that it becomes difficult to classify the disorder as progressive or not [29]. In this framework, we studied a group of children with generalized cerebellar atrophy and clinical evidence of slow progression during follow-up (SlowP). In other cases, the disease has a proven non-progressive course, as in Joubert syndrome, which is characterized by cerebellar hypoplasia limited to the vermis and peduncles [30]. Thus, we also considered the second group of children (non-progressive, NonP) affected by this kind of pathology. It is also important to note that the onset of the SlowP pathology is clinically indistinguishable from that of the NonP forms; therefore, practically, both diseases are present since birth.

In order to characterize quiet stance and gait initiation, we measured the classical posturographic parameters [31,32] and the first step length and velocity (as measures of performance). Besides, we also calculated the shifts of CoP and COM, as well as the horizontal distance (gap) between these two points in the imbalance and unloading phases, to highlight the net effect on COM dynamics. In order to document possible changes in muscular APAs, we evaluated the EMG activities that accompany the APA onset. Should we observe significant differences between each pathological group and healthy children, this would point out the involvement of the cerebellum also during the key phase of human maturation, in which the central nervous system learns gait initiation dynamics and how to optimize this motor process. Moreover, these findings would be fruitful in tailoring rehabilitation for such pathologies. Finally, a different motor pattern in children with SlowP vs. NonP would suggest possible compensation mechanisms. In particular, taking into consideration that children with SlowP suffer from generalized cerebellar damage, better motor behavior in SlowP vs. NonP could suggest the involvement of extracerebellar regions. On the other side, better behavior in children of the NonP group could as well stem from a compensatory involvement of the cerebellar hemispheres, which are unaffected by Joubert syndrome.

2. Materials and Methods

2.1. Participants

Thirteen participants with PCA were recruited at the "Istituto Neurologico Carlo Besta" of Milan: seven of them had radiological signs of generalized cerebellar atrophy and clinical evidence of slow progression during follow-up (SlowP, mean age: 12 ± 3 years), while the remaining six suffered from Joubert syndrome, i.e., a proven non-progressive pathology (NonP, mean age: 12 ± 3 years). All of them underwent clinical evaluation, including the administration of the Scale for the Assessment and Rating of Ataxia (SARA, [33]), an MRI scan for establishing the cerebellar alteration (atrophy and/or hypoplasia), and genetic screening. In particular, all children with Joubert syndrome showed a unique cerebellar and brainstem malformation known as the "molar tooth sign" [30]. The demographic and clinical data of each patient are reported in Table 1.

Seven healthy children, free from neurological or psychological pathologies and typically developing, were enrolled as a control group, from the primary school "FAES" in Milan (H, 4 males and 3 females, mean age: 10 ± 3 years). The experimental procedure was carried out in accordance with the standards of the Declaration of Helsinki. The Ethical Committee "Comitato Etico di Ateneo dell'Università degli Studi di Milano" approved the study and the written consent procedure, on 15 February 2016 (counsel 5/16). Before each acquisition, the child neuropsychiatrist and the experimenters explained the aim of the study and the details of the experimental procedure to the parents and to their child. Only those children who agreed with the study participated in the experiments. The parents of each participant, as her/his legal representatives, signed the consent procedure. All children were perfectly aware of the task since no one of them failed in accomplishing it.

Table 1. Demographic and clinical characteristics of children with PCA. SlowP: slowly progressive ataxia; NonP: non-progressive ataxia (Joubert syndrome); EXOSC3: exosome component 3; KCNC3: potassium voltage-gated channel subfamily C member 3; ADCK3: aarF domain-containing kinase 3; NPHP1: nephrocystin 1; AHI1: Abelson helper integration site 1; SUFU: negative regulator of hedgehog signaling; PCA: pediatric cerebellar ataxia. Details about a molecular diagnosis can be found at [34].

Patient	Age	Gender	Molecular Diagnosis	SARA
SlowP_01	9	M	mutation in a candidate gene	8
SlowP_02	8	M	mutation in a candidate gene	14
SlowP_03	12	M	EXOSC3: c.572G > A	15
SlowP_04	13	F	KCNC3: c.1268G > A	17
SlowP_05	13	F	to be evaluated	13
SlowP_06	16	F	ADCK3: c547C > T; c1042C > T	13
SlowP_07	17	M	to be evaluated	18
NonP_01	10	M	NPHP1: c.1358G > T; c.1438-4C > T	12
NonP_02	12	F	to be evaluated	15
NonP_03	18	M	to be evaluated	14
NonP_04	9	M	AHI1: c.1829G > C; c.2671C > T	15
NonP_05	12	F	SUFU: c.1217T > C	11
NonP_06	9	M	SUFU: c.1217T > C	15.5

2.2. Experimental Protocol

Subjects were asked to perform a gait initiation task several times. They were instructed to stand quietly on a force plate for 30 s and then to walk at their natural speed after a vocal prompt, self-selecting the leading limb [35]. After three to five steps, subjects stopped and returned to the initial position.

Each subject repeated the task until three *valid* trials were collected (i.e., trials in which the subject did not move the feet, arms, or head during the quiet stance preceding gait initiation). Subjects were allowed to rest 2 min before repeating the task. A maximum of nine trials was required to satisfy the above criterion; the average number of trials per subject was 6.6 ± 2.7. At the end of the resting periods between motor tasks, the experimenter asked the child, "Do you feel fatigued? Are you ready to start again?". Moreover, since the parents assisted at the trials, they could report to the experimenter any possible discomfort of their child. Neither children nor their parents complained about fatigue. The width of the base of support was self-selected by each subject in the first trial, then marked on the platform with adhesive tape and kept fixed for all further trials. The distance between lateral malleoli during quiet stance was comparable among groups (SlowP: median = 162.8 mm, range = 115.6 to 205.6 mm; NonP: 165.6 mm, 152.3 to 198.5 mm; H: 150.3 mm, 147.5 to 242.3 mm), with no significant differences (Kruskal–Wallis $p = 0.566$).

2.3. Recordings

Body kinematics was recorded by means of a six-cameras optoelectronic system (SMART-E, BTS, Milan, Italy) using a full-body marker set [36], which allowed estimating the Centre of Mass (COM) and its trajectory, according to Isaias et al. (2014) [18]. A dynamometric force plate (9286AA, KISTLER, Winterthur, Switzerland) was used to compute the Center of Pressure (CoP) position. Wireless probes (FREEEMG 1000, BTS, Milan, Italy) were employed bilaterally to record the surface electromyographic (EMG) activity of Tibialis Anterior (TA), Soleus (SOL), Rectus Femoris (RF), Biceps Femoris (BF), and Erector Spinae (ES). Electrodes were placed according to the Surface Electromyography for the Non-Invasive Assessment of Muscles (SENIAM) guidelines [37]. Synchronous data acquisition was accomplished by the SMART-E workstation; sampling rate being 60 Hz for optoelectronic cameras, 960 Hz for dynamometric signals, and 1000 Hz for EMGs.

2.4. Data Processing

During the 30 s quiet standing period, the statokinesigram, i.e., the trajectory of the CoP in the horizontal plane, was used to extract specific indexes of balance control. These indexes were: the area and the eccentricity of the ellipse containing 95% of CoP positions, the total length of CoP trajectory (CoP length), the average CoP velocity, and the peak-to-peak Mediolateral and Anteroposterior CoP displacements (ML and AP ranges, respectively) [31,32]. In particular, the ellipse area (A) and eccentricity (*e*) were calculated according to the following formulae:

$$A = a * b * \pi \tag{1}$$

$$e = \frac{\sqrt{|a^2 - b^2|}}{a} \tag{2}$$

in which *a* and *b* were the semi-major axis (i.e., half of the ellipse longest diameter) and the semi-minor axis (i.e., half of the shortest diameter), respectively.

Gait initiation was subdivided into three phases [13]: the imbalance phase, in which CoP moves backward and towards the future swing foot; the unloading phase, in which CoP moves laterally towards the stance foot, and the first swing, in which CoP moves forwards along the stance foot, from toe-off to heel-strike of the swing foot. The temporal events delimiting each phase were determined by visual inspection of the CoP trajectory; in particular, the onset of the CoP backward shift represented the APA onset.

For the *imbalance* and *unloading* phases, separately, we measured the phase duration, the length of the CoP trajectory, the maximum AP and ML shifts of both CoP and COM, and the distance from CoP to COM projection on the horizontal plane, at the end of the phase.

The *first swing* phase was evaluated by measuring the length of the first step, normalized to the lower limb length (*LL*), and its velocity (*v*), expressed in Froude number ($Fr = \frac{v}{\sqrt{g*LL}}$, *g* being gravity acceleration [38]); also these measurements were obtained from kinematic data.

For each subject, the kinematic and dynamometric variables were averaged over the three valid trials recorded. Data normality was evaluated by means of the Shapiro–Wilk test. Considering that data were not normally distributed, the differences among SlowP, NonP, and H groups were analyzed non-parametrically by using Kruskal–Wallis tests followed by Dunn post hoc, with Bonferroni adjustment. The level of significance was set to 0.05.

The analysis of EMG recordings regarded the timing of muscle activation or inhibition, surrounding the APA onset at gait initiation. Raw EMG data were high-pass filtered ($f_{cut} = 50$ Hz) with a zero-phase shift 6th-order elliptic filter, to remove movement artifacts, and then the signals were rectified. For each muscle, the traces of the recorded trials were time-aligned to the APA onset and averaged across trials. For each averaged EMG track, the period from 1 s to 0.5 s before the APA onset—where no EMG changes were observed—was assumed as reference. The signal was integrated (time constant = 11 ms), and the mean level in the reference period was subtracted. The onset of an excitatory or inhibitory EMG change was identified by a software algorithm, which searched the first time point in which the track fell above or below 2 SD of the reference signal (excitation or inhibition, respectively) and remained there for at least 50 ms. Whenever the criterion was met, the algorithm searched backward the point in which the signal started to deviate from the mean reference value [39]. No statistical analysis was performed on EMG timings because of the many cases in which no clear inhibitory or excitatory changes could be identified.

3. Results

3.1. Postural Parameters

The analysis of quiet stance (static posturography) highlighted alterations of postural control in subjects with cerebellar deficits. In fact, children of both the SlowP and NonP groups showed an ellipse area greater than H (Kruskal–Wallis p = 0.039), mainly due to a greater CoP displacement in the mediolateral direction (Kruskal–Wallis p = 0.022).

However, post hoc showed that this difference was statistically significant only in subjects with SlowP vs. H (Table 2). In particular, the latter group revealed an inversion of the normal ellipse configuration, with mediolateral oscillation as the preferred direction (Figure 1), which led to a reduction of the ellipse eccentricity.

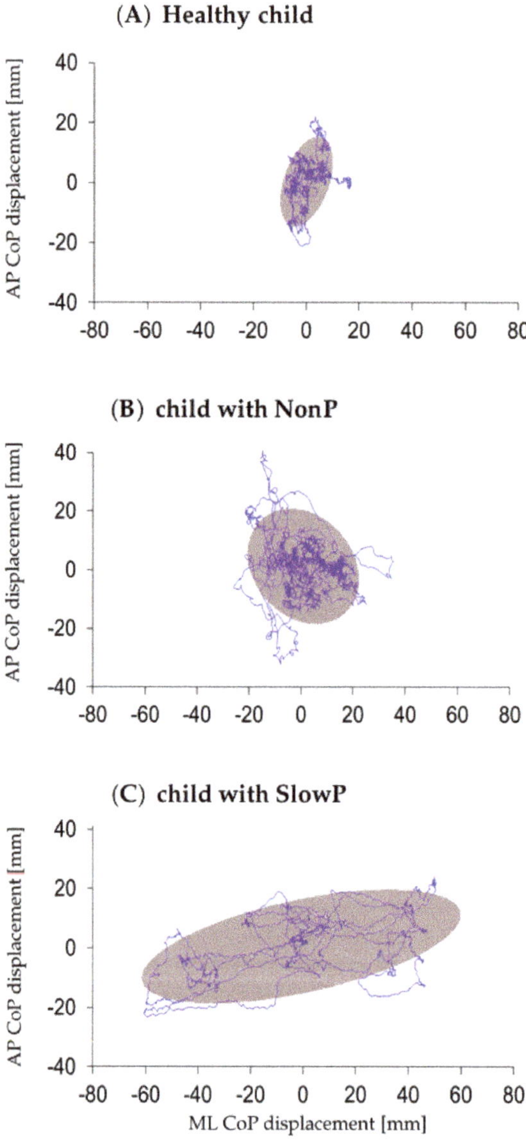

Figure 1. Statokinesigram, with 95% confidence ellipse, for representative children of the three groups: healthy (H, panel **A**), with non-progressive PCA (NonP, **B**), and with slow progressive PCA (SlowP, **C**). ML: mediolateral; AP: anteroposterior; PCA: pediatric cerebellar ataxia.

Table 2. Postural parameters during quiet stance. Data are shown as *median value (range)* for children with slow progressive PCA (SlowP), children with non-progressive PCA (NonP), and healthy children (H). ML: mediolateral; AP: anteroposterior; CoP: center of pressure. * Significant difference between SlowP and H groups, Dunn test $p < 0.05$.

	SlowP		NonP		H	
CoP length (mm)	1924	(1222 to 2983)	1481	(893 to 2911)	1594	(895 to 2309)
Average CoP velocity (mm/s)	64	(40 to 99)	49	(29 to 110)	53	(29 to 76)
ML range (mm)	* 37	(24 to 97)	29.72	(13 to 85)	* 19	(11 to 33)
AP range (mm)	39	(28 to 57)	31	(18.39 to 77.96)	26	(15 to 60)
Ellipse area (mm^2)	* 564	(354 to 2857)	447	(123 to 2075)	* 208	(54 to 609)
Ellipse eccentricity	0.68	(0.48 to 0.87)	0.73	(0.65 to 0.93)	0.79	(0.51 to 0.83)

3.2. Gait Initiation Parameters

The spatial and temporal parameters during the imbalance and unloading phases were not significantly affected by the pathology (Table 3). First step length and velocity were instead different among the three groups (Kruskal–Wallis, $p = 0.005$ for length and $p = 0.019$ for velocity). However, post hoc tests showed that such difference was statistically significant only in children with SlowP vs. H (Table 4).

Table 3. Postural parameters during the imbalance and unloading phases. Data are shown as *median value (range)* for children with slow progressive PCA (SlowP), children with non-progressive PCA (NonP), and healthy children (H). ML: mediolateral, positive towards the swing foot; AP: anteroposterior, positive when forwards; CoP→COM: horizontal distance from CoP to COM.

		SlowP		NonP		H	
	Phase duration (s)	0.41	(0.25 to 1.03)	0.42	(0.31 to 1.90)	0.31	(0.25 to 0.53)
	CoP length (mm)	34	(24 to 100)	70	(33 to 134)	42	(17 to 68)
	ML CoP shift (mm)	15.99	(13 to 28)	39.80	(−6 to 67)	25.95	(8 to 49)
IMBALANCE	**AP CoP shift (mm)**	−18	(−43 to −5)	−25	(−43 to −11)	−14	(−41 to −8)
	ML COM shift (mm)	−7.33	(−14.67 to −2)	−6	(−9.67 to 2)	−7	(−13.67 to −2.33)
	AP COM shift (mm)	4.33	(−0.67 to 20.33)	6	(2 to 8.67)	6	(2.33 to 11.67)
	ML CoP→ COM (mm)	−25	(−54.67 to −16.67)	−30.75	(−49.25 to 54)	−37.33	(−56.67 to −18)
	AP CoP → COM (mm)	24.67	(2.67 to 62.67)	28.83	(12 to 47.33)	21	(12 to 56)
	Phase duration (s)	0.46	(0.21 to 1.53)	0.33	(0.23 to 1.20)	0.41	(0.21 to 0.56)
	CoP length (mm)	172	(112 to 202)	134	(62 to 175)	143	(90 to 172)
	ML CoP shift (mm)	−104	(−147 to −48)	−127	(−168 to −9)	−121	(−152 to −76)
UNLOADING	**AP CoP shift (mm)**	9	(−12 to 28)	−1	(−26 to 17)	−9	(−59 to 9)
	ML COM shift (mm)	−34.67	(−66.67 to −21)	−34.37	(−49 to −17)	−35.67	(−53.33 to −22)
	AP COM shift (mm)	43	(21.67 to 47.33)	35.08	(12 to 55.5)	38.67	(28 to 55.33)
	ML CoP → COM (mm)	44.25	(33.67 to 90.33)	50.96	(−7 to 62.67)	46.67	(23 to 61.67)
	AP CoP → COM (mm)	51	(37.33 to 94.33)	67.75	(29.33 to 106)	80	(57.33 to 108.5)

Table 4. First swing parameters. Data are shown as *median value (range)* for children with slow progressive PCA (SlowP), children with non-progressive PCA (NonP), and healthy children (H). LL: lower limb length; Fr: Froude number; * significant difference between SlowP and H groups, Dunn test $p < 0.05$.

	SlowP		NonP		H	
First step length (%LL)	* 47.34	(38.39 to 58.70)	62.93	(51.83 to 72.67)	* 72.99	(47.46 to 83.61)
First step velocity (Fr)	* 0.24	(0.12 to 0.32)	0.31	(0.24 to 0.39)	* 0.39	(0.26 to 0.49)

3.3. EMG

Postural EMG changes accompanying APA onset, defined as the first CoP backward shift, could not be detected in all recorded muscles and for all subjects. A descriptive analysis of electromyographic recordings allowed appreciating the development of an inhibitory postural chain involving ES, BF, and SOL, followed by an excitatory chain in RF and TA. Such a general pattern was observed in both the stance and swing sides, irrespectively from the healthy or pathological status. Nevertheless, a different timing distribution of the muscular actions was found in the three groups (Table 5). In the stance limb side, healthy subjects showed a clear craniocaudal progression, for both the inhibitory (first ES and BF, then SOL) and excitatory (first RF, then TA) chains. Such a progression was lost both in children with NonP and in those affected by SlowP; moreover, in SlowP, the recruitment of the excitatory chain was delayed. On the contralateral side with respect to the swing limb, both chains had a caudocranial progression in the H group (first SOL, then BF and ES; first TA, then RF), while children with NonP displayed a disrupted progression of the inhibitory chain, where SOL de-activated after BF and ES, followed by an almost synchronous activation of RF and TA. Instead, in children with SlowP, the inhibitory chain still maintained a caudocranial progression but was overall delayed. Also, in this group, the excitatory actions in RF and TA were synchronous.

Table 5. Latencies (ms) of postural EMG changes with respect to the APA onset (time 0). Median, minimum, and maximum values, together with the number of subjects (n) in which APAs could be identified, for children with slow progressive PCA (SlowP), children with non-progressive PCA (NonP), and healthy children (H). EMG: electromyographic; APA: anticipatory postural adjustment.

			SlowP			NonP			H		
STANCE	Erector spinae	inhibition	−68	(−85 to −52)	n = 2	−79	(−109 to −49)	n = 2	−102	(−115 to −80)	n = 3
	Biceps femoris	inhibition	−70	(−70 to −70)	n = 4	−68	(−84 to −64)	n = 3	−100	(−187 to 119)	n = 7
	Soleus	inhibition	−72	(−199 to 63)	n = 6	−98	(−130 to −63)	n = 3	−72	(−146 to −25)	n = 5
	Rectus femoris	excitation	15	(−32 to 141)	n = 4	−35	(−112 to 22)	n = 5	−72	(−110 to 57)	n = 5
	Tibialis anterior	excitation	31	(−52 to 66)	n = 5	−41	(−49 to 26)	n = 5	−31	(−49 to 108)	n = 7
SWING	Erector spinae	inhibition	−55	(−99 to 2)	n = 3	−111	(−127 to −95)	n = 2	−90	(−263 to 100)	n = 4
	Biceps femoris	inhibition	−71	(−90 to −39)	n = 4	−136	(−160 to −113)	n = 2	−94	(−199 to 70)	n = 5
	Soleus	inhibition	−98	(−184 to −35)	n = 6	−53	(−100 to −24)	n = 4	−121	(−259 to −111)	n = 5
	Rectus femoris	excitation	3	(−40 to 24)	n = 3	−11	(−83 to 162)	n = 4	68	(−80 to 113)	n = 3
	Tibialis anterior	excitation	−2	(−51 to 145)	n = 4	6	(−67 to 45)	n = 4	−61	(−101 to 6)	n = 6

Of note, in the control group, the inhibition of the stance leg SOL started about 40 ms prior to TA excitation. While this timing was overall preserved in the children of the NonP group (about 60 ms), it was effectively increased in children with SlowP (about 100 ms, Figure 2). Similar changes were detected also for the swing leg.

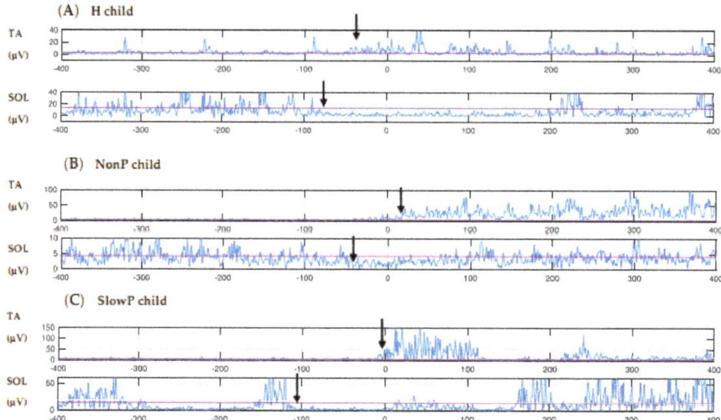

Figure 2. EMG (electromyographic) of shank muscles of the stance leg. Comparison among healthy children H, panel (**A**), children with non-progressive PCA NonP, (**B**) and children with slow progressive PCA SlowP, (**C**). One representative subject for each group. Time 0: APA (anticipatory postural adjustment) onset, defined as the first backward shift of the CoP (center of pressure). Black arrows show SOL (soleus) inhibition and the following TA (tibialis anterior) excitation. Note that the time delay between these two reciprocal actions gradually increased in children with NonP and SlowP with respect to H children.

4. Discussion

The aim of this study was to describe the postural control adopted by children with PCA during quiet stance and gait initiation, in order to draw considerations on the role of the cerebellum in the development of postural control. As a main result, we observed that in patients with slow progressive PCA, i.e., SlowP, both the static and dynamic components of postural control were disturbed, while the postural behavior of children with non-progressive PCA, i.e., NonP, was much similar to that of healthy children.

During the maintenance of upright posture, children with SlowP showed an increased ellipse area, mainly due to large mediolateral oscillations of the CoP. Considering CoP oscillations in anteroposterior direction too, this resulted in a general reduction of the ellipse eccentricity, outlining an omnidirectional decrease of stability. This finding was in agreement with the results described in adults with cerebellar lesions [26]. No statistical posturographic differences were instead found between children with NonP and H participants.

Gait initiation parameters during the *imbalance* and *unloading* phases remained substantially unchanged in patients of both pathological groups compared to H controls. Also, this observation fitted with previous results obtained in adults with cerebellar ataxia [21,40]. First step length and velocity showed instead a marked reduction in children with SlowP with respect to H, possibly reflecting a compensatory strategy for their poor balance control, and in agreement with what has been previously described in adults [21].

Electromyographic data, despite the roughness of the descriptive approach, suggested that patients with NonP and SlowP suffered more alterations in the temporal (when) than in the spatial distribution (to what muscle) and in the sign of activity (how, i.e., excitation or inhibition). This aspect agreed with the general view that assigns to the cerebellum the role of a "timing-machine" [41–45] and leaves the pattern selection to other brain structures, like the basal ganglia. Such a perspective has been confirmed also for what regards APAs in adults [46,47].

A short comment also deserves TA and SOL reciprocal activation. Despite our analysis was descriptive, patients with PCA and healthy children displayed the classical anticipatory postural pattern, characterized by SOL inhibition followed by TA activation of the stance limb (see Introduction). However, the latency between SOL and TA activity in the healthy group (about 40 ms) was consistent with what reported by Isaias et al. (2014) [18] and much lower than what has been found in adults (about 100 ms, [7]). This observation supported the choice of devoting a paper to gait initiation in children and, at the same time, suggested that the present H group, despite scarce, well represents the underlying population. On the contrary, indications of altered timing were observed in patients with PCA, in which the SOL-TA latency slightly increased to about 60 ms in children with NonP and attained about 100 ms in children with SlowP. This suggested a framework of abnormal feed-forward muscle synergies [21,24].

In summary, the reported differences in postural behavior between children with typical development and children affected by PCA support our hypothesis that the cerebellum plays a role also during the key phase of human maturation, in particular, in building internal models of gait initiation dynamics. This finding further stresses the importance of including postural training exercises in the rehabilitation programs for these pathologies [48]. Moreover, the observation that the gait initiation protocol allowed distinguishing motor deficits in children with SlowP vs. NonP, although the corresponding SARA scores were comparable, suggests that such protocol may be useful to monitor the evolution of motor deficits over time. The following two subsections are devoted to discussing the putative reasons for the different motor patterns we observed in patients of the two groups, as well as the resulting clues about the compensation mechanisms.

4.1. Disease Progression and Postural Behavior

When looking to the present results as a whole, children with SlowP seem to have a worse postural behavior with respect to both children of the NonP and H groups. This result is unlikely related to the severity of the pathology since all patients had a homogeneous SARA score, which, in turn, indicates comparable motor deficits in clinical terms. Therefore, the difference might stem either from the kind of cerebellar lesion (generalized atrophy vs. vermian hypoplasia) or from the progressive or non-progressive nature of the pathology. In this regard, children with SlowP suffered from generalized cerebellar atrophy, which represents macroscopic neuronal death, and received an ascertained clinical and/or radiological diagnosis of slow progression. On the contrary, Joubert syndrome, affecting children of the NonP group, is a congenital malformation that causes anomalous organogenesis of both the cerebellar vermis and peduncles. Therefore, it has an intrinsically stable nature along with the growth of the subject. In fact, once the organogenesis is completed, the vermian hypoplasia remains stable throughout the patient's lifetime.

It could be argued that our observation of worse postural control in children with SlowP may be related to the larger extent of their cerebellar compromission (generalized atrophy vs. vermian hypoplasia). However, literature reports an emblematic case that contrasts with this interpretation. In fact, Titomanlio et al. (2005) [49] published a case report in which a 17-year-old subject with complete cerebellar agenesis showed only mild ataxia with slight dysmetria and moderate mental retardation, but no difficulty in attaining very complex motor tasks. Such evident functional compensation could be explained only through the plasticity of the remaining brain areas, which had to cope with a lesion that is stable since embryogenesis. This report suggests restricting the hypothesis to the progressive nature of the pathology.

Returning to the present study, we envisage that children with NonP could use the plasticity of their intact brain areas, which may include the cerebellar hemispheres, to effectively compensate for their stable lesion and attain an almost normal psychomotor development. On the contrary, children with SlowP suffer from a continuous cerebellar degeneration, which conflicts with the consolidation of compensatory functional strategies. This perspective not only fits with the gradual worsening of postural deficits we documented here when passing from healthy children to patients with NonP and to patients with SlowP but would also explain why patients with adult-onset cerebellar lesions show even more pronounced postural deficits [46]. Indeed, neural plasticity gradually but consistently decreases over the lifespan [50].

4.2. Putative Compensatory Network

Finally, it remains to figure out which neural substrate could be involved in functional compensation. In this regard, it is interesting to highlight recent evidence showing subcortical bidirectional connections between the basal ganglia and the cerebellum [51–53].

The functional role of the basal ganglia to cerebellum connections has been deeply investigated. Indeed, it has been observed that patients with Parkinson's disease (PD) show abnormal functioning also in the cerebellum [54,55]. A SPECT study in patients with PD confirms an increased cerebellar activity when the effect of the anti-parkinsonian drug extinguishes [56]. Considering the reciprocal connectivity, it is of interest that functional MRI has shown increased putamen-cerebellar activity in patients with PD performing simple motor tasks and that greater putamen-cerebellar connectivity is significantly correlated with better motor performance. On the contrary, the administration of levodopa, which compensates the low endogenous dopamine production in patients with PD, has reduced this connectivity, relieving the cerebellum from its compensatory task [57]. It has also been observed that the compensatory role of the cerebellum contributes to preventing the full manifestation of the typical motor symptoms during the initial stage of PD; this compensatory ability saturates with time, leading these patients to develop cerebellar symptoms too [58].

These pieces of evidence allow arguing that, reciprocally, intact basal ganglia may compensate for cerebellar deficits. This hypothesis is still to be demonstrated, but should it be proved, it would provide a straightforward explanation for the graded postural impairments we found in children affected by PCA, as well as for the worse impairments reported in adult patients with cerebellar ataxia [46]. Evidence in this regard might come from functional MRI and diffusion tensor imaging techniques.

4.3. Limitations of the Study

The two main limits of the present study were the small number of participants and the low number of valid trials recorded in each of them. In particular, we could not observe APAs in all subjects, and this prevented a statistical analysis of EMG data. While it could be feasible to recruit more H subjects, the rarity of PCAs limited the number of children that precisely fall within the SlowP or NonP groups. With regard to the low number of valid trials, it could be increased only by prolonging the experimental session, which would quickly become burdensome for children, and especially for those affected by PCA.

5. Conclusions

Although all children with PCA showed clinically similar motor impairments, only children with SlowP were less stable in standing and showed a significantly shorter and slower first step than healthy children. Also, the descriptive EMG analysis in lower limb and back muscles pointed to a worse postural control in children of the SlowP group. On the basis of recent literature, we proposed that such different behavior stems from the disease progression, which interferes with the consolidation of compensatory strategies in children with SlowP but not in those affected by NonP.

Author Contributions: Conceptualization, funding acquisition, and supervision, P.C.; formal analysis, V.F., C.P. (Chiara Palmisano), S.M.M., C.M.M.S., and R.E.; investigation V.F., S.M.M., and C.M.M.S.; software, V.F., C.P. (Chiara Palmisano), and R.E.; clinical resources, S.D., C.P. (Chiara Pantaleoni), A.A., and N.N.; writing—original draft preparation, V.F., S.M.M., and C.P. (Chiara Palmisano); writing—review and editing, V.F., C.P. (Chiara Palmisano), S.M.M., R.E., P.C., S.D., C.P. (Chiara Pantaleoni), A.A., and N.N. All authors have read and agreed to the published version of the manuscript.

Funding: This study was supported by Fondazione Mariani for Child Neurology.

Conflicts of Interest: The authors declare no conflict of interest.

References

1. Aruin, A.S.; Latash, M.L. The role of motor action in anticipatory postural adjustments studied with self-induced and externally triggered perturbations. *Exp. Brain Res.* **1995**, *106*, 291–300. [CrossRef] [PubMed]

2. Ito, T.; Azuma, T.; Yamashita, N. Anticipatory control related to the upward propulsive force during the rising on tiptoe from an upright standing position. *Eur. J. Appl. Physiol.* **2004**, *92*, 10–186. [CrossRef] [PubMed]

3. Bouisset, S.; Zattara, M. Biomechanical study of the programming of anticipatory postural adjustments associated with voluntary movement. *J. Biomech.* **1987**, *20*, 735–742. [CrossRef]

4. Cordo, P.J.; Nashner, L.M. Properties of postural adjustments associated with rapid arm movements. *J. Neurophysiol.* **1982**, *47*, 287–302. [CrossRef]

5. Marsden, C.D.; Merton, P.A.; Morton, H.B. Human postural responses. *Brain* **1981**, *104*, 513–534. [CrossRef]

6. Massion, J. Movement, posture and equilibrium: Interaction and coordination. *Prog. Neurobiol.* **1992**, 35–56. [CrossRef]

7. Crenna, P.; Frigo, C. A motor programme for the initiation of forward-oriented movements in humans. *J. Physiol.* **1991**, *437*, 635–653. [CrossRef]

8. Nissan, M.; Whittle, M.W. Initiation of gait in normal subjects: A preliminary study. *J. Biomed. Eng.* **1990**, *12*, 165–171. [CrossRef]

9. Burleigh, A.L.; Horak, F.B.; Malouin, F. Modification of postural responses and step initiation: Evidence for goal-directed postural interactions. *J. Neurophysiol.* **1994**, *72*, 2892–2902. [CrossRef]

10. Burleigh, A.; Horak, F. Influence of instruction, prediction, and afferent sensory information on the postural organization of step initiation. *J. Neurophysiol.* **1996**, *75*, 1619–1628. [CrossRef]

11. McIlroy, W.E.; Maki, B.E. Do anticipatory postural adjustments precede compensatory stepping reactions evoked by perturbation? *Neurosci. Lett.* **1993**, *164*, 199–202. [CrossRef]

12. Yiou, E.; Caderby, T. Balance control during gait initiation: State-of-the-art and research perspectives. *World J. Orthop.* **2017**, *8*, 815–828. [CrossRef] [PubMed]

13. Crenna, P.; Carpinella, I.; Rabuffetti, M.; Rizzone, M.; Lopiano, L.; Lanotte, M.; Ferrarin, M. Impact of subthalamic nucleus stimulation on the initiation of gait in Parkinson's disease. *Exp. Brain Res.* **2006**, *172*, 519–532. [CrossRef] [PubMed]

14. Wang, Y.; Zatsiorsky, V.M.; Latash, M.L. Muscle synergies involved in shifting the center of pressure while making a first step. *Exp. Brain Res.* **2005**, *167*, 196–210. [CrossRef] [PubMed]

15. Ledebt, A.; Bril, B.; Brenière, Y. The build-up of anticipatory behaviour: An analysis of the development of gait initiation in children. *Exp. Brain Res.* **1998**, *120*, 9–17. [CrossRef] [PubMed]

16. Assaiante, C.; Woollacott, M.; Amblard, B. Development of postural adjustment during gait initiation: Kinematic and EMG analysis. *J. Mot. Behav.* **2009**, *32*, 211–226. [CrossRef]

17. Malouin, F.; Richards, C.L. Preparatory adjustments during gait initiation in 4–6-year-old children. *Gait Posture* **2000**, *11*, 239–253. [CrossRef]

18. Isaias, I.U.; Dipaola, M.; Michi, M.; Marzegan, A.; Volkmann, J.; Roidi, M.L.R.; Frigo, C.A.; Cavallari, P. Gait initiation in children with rett syndrome. *PLoS ONE* **2014**, *9*, e92736. [CrossRef]

19. Dipaola, M.; Frigo, C.A.; Cavallari, P.; Iasias, I.U. Alterations of load transfer mechanism during gait initiation in Parkinson's disease. In Proceedings of the EHB 2017 IEEE International Conference on e-Health and Bioengineering, Sinaia, Romania, 22–24 June 2017; pp. 579–582. [CrossRef]

20. Timmann, D.; Horak, F.B. Perturbed step initiation in cerebellar subjects 1. Modifications of postural responses. *Exp. Brain Res.* **1998**, *119*, 73–84. [CrossRef]

21. Timmann, D.; Horak, F.B. Perturbed step initiation in cerebellar subjects: 2. Modification of anticipatory postural adjustments. *Exp. Brain Res.* **2001**, *141*, 110–120. [CrossRef]

22. Martino, G.; Ivanenko, Y.P.; Serrao, M.; Ranavolo, A.; D'Avella, A.; Draicchio, F.; Conte, C.; Casali, C.; Lacquaniti, F. Locomotor patterns in cerebellar ataxia. *J. Neurophysiol.* **2014**, *112*, 2810–2821. [CrossRef] [PubMed]

23. Ilg, W.; Golla, H.; Thier, P.; Giese, M.A. Specific influences of cerebellar dysfunctions on gait. *Brain* **2007**, *130*, 786–798. [CrossRef] [PubMed]

24. Asaka, T.; Wang, Y. Feedforward postural muscle modes and multi-mode coordination in mild cerebellar ataxia. *Exp. Brain Res.* **2011**, *210*, 153–163. [CrossRef] [PubMed]

25. Serrao, M.; Pierelli, F.; Ranavolo, A.; Draicchio, F.; Conte, C.; Don, R.; Di Fabio, R.; Lerose, M.; Padua, L.; Sandrini, G.; et al. Gait pattern in inherited cerebellar ataxias. *Cerebellum* **2012**, *11*, 194–211. [CrossRef] [PubMed]

26. Morton, S.M.; Bastian, A.J. Cerebellar control of balance and locomotion. *Neuroscientist* **2004**, *10*, 247–259. [CrossRef]

27. Bastian, A.J. Learning to predict the future: The cerebellum adapts feedforward movement control. *Curr. Opin. Neurobiol.* **2006**, 645–649. [CrossRef]

28. Stoodley, C.J. The cerebellum and neurodevelopmental disorders. *The Cerebellum* **2017**, *15*, 34–37. [CrossRef]

29. Valence, S.; Cochet, E.; Rougeot, C.; Garel, C.; Chantot-Bastaraud, S.; Lainey, E.; Afenjar, A.; Barthez, M.A.; Bednarek, N.; Doummar, D.; et al. Exome sequencing in congenital ataxia identifies two new candidate genes and highlights a pathophysiological link between some congenital ataxias and early infantile epileptic encephalopathies. *Genet. Med.* **2019**, *21*, 553–563. [CrossRef]

30. Romani, M.; Micalizzi, A.; Valente, E.M. Joubert syndrome: Congenital cerebellar ataxia with the molar tooth. *Lancet Neurol.* **2013**, *12*, 894–905. [CrossRef]

31. Prieto, T.E.; Myklebust, J.B.; Hoffmann, R.G.; Lovett, E.G.; Myklebust, B.M. Measures of postural steadiness: Differences between healthy young and elderly adults. *IEEE Trans. Biomed. Eng.* **1996**, *43*, 956–966. [CrossRef]

32. Rocchi, L.; Chiari, L.; Horak, F.B. The effects of deep brain stimulation and levodopa on postural sway in subjects with Parkinson's disease. *J. Neurol. Neurosurg. Psychiatry* **2002**, *73*, 240. [CrossRef] [PubMed]

33. Schmitz-Hübsch, T.; Du Montcel, S.T.; Baliko, L.; Berciano, J.; Boesch, S.; Depondt, C.; Giunti, P.; Globas, C.; Infante, J.; Kang, J.S.; et al. Scale for the assessment and rating of ataxia: Development of a new clinical scale. *Neurology* **2006**, *66*, 1717–1720. [CrossRef] [PubMed]

34. Orphanet. Available online: https://www.orpha.net (accessed on 27 January 2020).

35. Hiraoka, K.; Hatanaka, R.; Nikaido, Y.; Jono, Y.; Nomura, Y.; Tani, K.; Chujo, Y. Asymmetry of anticipatory postural adjustment during gait initiation. *J. Hum. Kinet.* **2014**, *42*, 7–14. [CrossRef]

36. Ferrari, A.; Benedetti, M.G.; Pavan, E.; Frigo, C.; Bettinelli, D.; Rabuffetti, M.; Crenna, P.; Leardini, A. Quantitative comparison of five current protocols in gait analysis. *Gait Posture* **2008**, *28*, 207–216. [CrossRef] [PubMed]

Appl. Sci. **2020**, *10*, 1606

37. Hermens, H.J.; Freriks, B.; Disselhorst-Klug, C.; Rau, G. Development of recommendations for SEMG sensors and sensor placement procedures. *J. Electromyogr. Kinesiol.* **2000**, *10*, 361–374. [CrossRef]

38. Hof, A.L. Scaling gait data to body size. *Gait Posture* **1996**, 222–223. [CrossRef]

39. Marchese, S.M.; Esposti, R.; Bolzoni, F.; Cavallari, P. Transcranial Direct Current Stimulation on Parietal Operculum Contralateral to the Moving Limb Does Not Affect the Programming of Intra-Limb Anticipatory Postural Adjustments. *Front. Physiol.* **2019**, *10*, 1–9. [CrossRef]

40. Mummel, P.; Timmann, D.; Krause, U.W.H.; Boering, D.; Thilmann, A.F.; Diener, H.C.; Horak, F.B. Postural responses to changing task conditions in patients with cerebellar lesions. *J. Neurol. Neurosurg. Psychiatry* **1998**, *65*, 734–742. [CrossRef]

41. Ivry, R.B.; Keele, S.W. Timing functions of the cerebellum. *J. Cogn. Neurosci.* **1989**, *1*, 136–152. [CrossRef]

42. Ivry, R.B.; Spencer, R.M.; Zelaznik, H.N.; Diedrichsen, J. The cerebellum and event timing. *Ann. N. Y. Acad. Sci.* **2002**, 302–317. [CrossRef]

43. Cerri, G.; Esposti, R.; Locatelli, M.; Cavallari, P. Coupling of hand and foot voluntary oscillations in patients suffering cerebellar ataxia: Different effect of lateral or medial lesions on coordination. *Prog. Brain Res.* **2005**, *148*, 227–241. [CrossRef] [PubMed]

44. D'Angelo, E. Rebuilding cerebellar network computations from cellular neurophysiology. *Front. Cell. Neurosci.* **2010**, *4*, 131. [CrossRef]

45. Bareš, M.; Apps, R.; Avanzino, L.; Breska, A.; D'Angelo, E.; Filip, P.; Gerwig, M.; Ivry, R.B.; Lawrenson, C.L.; Louis, E.D.; et al. Consensus paper: Decoding the contributions of the cerebellum as a time machine. from neurons to clinical applications. *Cerebellum* **2019**, *18*, 266–286. [CrossRef] [PubMed]

46. Bruttini, C.; Esposti, R.; Bolzoni, F.; Vanotti, A.; Mariotti, C.; Cavallari, P. Temporal disruption of upper-limb anticipatory postural adjustments in cerebellar ataxic patients. *Exp. Brain Res.* **2015**, *233*, 197–203. [CrossRef] [PubMed]

47. Bolzoni, F.; Esposti, R.; Marchese, S.M.; Pozzi, N.G.; Ramirez-Pasos, U.E.; Isaias, I.U.; Cavallari, P. Disrupt of intra-limb APA pattern in Parkinsonian patients performing index-finger flexion. *Front. Physiol.* **2018**, *9*, 1745. [CrossRef] [PubMed]

48. Marquer, A.; Barbieri, G.; Pérennou, D. The assessment and treatment of postural disorders in cerebellar ataxia: A systematic review. *Ann. Phys. Rehabil. Med.* **2014**, *57*, 67–78. [CrossRef]

49. Titomanlio, L.; Romaniello, R.; Borgatti, R. Cerebellar agenesis. *Handb. Cereb. Cereb. Disord.* **2005**, 1855–1872. [CrossRef]

50. Lu, C.; Huffmaster, S.L.A.; Harvey, J.C.; MacKinnon, C.D. Anticipatory postural adjustment patterns during gait initiation across the adult lifespan. *Gait Posture* **2017**, *57*, 182–187. [CrossRef]

51. Bostan, A.C.; Dum, R.P.; Strick, P.L. The basal ganglia communicate with the cerebellum. *Proc. Natl. Acad. Sci. USA* **2010**, *107*, 8452–8456. [CrossRef]

52. Bostan, A.C.; Dum, R.P.; Strick, P.L. Cerebellar networks with the cerebral cortex and basal ganglia. *Trends Cogn. Sci.* **2013**, *17*, 241–254. [CrossRef]

53. Hoshi, E.; Tremblay, L.; Féger, J.; Carras, P.L.; Strick, P.L. The cerebellum communicates with the basal ganglia. *Nat. Neurosci.* **2005**, *8*, 1491–1493. [CrossRef] [PubMed]

54. Wu, T.; Hallett, M. The cerebellum in Parkinson's disease. *Brain* **2013**, *136*, 696–709. [CrossRef] [PubMed]

55. Yu, H.; Sternad, D.; Corcos, D.M.; Vaillancourt, D.E. Role of hyperactive cerebellum and motor cortex in Parkinson's disease. *Neuroimage* **2007**, *35*, 222–233. [CrossRef] [PubMed]

56. Rascol, O.; Sabatini, U.; Fabre, N.; Brefel, C.; Loubinoux, I.; Celsis, P.; Senard, J.M.; Montastruc, J.L.; Chollet, F. The ipsilateral cerebellar hemisphere is overactive during hand movements in akinetic parkinsonian patients. *Brain* **1997**, *120*, 103–110. [CrossRef]

57. Simioni, A.C.; Dagher, A.; Fellows, L.K. Compensatory striatal-cerebellar connectivity in mild-moderate Parkinson's disease. *NeuroImage Clin.* **2016**, *10*, 54–62. [CrossRef]

58. Martinu, K.; Monchi, O. Cortico-basal ganglia and cortico-cerebellar circuits in Parkinson's disease: Pathophysiology or compensation? *Behav. Neurosci.* **2013**, *127*, 222–236. [CrossRef]

Review

Balance Control in Obese Subjects during Quiet Stance: A State-of-the Art

Veronica Cimolin [1,*], Nicola Cau [2], Manuela Galli [1] and Paolo Capodaglio [2]

[1] Department of Electronics, Information and Bioengineering (DEIB) Politecnico di Milano; Piazza Leonardo da Vinci 32, 20133 Milano, Italy; manuela.galli@polimi.it

[2] Ospedale San Giuseppe, Istituto Auxologico Italiano, IRCCS, Strada Luigi Cadorna 90, 28824 Piancavallo (VB), Italy; nicola.cau@gmail.com (N.C.); p.capodaglio@auxologico.it (P.C.)

* Correspondence: veronica.cimolin@polimi.it

Received: 3 February 2020; Accepted: 4 March 2020; Published: 7 March 2020

Abstract: Obese individuals are characterized by a reduced balance which has a significant effect on a variety of daily and occupational tasks. The presence of excessive adipose tissue and weight gain could increase the risk of falls; for this reason, obese individuals are at greater risk of falls than normal weight subjects in the presence of postural stress and disturbances. The quality of balance control could be measured with different methods and generally in clinics its integrity is generally assessed using platform stabilometry. The aim of this narrative review is to present an overview on the state of art on balance control in obese individuals during quiet stance. A summary of knowledge about static postural control in obese individuals and its limitations is important clinically, as it could give indications and suggestions to improve and personalize the development of specific clinical programs.

Keywords: posture; stability; force platform; stabilometry; posturography; obesity

1. Introduction

Postural stability can be explained as the ability of one's motor control to maintain the standing posture, even during external perturbations. It is often described as static (quiet standing) or dynamic (maintaining a stable position while undertake a prescribed movement) [1].

Postural stability is achieved when a subject can regain equilibrium after being disturbed. Postural control requires the integration of inputs from various sensory systems, which are: (a) the proprioceptive system (receptors in joints, muscles, tendons, skin) [2], (b) the visual system [3] and (c) the vestibular system (semicircular canals and macular otoliths) [4]. Sensorimotor integration and balance regulation rely on the cerebellum; most of the evidence of the function of the cerebellum in humans comes from patients with cerebellar damage who display balance abnormalities and gait ataxia [5].

This multimodal (integrating visual, vestibular, and proprioceptive inputs) and redundant control causes chaotic center of mass (CoM) oscillations known as postural sway [6–8]. In clinics the quality and integrity of balance control is generally assessed using various measures of postural sway [6].

Instrumental evaluation with platform stabilometry or static posturography is a technique where CoM fluctuations are represented by the center of foot pressure (CoP) displacements [9] and it consists in the measurement of forces exerted against a force platform during quiet stance. This method is widely used in clinical settings to evaluate the integrity of the postural control system and to obtain functional markers on fine competencies and their development in different testing conditions (i.e., eyes open vs. eyes closed, feet position, and presence of external stimuli) [10]. Commonly, the study of properties of the CoP trajectory in clinics is performed using traditional time series, in particular analyzing length, displacement excursion, velocity and frequency analysis. The main advantages of these analysis techniques are the simplicity of the experimental set-ups and the safety for the subjects evaluated, a particularly important consideration in pathological conditions [11]. However, some limitations are present: (a) the lack of a normal pattern due to intra- and inter-individuals variability; (b) the difficulties inherent to standardizing the measurement experimental set-up (reproducibility of the experimental protocol, environmental conditions, random errors, signal processing); and (c) the presence of highly coupled parameters measured by means of the force platform [12,13]. Due to the presence of these limitations, it is necessary to have reliable methods available to extract physiologically significant information from the platform data [11,14]. Alongside the study of properties of the CoP trajectory using traditional time series, some advanced mathematical methods have been developed applying a dynamic approach, such as entropy and fractal dimension (FD) analysis [11,15,16]. Even if previous studies [14,17,18] suggested that these methods could represent reliable techniques complementary to the analysis in time and frequency domain in order to evaluate postural control, their use in clinical setting is not yet widespread.

Balance is a fundamental element in carrying out daily life activities. Aging and various pathologies increase the postural instability which can lead to falls [19]. In particular, some anthropometric measurements (height and weight) are demonstrated to significantly influence both postural sway and postural stability [12,20,21]. The excessive presence of fat adds mass to different regions and modifies the body geometry [22]: in this way the altered CoM position leads to balance impairments [23].

It is estimated that in 2016, more than 1.9 billion adults aged 18 years and older were overweight; of these over 650 million were obese [24]. A recent paper suggests that by 2030 nearly one in two adults in the United States will be obese and nearly one in four adults will have severe obesity [25]. Patients with obesity have intrinsically reduced postural stability and balance as compared with their normal-weight counterparts. This reduced stability increases linearly with body mass index (BMI). They are at increased risk of falling at any age [26]. Although many studies investigated the risk of falling in obese individuals, balance exercises in these individuals are often not implemented in the rehabilitation program. After implementing a rehabilitation program with specific balance exercises, patients with obese have shown to improve their balance control. Maffiuletti et al. [19] showed that just 4-min of specific balance training incorporated into the physical exercise routine improved postural stability in patients with severe obesity. The recommended exercise program for patient with obesity is a multicomponent 90-min exercise program that includes 15-min balance training, 15-min flexibility, 30-min aerobic exercise and 30-min high-intensity resistance training [27]. Different aspects of balance control can be addressed:

1. Treatment of biomechanical constraints (weakness, reduced range of motion, reduced flexibility, and improper postural alignment).
2. Weight shifting exercise to treat reduced limits of stability.
3. Sensory retraining of balance control.
4. Training of anticipatory postural adjustments focused on improving postural preparation for transition from one position to another (sit-to-stand single-leg-stance, step initiation, and compensatory forward stepping).
5. Training postural responses to perturbations.
6. Dynamic stability during gait (i.e., walking in different directions and environments).

The consensus is that obese individuals exhibit poor postural stability [28,29]; the aim of this review will be to present an overview on the state of art on balance control in obese individuals during quiet stance. A summary of our current knowledge about static postural control in obese individuals is important from a clinical perspective, as it could allow for the development of clinical programs that are more oriented towards the needs of this population, even avoiding the risk of comorbidity due to falls.

2. Methods

We conducted an extensive search of the relevant literature, with a focus on studies/articles published in the last 15 years (2004–2019). The literature search was performed in November 2019 on the following electronic databases: Web of Science, PubMed MEDLINE, Scopus, and Mendeley. Customized queries including keywords and Boolean logic with AND/OR operators were entered in this form: "(balance OR posture OR stability) AND (posturography) AND (obesity)", with document type set to "Article". The search was limited to full original articles written in English. Bibliographies of identified papers were hand searched for supplemental relevant items.

All included studies had to meet the following criteria: overweight and obese adults (18–75 years; with body mass index (BMI) $\geq$ 25 Kg/m^2); studies assessing obese patients with genetic obesity; evaluations performed with force platform. We excluded studies that were not primarily focused on the evaluation of quiet standing evaluation (dynamic balance) in obese subjects, or non-adult participants and elderly subjects.

3. Results

A total of 27 records were retrieved from the electronic databases. Eight items were added by visual inspection of reference lists and review articles. After removing seven duplicates, titles and abstracts screening led to exclude four papers. Out of the remaining 24 articles, 12 failed to meet inclusion criteria. The main reasons for exclusion were as follows: studies assessing elderly or non-adult patients; patients with neuropathy; the presence of other pathologies (multiple sclerosis). The selection process is summarised in Figure 1.

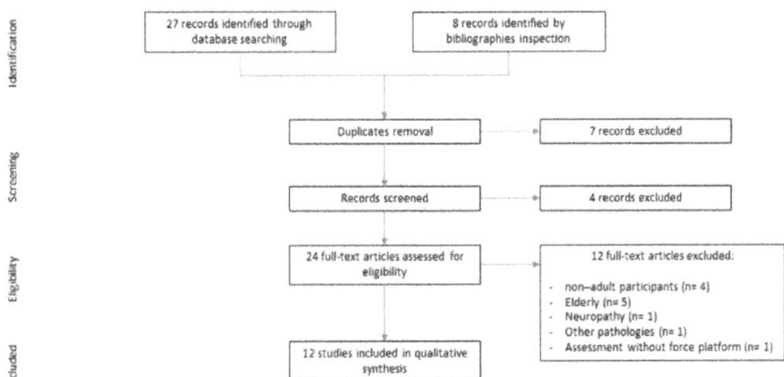

Figure 1. Diagram of study selection.

Table 1 presents a summary of the papers on postural ability in obese individuals with the demographic characteristics of the evaluated individuals and some details of the experimental set-up (trial duration, conditions, feet position, analyzed parameters).

Table 1. Summary of the main details of the reviewed studies.

Source	Year	Country	# Participants and Gender (# M/F)	Age (Yrs.)	BMI (Kg/m²) (the Weight (kg) Is Reported When BMI Is Not Present)	Parameters	Trial Duration	Conditions	Foot Position
Hue et al. [30]	2007	Canada	Total: 59 (M)	40.5 ± 9.5	35.2 ± 11.7	CoP RMS (AP and ML) CoP Range (AP and ML) Sway area Mean CoP velocity (AP and ML) RMS of CoP velocity	35 s (last 30 s used for the analysis)	Eyes open/eyes closed	Feet 10 cm apart
Teasdale et al. [31]	2007	Canada	Total: 44 (M) Divided into Control group: 16 Obese group: 14 Morbid group: 14	Control group: 38.6 ± 9.4 Obese group: 37.9 ± 7.7 Morbid group: 44.4 ± 8.9	Control group: 22.7 ± 2.2 Obese group: 33 ± 3 Morbid group: 50.5 ± 6.8	CoP RMS (AP and ML) CoP Range (AP and ML) Mean CoP velocity (AP and ML)RMS of CoP velocity	35 s (last 30 s used for the analysis)	Eyes open/Eyes closed	Feet together
Menegoni et al. [32]	2009	Italy	Total: 44 M: 22/F: 22	19–58	M:41.1 ± 4.1 F:40.2 ± 5	CoP RMS (AP and ML) CoP Range (AP and ML) Mean CoP velocity (AP and ML)	60 s	Eyes open	Standardised (distance between heels approx. 8 cm and angle between feet of 30°)
Blaszczyk et al. [33]	2009	Poland	Total. 100 (F)	18–53	37.2 ± 5.2	CoP range (AP and ML) CoP length	30 s	Eyes open/eyes closed	Feet apart and slightly turned out
Cimolin et al. [17]	2011	Italy	Total: 11 (PWS) M: 5/F: 6	34.4 ± 3.7	41.4 ± 8.1	CoP Range (AP and ML) CoP length peak of the spectrum (AP and ML) FD	30 s	Eyes open	Feet at a 30° angle
Cruz-Gomez et al. [34]	2011	Mexico	Total: 180 M: 90/F: 90 Divided into Lean (%): M: 41; F: 37 Overweight (%): M: 48; F: 33 Obese (%): M: 11; F: 30	12–67 (M:34.9 ± 12.27; F:36.76 ± 12.02)	M:25.97 ± 3.73 F:26.83 ± 4.77	CoP length CoP area CoP velocity	25.6 s	1. Hard surface and eyes open. 2. Hard surface and eyes closed. 3. Soft surface and eyes open. 4. Soft surface and eyes closed	According to the manufacturer reference
Rigoldi et al. [35]	2011	Italy	Total: 45 (DS) (gender not detailed)	22–46	57.9 + 10.8 Kg	CoP Range (AP and ML) CoP length peak of the spectrum (AP and ML)	30 s	Eyes open/eyes closed	Feet at a 30° angle

Table 1. *Cont.*

Source	Year	Country	# Participants and Gender (# M/F)	Age (Yrs.)	BMI (Kg/m^2) (the Weight (kg) Is Reported When BMI Is Not Present)	Parameters	Trial Duration	Conditions	Foot Position
Hita-Contreras et al. [36]	2013	Spain	Total: 100 (F)	57.51 ± 3.99	27.10 ± 4.71	CoP RMS (AP and ML) Sway area Mean CoP velocity (AP and ML)	30 s	1. Eyes open 2. Eyes closed 3. Foam surface and eyes open 4. Foam surface and eyes closed	Feet at a 30° angle
Cimolin et al. [37]	2014	Italy	Total 59 M: 15; F: 11 M: 6; F: 7 (PWS) M: 11; F: 9 (DS)	34.2 ± 10.7 32.4 ± 4.2 (PWS) 29.1 ± 8.1 (DS)	40.6 ± 4.6 40.3 ± 6.6 (PWS) 35.8 ± 6.2 (DS)	CoP Range (AP and ML) CoP length peak of the spectrum (AP and ML) FD	30 s	Eyes open	Feet at a 30° angle
Cieslinska-Swider et al. [38]	2017	Poland	Total: 80 (F) Divided into Group A: 40 with android type of obesity Group G: 40 with gynoid type of obesity	Group A: 38 ± 12 Group G: 36 ± 11	Group A: 37.6 ± 5.5 Group G: 36.9 ± 5.1	CoP range (AP and ML) CoP mean velocity CoP peak velocity	30 s	Eyes open/eyes closed	Feet apart and slightly turned out
Hirjakova et al. [39]	2018	Slovakia	Total: 22 M: 13/F: 9	32.5 ± 1.3	32.0 ± 0.9	CoP Range (AP and ML) Mean CoP velocity (AP and ML)	50 s	Eyes open	Self-selected stance width
Cieslinska-Swider et al. [8]	2019	Poland	Total: 32 (F)	35.9 ± 9.8	36.4 ± 5.2	CoP range (AP and ML) CoP mean velocity CoP peak velocity	30 s	Eyes open/eyes closed	Feet apart and slightly turned out

Abbreviations: M: male; F: female; PWS: Prader–Willi syndrome; DS: Down syndrome; BMI: body mass index; ROM: range Of motion; AP: antero-posterior; ML: medio-lateral; FD: fractal dimension.

3.1. Study Population

The sample size of participants ranged from n = 11 to n = 180 (mean age range: 19 years to 60 years). A total of 860 subjects were involved (approximately 60% females and 40% males), but we could not exclude subjects overlapping in studies conducted by the same research group.

Experimental Set-up and Parameters

In terms of the experimental set-up, the common postural techniques are used in terms of trial duration (30 s or 60 s) and conditions (eyes open and eyes closed). On the contrary, some differences were found as for the feet position; this could represent a bias as it is demonstrated that anthropometric measurements (e.g., maximum foot width), and the foot position could influence balance [12]. Almost all the studies have been conducted using parameters obtained by time-domain analysis of the CoP trajectory (length, displacement excursion, velocity).

To the best of our knowledge, most studies in obese individuals have adopted the traditional time-domain approach. However, a growing number of studies have been designed to explore other approaches for the analysis of the CoP trajectories during quiet standing and to characterize the preferential involvement of specific neuronal loops in postural regulation (frequency), its irregularity and complexity (entropy), its chaotic pattern (fractal dimension method), etc. [40].

Nevertheless, these alternative methodologies have not been used in obese patients. To our knowledge, only few researches [17,37] quantified posture using the FD approaches in addition to the time and frequency domain in genetic obese patients (Prader-Willi syndrome and Down syndrome). The results showed that patients presented higher CoP fluctuations in both AP and ML directions with longer CoP trajectory than the control group [35,41–43] and similar values as for the frequency analysis. Furthermore, the pathological individuals presented larger excursions of CoP with the same velocity of oscillation if compared to normal weight individuals. In addition, patients with genetic obesity exhibited higher FD values, which were higher in patients with Down syndrome. According to these results, the time domain parameters do not seem to clearly highlight early anomalies in postural maintenance: the traditional method does not investigate the chaotic fluctuations of CoP trajectories. Frequency analysis and dynamical system theory could identify early changes and may match time frequency analysis to assess balance. However, the clinical interpretation of the results is not completely clear and requires further in-depth investigation.

3.2. Gender Effects

Gender differences (considering in particular gynoid and android shape) are observed in terms of body mass distribution, even if android fat distribution is also found in females, especially in postmenopausal women: nevertheless, the research for possible gender-specific differences in balance has reported contradictory findings. Some studies were conducted just on a specific gender (males or females).

3.2.1. Males

Hue et al. [30] observed that obese men presented higher instability, in both the antero-posterior (AP) and medio-lateral (ML) CoP excursions, than normal weight individuals; an improvement was found in severely obese men after weight loss [31] and specific balance training [19]. This result supports the suggestion that body weight is an important predictor of postural stability [30].

3.2.2. Females

Blaszczyk et al. [33] observed a significant postural instability in all obese patients; CoP fluctuations were higher in patients with the highest body mass index (>40 kg/m^2). In another study, Cienliska-Swider et al. [38] quantified CoP characteristics in a group of obese women with android type of obesity as compared with a group of obese women with gynoid type of obesity, standing with eyes open and closed. They found that women with abdominal obesity showed a larger sway range in the AP direction under both conditions and a greater maximal CoP velocity than subjects with gynoidal obese type under the eyes closed condition. Women with abdominal obesity seem to exhibit greater postural instability in comparison with women with gynoid fat distribution. In another study conducted by the same authors [8] on young obese women, they found that participants exhibited sagittal plane postural instability only with their eyes closed. After body weight reduction, ML static stability decreased, directly connected to a change in the base of support. Recently, Hita-Contreras [36] analyzed the relationship between body weight/body fat distribution and postural balance and their correlation with falls in postmenopausal women, which are characterized by weight gain and increased central adiposity. In the obese group, higher excursions were found in the antero-posterior direction under both eyes-open and eyes-closed conditions, as well as for the CoP velocity. In particular, in overweight and obese individuals with an android body fat distribution a good correlation with the risk of falling was found. Similar results were found in obese older women, suggesting that obesity has a negative impact on the capacity of older woman to adequately use proprioceptive information for posture control [44].

3.2.3. Males vs. Females

If instability was generally observed both in obese males and female, it is not clear whether gender could influence balance. Cruz-Gomez et al. [34] assessed the influence of BMI group (lean/overweight/obese) and gender on the postural sway of adolescents and adults during quiet upright stance in four conditions (eyes open/closed on hard/soft surface). The postural stability of obese subjects decreases with eyes closed, if compared to normal weight individuals, with no influence of the gender. Thus, they found no interactions between the BMI group and the gender, independently of the age of the subjects. On the contrary, in another study, Menegoni et al. [32] identified a gender-specific effect in obese individuals: both genders displayed instability in AP direction, while only males presented ML destabilization. The difference in male and female body tissue distribution could potentially offer two related explanations to these outcomes. While males typically store more fatty tissue around their abdomen, (android shape), females usually carry fat around the hips and the upper portion of their legs (gynoid shape) [45]. AP instability is exacerbated in both populations by greater overall mass which leads to higher ankle torque. The male distribution of mass, however, results in increased loading of the pelvic girdle which, in turn may contribute to greater ML CoP excursion. Similar, yet alternatively, the anatomically lower position of fat storage associated with the female gynoid shape has a consequently lower CoM than the male android shape.

3.3. General Considerations

In general, body weight is considered a predictor of postural instability [30], from adolescence onwards [26]. In obese individuals there is significantly greater forward CoP displacement during dynamic standing balance activities [46]. Excessive body weight affects posture linearly with the increase of BMI ($0.39 < rho < 0.60$, $p < 0.05$) [47,48], similarly to the later stages of pregnancy [31], the center of gravity shifts forward, lumbar lordosis increases together with the pelvic forward tilt, dorsal kyphosis and secondary cervical lordosis become more pronounced [49]. Otherwise, increased CoP parameter values and therefore increased postural instability during quiet stance has been reported in both morbidly and slightly obese subjects [39]. Two hypotheses have been proposed to explain the presence of higher oscillations in obese individuals in comparison with normal weight subjects: a) the reduction of plantar sensitivity due to the hyper activation of the plantar mechanoreceptors for the continuous pressure of supporting the large mass; b) the presence of high mechanical request in obese subjects due to a whole body center of mass further away from the axis of rotation causing a greater gravitational torque [30].

4. Conclusions

According to this narrative review, the research investigating the effect of adiposity on postural balance is limited and, to date, has primarily focused on parameters related to time-domain approach during bipedal stance.

Because of the elevated mass to height ratio of excessively muscular people, BMI becomes a relatively ineffective method for differentiating between highly overweight subjects and bulky, yet fit, muscular subjects. This could be a potential explanation for the discrepancies found in studies using BMI as their sole classification metric. The relationship between excessive adiposity and balance may be better explained using different measures of fat percentage. Rather, by including the ratio of circumference between waist and hip, researchers could add a quantitative parameter (in addition to the traditional BMI) which characterizes the shape of a subject and subsequently ensure better homogeneity amongst groups.

Considering that the prevalence of obesity is increasing at a rapid rate all over the world, health systems will have to face the growing problems directly related to obesity. It is now known that an increase in BMI is associated with an increase in functional limitation, decreased stability and an increased risk of falls. Complications associated with falls are often more difficult to treat in obese individuals in comparison with normal-weight subjects. Targeted physical activity and rehabilitation would seem to lead to improvements in terms of balance in obese people; it is still unclear whether a regular exercise program and weight loss could be the first steps in countering the reduced balance related to obesity. Balance exercises are often neglected in rehabilitation programs; evidence exists that majority of patients with obesity complain of dizziness, but they seem to underestimate the risk of fall; considering the increased risk of fall, balance exercises for the patient with obesity should be implemented even in the absence of specific balance disorders.

The main limitation of this review is mainly related to the age of the evaluated patients. A more extensive review could be conducted evaluating also the effects of age (elderly) and non-adult participant's (children and adolescents). In addition, it could be interesting to consider dynamic balance training in these patients, which has been proven improving balance performance and decreases falls in these subjects.

Author Contributions: V.C. contributed to conceptualization, Writing—original draft, Writing—review & editing; N.C. Data curation, Formal analysis, Writing—original draft; M.G. contributed to conceptualization, Supervision, Writing—review & editing; P.C. Project administration, Supervision, Writing–review & editing. All authors have read and agreed to the published version of the manuscript.

Funding: This research received no external funding.

Conflicts of Interest: The authors declare no conflict of interest.

References

1. Karimi, M.T.; Solomonidis, S. The relationship between parameters of static and dynamic stability tests. *J. Res. Med. Sci. Off. J. Isfahan Univ. Med. Sci.* **2011**, *16*, 530–535.

2. Maranesi, E.; Fioretti, S.; Ghetti, G.G.; Rabini, R.A.; Burattini, L.; Mercante, O.; Di Nardo, F. The surface electromyographic evaluation of the Functional Reach in elderly subjects. *J. Electromyogr. Kinesiol.* **2016**, *26*, 102–110. [CrossRef]

3. Alghadir, A.H.; Alotaibi, A.Z.; Iqbal, Z.A. Postural stability in people with visual impairment. *Brain Behav.* **2019**, *9*, e01436. [CrossRef] [PubMed]

4. Appiah-Kubi, K.; Wright, W. Vestibular training promotes adaptation of multisensory integration in postural control. *Gait Posture* **2019**, *73*, 215–220. [CrossRef] [PubMed]

5. Yanagihara, D. Role of the cerebellum in postural control. *J. Phys. Fit. Sports Med.* **2014**, *3*, 169–172. [CrossRef]

6. Błaszczyk, J.W. The use of force-plate posturography in the assessment of postural instability. *Gait Posture* **2016**, *44*, 1–6. [CrossRef] [PubMed]

7. Maurer, C.; Peterka, R.J. A New Interpretation of Spontaneous Sway Measures Based on a Simple Model of Human Postural Control. *J. Neurophysiol.* **2005**, *93*, 189–200. [CrossRef] [PubMed]

8. Cieślińska-Świder, J.M.; Błaszczyk, J.W. Posturographic characteristics of the standing posture and the effects of the treatment of obesity on obese young women. *PLOS ONE* **2019**, *14*, e0220962. [CrossRef]

9. Winter, D. Human balance and posture control during standing and walking. *Gait Posture* **1995**, *3*, 193–214. [CrossRef]

10. Maranesi, E.; Merlo, A.; Fioretti, S.; Zemp, D.D.; Campanini, I.; Quadri, P. A statistical approach to discriminate between non-fallers, rare fallers and frequent fallers in older adults based on posturographic data. *Clin. Biomech.* **2016**, *32*, 8–13. [CrossRef]

11. Sabatini, A.M. Analysis of postural sway using entropy measures of signal complexity. *Med. Biol. Eng. Comput.* **2000**, *38*, 617–624. [CrossRef] [PubMed]

12. Chiari, L.; Rocchi, L.; Cappello, A. Stabilometric parameters are affected by anthropometry and foot placement. *Clin. Biomech.* **2002**, *17*, 666–677. [CrossRef]

13. Kim, Y.; Morshed, S.; Joseph, T.; Bozic, K.; Ries, M.D. Clinical Impact of Obesity on Stability Following Revision Total Hip Arthroplasty. *Clin. Orthop.* **2006**, *453*, 142–146. [CrossRef] [PubMed]

14. Capodaglio, P.; Menegoni, F.; Vismara, L.; Cimolin, V.; Grugni, G.; Galli, M. Characterisation of balance capacity in Prader–Willi patients. *Res. Dev. Disabil.* **2011**, *32*, 81–86. [CrossRef]

15. Doyle, T.L.; Newton, R.U.; Burnett, A.F. Reliability of Traditional and Fractal Dimension Measures of Quiet Stance Center of Pressure in Young, Healthy People. *Arch. Phys. Med. Rehabil.* **2005**, *86*, 2034–2040. [CrossRef]

16. Galli, M.; Rigoldi, C.; Celletti, C.; Mainardi, L.; Tenore, N.; Albertini, G.; Camerota, F. Postural analysis in time and frequency domains in patients with Ehlers-Danlos syndrome. *Res. Dev. Disabil.* **2011**, *32*, 322–325. [CrossRef]

17. Cimolin, V.; Galli, M.; Rigoldi, C.; Grugni, G.; Vismara, L.; Mainardi, L.; Capodaglio, P. Fractal dimension approach in postural control of subjects with Prader-Willi Syndrome. *J. NeuroEngineering Rehabil.* **2011**, *8*, 45. [CrossRef]

18. Goldberger, A.L.; Amaral, L.A.N.; Hausdorff, J.M.; Ivanov, P.C.; Peng, C.-K.; Stanley, H.E. Fractal dynamics in physiology: Alterations with disease and aging. *Proc. Natl. Acad. Sci. USA* **2002**, *99*, 2466–2472. [CrossRef]

19. Maffiuletti, N.A.; Agosti, F.; Proietti, M.; Riva, D.; Resnik, M.; Lafortuna, C.L.; Sartorio, A. Postural instability of extremely obese individuals improves after a body weight reduction program entailing specific balance training. *J. Endocrinol. Invest.* **2005**, *28*, 2–7. [CrossRef]

20. Alonso, A.C.; Mochizuki, L.; Silva Luna, N.M.; Ayama, S.; Canonica, A.C.; Greve, J.M.D.A. Relation between the Sensory and Anthropometric Variables in the Quiet Standing Postural Control: Is the Inverted Pendulum Important for the Static Balance Control? *BioMed Res. Int.* **2015**, *2015*, 1–5. [CrossRef]

21. Cau, N.; Cimolin, V.; Galli, M.; Precilios, H.; Tacchini, E.; Santovito, C.; Capodaglio, P. Center of pressure displacements during gait initiation in individuals with obesity. *J. NeuroEngineering Rehabil.* **2014**, *11*, 82. [CrossRef] [PubMed]

22. de Souza, S.A.F.; Faintuch, J.; Valezi, A.C.; Sant' Anna, A.F.; Gama-Rodrigues, J.J.; de Batista Fonseca, I.C.; Souza, R.B.; Senhorini, R.C. Gait Cinematic Analysis in Morbidly Obese Patients. *Obes. Surg.* **2005**, *15*, 1238–1242. [CrossRef] [PubMed]

23. Li, X.; Aruin, A.S. The effect of short-term changes in body mass distribution on feed-forward postural control. *J. Electromyogr. Kinesiol.* **2009**, *19*, 931–941. [CrossRef]

24. World Health Organisation Obesity and overweigth. Available online: https://www.who.int/news-room/fact-sheets/detail/obesity-and-overweight(accessed on 6 March 2020).

25. Ward, Z.J.; Bleich, S.N.; Cradock, A.L.; Barrett, J.L.; Giles, C.M.; Flax, C.; Long, M.W.; Gortmaker, S.L. Projected U.S. State-Level Prevalence of Adult Obesity and Severe Obesity. *N. Engl. J. Med.* **2019**, *381*, 2440–2450. [CrossRef]

26. McGraw, B.; McClenaghan, B.A.; Williams, H.G.; Dickerson, J.; Ward, D.S. Gait and postural stability in obese and nonobese prepubertal boys. *Arch. Phys. Med. Rehabil.* **2000**, *81*, 484–489. [CrossRef] [PubMed]

27. Mathus-Vliegen, E.M.H.; Basdevant, A.; Finer, N.; Hainer, V.; Hauner, H.; Micic, D.; Maislos, M.; Roman, G.; Schutz, Y.; Tsigos, C.; et al. Prevalence, Pathophysiology, Health Consequences and Treatment Options of Obesity in the Elderly: A Guideline. *Obes. Facts* **2012**, *5*, 460–483. [CrossRef]

28. Menegoni, F.; Milano, E.; Trotti, C.; Galli, M.; Bigoni, M.; Baudo, S.; Mauro, A. Quantitative evaluation of functional limitation of upper limb movements in subjects affected by ataxia. *Eur. J. Neurol.* **2009**, *16*, 232–239. [CrossRef]

29. Son, S.M. Influence of Obesity on Postural Stability in Young Adults. *Osong Public Health Res. Perspect.* **2016**, *7*, 378–381. [CrossRef]

30. Hue, O.; Simoneau, M.; Marcotte, J.; Berrigan, F.; Doré, J.; Marceau, P.; Marceau, S.; Tremblay, A.; Teasdale, N. Body weight is a strong predictor of postural stability. *Gait Posture* **2007**, *26*, 32–38. [CrossRef]

31. Teasdale, N.; Hue, O.; Marcotte, J.; Berrigan, F.; Simoneau, M.; Doré, J.; Marceau, P.; Marceau, S.; Tremblay, A. Reducing weight increases postural stability in obese and morbid obese men. *Int. J. Obes.* **2007**, *31*, 153–160. [CrossRef]

32. Menegoni, F.; Galli, M.; Tacchini, E.; Vismara, L.; Cavigioli, M.; Capodaglio, P. Gender-specific Effect of Obesity on Balance. *Obesity* **2009**, *17*, 1951–1956. [CrossRef] [PubMed]

33. Błaszczyk, J.W.; Cieślinska-Świder, J.; Plewa, M.; Zahorska-Markiewicz, B.; Markiewicz, A. Effects of excessive body weight on postural control. *J. Biomech.* **2009**, *42*, 1295–1300. [CrossRef] [PubMed]

34. Cruz-Gómez, N.S.; Plascencia, G.; Villanueva-Padrón, L.A.; Jáuregui-Renaud, K. Influence of Obesity and Gender on the Postural Stability during Upright Stance. *Obes. Facts* **2011**, *4*, 212–217. [CrossRef] [PubMed]

35. Rigoldi, C.; Galli, M.; Mainardi, L.; Crivellini, M.; Albertini, G. Postural control in children, teenagers and adults with Down syndrome. *Res. Dev. Disabil.* **2011**, *32*, 170–175. [CrossRef]

36. Hita-Contreras, F.; Martínez-Amat, A.; Lomas-Vega, R.; Álvarez, P.; Mendoza, N.; Romero-Franco, N.; Aránega, A. Relationship of body mass index and body fat distribution with postural balance and risk of falls in Spanish postmenopausal women. *Menopause* **2013**, *20*, 202–208. [CrossRef]

37. Cimolin, V.; Galli, M.; Rigoldi, C.; Grugni, G.; Vismara, L.; de Souza, S.A.F.; Mainardi, L.; Albertini, G.; Capodaglio, P. The fractal dimension approach in posture: a comparison between Down and Prader–Willi syndrome patients. *Comput. Methods Biomech. Biomed. Engin.* **2014**, *17*, 1535–1541. [CrossRef]

38. Cieślińska-Świder, J.; Furmanek, M.P.; Błaszczyk, J.W. The influence of adipose tissue location on postural control. *J. Biomech.* **2017**, *60*, 162–169. [CrossRef]

39. Hirjaková, Z.; Šuttová, K.; Kimijanová, J.; Bzdúšková, D.; Hlavačka, F. Postural Changes During Quiet Stance and Gait Initiation in Slightly Obese Adults. *Physiol. Res.* **2018**, 985–992. [CrossRef]

40. Paillard, T.; Noé, F. Techniques and Methods for Testing the Postural Function in Healthy and Pathological Subjects. *BioMed Res. Int.* **2015**, *2015*, 1–15. [CrossRef]

41. Cimolin, V.; Galli, M.; Grugni, G.; Vismara, L.; Precilios, H.; Albertini, G.; Rigoldi, C.; Capodaglio, P. Postural strategies in Prader–Willi and Down syndrome patients. *Res. Dev. Disabil.* **2011**, *32*, 669–673. [CrossRef]

42. Galli, M.; Rigoldi, C.; Mainardi, L.; Tenore, N.; Onorati, P.; Albertini, G. Postural control in patients with Down syndrome. *Disabil. Rehabil.* **2008**, *30*, 1274–1278. [CrossRef] [PubMed]

43. Galli, M.; Cimolin, V.; Vismara, L.; Grugni, G.; Camerota, F.; Celletti, C.; Albertini, G.; Rigoldi, C.; Capodaglio, P. The effects of muscle hypotonia and weakness on balance: A study on Prader–Willi and Ehlers–Danlos syndrome patients. *Res. Dev. Disabil.* **2011**, *32*, 1117–1121. [CrossRef] [PubMed]

44. Dutil, M.; Handrigan, G.A.; Corbeil, P.; Cantin, V.; Simoneau, M.; Teasdale, N.; Hue, O. The impact of obesity on balance control in community-dwelling older women. *AGE* **2013**, *35*, 883–890. [CrossRef] [PubMed]

45. Clark, C.C.T.; Barnes, C.M.; Holton, M.; Summers, H.D.; Stratton, G. Profiling movement quality and gait characteristics according to body-mass index in children (9–11 y). *Hum. Mov. Sci.* **2016**, *49*, 291–300. [CrossRef] [PubMed]

46. Berrigan, F.; Simoneau, M.; Tremblay, A.; Hue, O.; Teasdale, N. Influence of obesity on accurate and rapid arm movement performed from a standing posture. *Int. J. Obes.* **2006**, *30*, 1750–1757. [CrossRef] [PubMed]

47. Gilleard, W.; Smith, T. Effect of obesity on posture and hip joint moments during a standing task, and trunk forward flexion motion. *Int. J. Obes.* **2007**, *31*, 267–271. [CrossRef]

48. Corbeil, P.; Simoneau, M.; Rancourt, D.; Tremblay, A.; Teasdale, N. Increased risk for falling associated with obesity: mathematical modeling of postural control. *IEEE Trans. Neural Syst. Rehabil. Eng.* **2001**, *9*, 126–136. [CrossRef]

49. Rodacki, A.L.F.; Fowler, N.E.; Provensi, C.L.G.; Rodacki, C.D.L.N.; Dezan, V.H. Body mass as a factor in stature change. *Clin. Biomech.* **2005**, *20*, 799–805. [CrossRef]

 *applied*
sciences

Article

Estimation of Transition Frequency during Continuous Translation Surface Perturbation

Nur Fatin Fatina Mohd Ramli [1,*], Mohd Azuwan Mat Dzahir [1,2] and Shin-Ichiroh Yamamoto [1]

[1] Department of Bioscience Engineering, Shibaura Institute of Technology, 307 Fukasaku, Minuma-ku, Saitama-City, Saitama 337-8570, Japan; azuwan@utm.my (M.A.M.D.); yamashin@se.shibaura-it.ac.jp (S.-I.Y.)
[2] School of Mechanical Engineering, Faculty of Engineering, Universiti of Teknologi Malaysia, Skudai 81310, Johor Bahru, Malaysia
* Correspondence: nb16106@shibaura-it.ac.jp; Tel.: +81-80-4084-2365

Received: 2 October 2019; Accepted: 11 November 2019; Published: 14 November 2019

Abstract: Depending on task requirements, a human is able to select distinct strategies such as the use of an ankle strategy and hip strategy to maintain their balance. Postural control actions often co-occur with other movements, and such movements may bring about a change from one type of postural coordination to another. The selection of a postural control strategy has typically been investigated by the transition of the center of mass (COM), center of pressure (COP), and in between angle joint motion along with their characteristics. In this paper, we proposed a method using the logistic function of the sigmoid model based on cross-correlation coefficient (CCF) data for investigating and observing the transition of postural control strategies of COM–COP and ankle-hip angles towards anterior–posterior (AP) continuous translation perturbation. Subjects were required to stand on the motion base platform where perturbations with an increasing frequency (0.2 Hz to 0.8 Hz) and decreasing frequency (0.8 Hz to 0.2 Hz) in steps of 0.02 Hz, were induced. As the frequency increased, the COM and COP displacements were decreased, with the opposite trend observable with decreasing frequency. This pattern was also observed at the head peak-to-peak amplitude. Meanwhile, ankle and hip angular displacements were increased during increasing frequency and decreased during decreasing frequency. In this paper, the proposed sigmoid model could identify the transition frequency of COM–COP and ankle–hip transition. The mean transition frequency of COM–COP during increasing frequency was 0.44 Hz, and the ankle–hip transition frequency was 0.42 Hz. Meanwhile, for decreasing frequency, the COM–COP transition frequency was 0.55 Hz, and for the ankle–hip transition the frequency was 0.56 Hz. With frequencies, both increasing and decreasing, the COM–COP and ankle–hip transition frequencies occurred almost at the same frequency. Furthermore, the transition occurred at a lower time scale during increasing frequency compared to decreasing frequency. In conclusion, the continuous translation surface perturbation provided information on the behavior of postural control strategies. A sudden change in 'phase angle' was observed, where either an ankle or hip strategy was implemented to maintain balance. Besides, the transition frequency of postural control strategies could be determined to occur between 0.4 Hz and 0.6 Hz, based on the average value, for healthy young subjects in the AP plane. Furthermore, the proposed sigmoid model was believed to be able to be used in the determination of transition frequency in postural control strategies.

Keywords: postural control strategies; sigmoid model; cross-correlation coefficient; continuous support surface translation perturbation; kinematics; kinetics; transition

---

1. Introduction

The ability to maintain balance in an upright position during quiet standing and dynamic task conditions is necessary for successful performance in daily life tasks [1,2]. Human postural control tends to initiate and constraint joint movement so that the center of mass (COM) is positioned over the base of support (BOS) and aligned with the center of pressure (COP), which is known as an equilibrium state. However, any external perturbation that is induced to those parts of the body will result in the shift of COM closer to the BOS border and interrupt alignment between COM and COP, which will cause a non-equilibrium state [3]. However, the balance control systems degenerate as we grow older, and this factor becomes the main reason for the high risk of falls among elderly people. Therefore, a better and deeper understanding of postural control strategies needs to be understood first before the development of training and rehabilitation tools.

Depending on task requirements, a human is able to select distinct strategies. Two primary postural control strategies were identified in upright bipeds, which are an ankle strategy and hip strategy [4,5]. The ankle strategy is viewed as an inverted pendulum with the motion around the ankle joint. Meanwhile, the hip strategy is used when the postural stability creates a particular constraint on the posture when the motion is around the hip joint. With the presence of a hip strategy, Buchanan and Horak suggested that a single link inverted pendulum will split into a multi-link model [6]. Besides these two primary postural control strategies, an additional strategy mode has also been identified that incorporates the ankle and hip strategies into a coordinated strategy. However, how these strategies are selected to maintain balance is still unknown and needs further investigation. In this study, we only observed the ankle and hip strategies since these are the primary strategies used in postural control.

Postural control actions often co-occur with other movements, and such movements may bring about a change from one type of postural coordination to another. This can be caused by the basic patterns of postural coordination being centrally represented by a set of motor programs, and postural transitions being behavioral consequences of changes between programs operating at the level of the central nervous system (CNS) [7,8]. These, as well as the changes between postural states, are consequences of the self-organized nature of the postural system, which exhibit properties of non-equilibrium phase transitions between attractors [9]. Typically, continuous relative phase (CRP) [10] or point estimate relative phase (PRP) [11] have been applied to investigate the coordination relation of two variables. The in-phase mode ($\sim0°$) indicates that two stationary signals move in the same direction, whereas the anti-phase mode ($\sim180°$) indicates that the signals are moving in the opposite direction. However, until now, there have been no studies that have mentioned the exact frequency at which the postural strategy changes happens. Other methods can also be considered in defining the coordination relation of two variables, such as cross correlation function (CCF) analysis. This analysis is a powerful, efficient, and relatively easy-to-apply tool for quantifying associations between variables. It also can be very useful for investigating spatial and temporal relationships between time-varying signals and can provide further insight into the coordination of movement, muscle activation patterns, and isolation noise within a signal. Furthermore, since the CCF coefficient data shows a resemblance to a sigmoidal pattern, we proposed that the transition frequency could be identified by using the logistic function of the sigmoid model. This is because the sigmoid model provides a good example of non-linear and quickly increasing functions of the probability of disclosure and makes computation easier [12]. While most traditional nonlinear activation functions are bounded, simpler piecewise linear activation functions have become popular because of their computational efficiency and robustness in preventing saturation [13]. Therefore, the sigmoid model is a reliable analysis technique, which mimics the physiologically based prediction of the input/output relation [14].

There have been several studies into the coordination of the body that affords stability in dynamic postural balance under both discrete [15,16] and continuous oscillation [17,18] of motion base support. Many researchers have reported the use of support surface perturbation as an experimental method to investigate different patterns of postural strategies [19–21]. According to previous studies, lower extremities play an important role in balance. The ascending sensory pathway from the sole then regulates muscle activation to initiate ankle motion, then activating hip motion, and finally the upper body, including trunk and head [20]. Without sufficient activation or response from the lower extremities, it is difficult for nervous systems to provide feedback to generate an effective strategy and prevent falling. Therefore, it is believed that support surface perturbation, which acts as an external perturbation, is enough to induce postural control strategy patterns. Besides, this perturbation can replicate daily life activities such as slipping, tripping, and also stepping. Both increasing frequency and decreasing frequency were implemented in this paper. In previous studies, large step size changes of continuous translation frequency were used [10,21]. Motor sensory perception could sense slight changes of the translation frequency, which could lead to sudden changes in postural control strategies and the possibility of the subjects not reacting naturally. Therefore, a small step size of translation frequency was implemented in this paper to provide a more precise transition frequency.

This paper aims to investigate and observe the transition frequency of COM–COP and ankle–hip transition towards continuous translation frequency perturbation. The logistic function of the sigmoid model based on CCF coefficient data was used to determine the transition frequency. Besides, the effect of frequency order perturbation towards the transition frequency was also observed. Moreover, this paper is the first paper to introduce the sigmoid model based on CCF coefficient data for determining the transition frequency of postural control strategies. We examined the hypotheses that (1) a sudden shift from in-phase to anti-phase happens as effect of translation frequency perturbation; (2) the transition frequency range decreases with the implementation of a small step size in continuous translation perturbation; (3) the different order of translation frequencies affects the transition frequency; and (4) the sigmoid model based on CCF coefficient data can be used in determining the transition frequency of postural control strategies. The findings from this study will contribute to further understanding of human postural control strategies.

2. Methods

2.1. Participants

In this study, 20 healthy young male subjects participated (25.70 ± 4.42 years old, 65.85 ± 9.53 kg, 170.43 ± 6.63 cm). The subject sample size (n) was confirmed with sample size calculation. The minimum size for every parameter (increasing frequency (COM–COP, $n = 15$; ankle-hip, $n = 9$), and decreasing frequency (COM–COP, $n = 13$; ankle–hip, $n = 9$)) was obtained. From the calculation, it was confirmed that the current sample size was sufficient for conducting the experiment. Young and healthy subjects were selected for the determination of transition frequency due to their agility and ability for a better control strategy. All healthy subjects were free from neurological disorders, balance disorders, and vestibular function disorders. They had no history of neurological impairment. Information regarding the subjects' history of falls and physical condition were recorded as references. The experimental procedure was approved by the ethical committee of our research institute.

2.2. Experimental Procedures

Subjects were requested to stay standing quietly for 10 s to record quiet standing data. Then, subjects were exposed to external surface continuous translation perturbation with two types of frequency orders, which increased (0.2 Hz to 0.8 Hz) and decreased (0.8 Hz to 0.2 Hz) in 0.02 Hz steps with a displacement of 100 mm peak-to-peak. Buchanan and Horak stated that the transition frequency occurs at a frequency of 0.5 Hz and above [6]. Therefore, the selection of these frequencies, which is from 0.2 H to 0.8 Hz, with 0.5 Hz in between, was used in the determination of transition frequency. A 0.02 Hz step of frequency was implemented in order for subjects not to notice the slight changes of the translation frequency. Besides, this approach can avoid sudden changes, and the subjects can react naturally in executing postural control strategies. Each frequency changed after five oscillation cycles. The subjects were instructed to maintain their postural balance while the continuous translation perturbations were induced as they were barefoot, with eyes open, and focused at a mark that was set at eye level. The total duration of the experiment was about 373 s for each subject in both increasing and decreasing frequencies. The subjects' vision and vestibular sense were not manipulated.

2.3. Experimental Apparatus

An external continuous translation perturbation in an anterior–posterior (AP) direction was produced by a 6-axis movable platform (MB-150, Cosmate, Tokyo, Japan). A force platform (9286A, Kistler, Yokohama, Japan) was used to derive the displacement of the body's COP and mounted on the moving platform. A 3-D motion capture analysis system with ten high-precision infrared cameras (Kestrel camera, Motion Analysis Corp., CA, USA) was used to record the motion of the passive marker attached over the joints of the subjects. Eighteen fixed reflective markers (placed at the 3rd metatarsal, lateral malleolus, lateral condyle, trochanter of the femur, iliac crest, acromion of spacula, top of the head, and four markers on the force plate) were attached over the subjects' joints. Both right and left knees were locked to prevent bias movement from the knees. This study focused on two segmental links of human postural strategies, which are ankle and hip strategies. Therefore, the absence from the knee joint was necessary to purely obtain the ankle and hip joint data. The method used to lock the knee joint was by using a pair of wood splints sandwiched around the knee joints in account of the subject comfort.

2.4. Data Collection

Motion capture data, which are marker data, underwent their own post-processing in order to create and gather the marker names and coordinates. All disconnected frames of marker positions needed to be corrected and smoothed to eliminate noise. Furthermore, force plate data were also filtered with a 2nd Butterworth filter with a cut off of 6 Hz to eliminate noise, especially from the power line and movement. Each data point was then resampled to a sampling frequency of 200 Hz. The coordinated data from motion capture were collected as raw data for the calculation process. The COM was calculated from the eight-segment model. The total body COM position was obtained by the weighted summation of the individual segment COM position, as mentioned in Equation (1). Constant value for each segment was shown as in Table 1 [22]. The ground reaction force (F_v) was obtained by summing up all the vertical forces from the strange gauge (i.e., f_{z1}, f_{z2}, f_{z3}, and f_{z4}) of the force plate, as shown in Figure 1. The force plate moment in the x-axis (M_x) was generated by these vertical forces, as shown in Equation (2). The joint movement coordinates (x, y, z) obtained from motion analysis systems with a 1 kHz sample rate were used to measure joint angle displacement (θ_{ankle} and θ_{hip}) and body segment length (h_{ankle}, h_{hip}, and h_{seg}) for segmental COM location. The COP displacement was determined from Equation (3) below where d_z was the distance from the surface to the platform origin. The ankle and hip angle displacement were calculated with Equation (4) by using 2 vectors, as shown in Figure 2. The average of head peak-to-peak amplitude for each frequency was also calculated.

Table 1. Constant value for each segment [22].

Segment	Segment Weight/Total Body Weight
Head, arm, trunk (HAT)	0.536
Pelvis	0.142
Left/Right Thigh	0.1
Left/Right Leg	0.0465
Left/Right Foot	0.0145

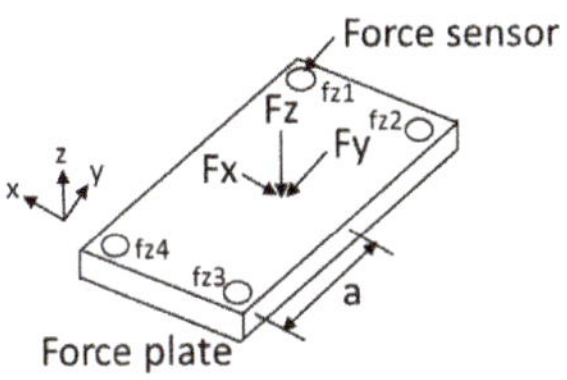

Figure 1. Force distribution on force plate.

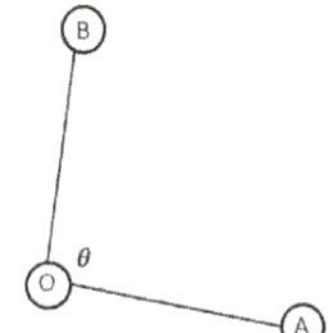

Figure 2. Joint angle.

The COM and COP displacement in the AP direction was as follows:

$$COM_{AP}[mm] = \sum \left(COM_{each\ seg} \times constant\ value_{each\ seg} \right) \tag{1}$$

where the constant value is in the table below.

$$M_x = a(f_{z1} + f_{z2} + f_{z3} + f_{z4}) \tag{2}$$

$$COP_{AP}[mm] = \frac{M_x - \left(F_y \cdot d_z \right)}{F_v} \tag{3}$$

where a is the sensor offset value and F_y is the horizontal component in y-direction force.

The ankle and hip angle displacement in the AP direction were as follows:

$$\cos \theta (deg) = \frac{\vec{A} \cdot \vec{B}}{|A||B|} \tag{4}$$

2.5. Data Analysis

2.5.1. Cross-Correlation Coefficient

This study focused on observing the changes of postural control strategies, especially regarding the transition frequency of COM–COP and ankle–hip angle toward increasing and decreasing frequency translation perturbation. In this paper, CCF analysis was used to observe the correlation between COM–COP and ankle–hip angle. This analysis is a powerful, efficient, and relatively easy-to-apply tool for quantifying associations between variables. It also can be very useful for investigating spatial and temporal relationships between time-varying signals and can provide further insight into the coordination of movement, muscle activation patterns, and isolation noise within a signal [23]. The coefficient value range was between −1.0 and 1.0. A CCF coefficient of −1.0 shows a perfect negative correlation, while a coefficient of 1.0 shows a perfect positive correlation. A calculated value greater than 1.0 or less than −1.0 means that there was an error in the experimental, data collection, and processing measurement. Cross-correlation involves correlating two different time-varying signals against each other. It was calculated using Equation (5), where x and y are the sample data:

$$r = \frac{n(\sum xy) - (\sum x)(\sum y)}{\sqrt{\left[n \sum x^2 - (\sum x)^2\right]\left[n \sum y^2 - (\sum y)^2\right]}} \tag{5}$$

There was a total of 31 frequencies, from 0.2 Hz to 0.8 Hz and 0.8 Hz to 0.2 Hz with 0.02 Hz steps. Each frequency comprised 5 cycles, which gave a total of 155 cycles except 0.2 Hz for increasing frequency and 0.8 Hz for decreasing frequency translation perturbation, which comprised 6 cycles. The first cycle of 0.2 Hz and 0.8 Hz was eliminated in order to cut out unnecessary movement at the beginning of the experiment. An average of 5 cycles from every frequency was calculated. Then, by using the value of average cycle, CCF coefficients were calculated for COM–COP and ankle–hip angle in order to observe the correlation of these two cycles.

2.5.2. Logistic Function (Sigmoid Model)

The sigmoid model provides a good example of non-linear and quickly increasing and decreasing functions of the probability of disclosure and makes computation easier [12]. Besides, the sigmoid model is a reliable analysis technique that mimics the physiologically based prediction. In this paper, the CCF coefficient data distribution graph exhibits a sigmoidal pattern at increasing and decreasing stimulation intensities. Therefore, we proposed the sigmoid model in order to determine the transition phase of human postural control strategies. A general least square model similar to one developed in the investigation of the ascending limb of all recruitment curves and transcranial magnetic stimulation (TMS) research was used [13]. In this paper, a custom four-parameter sigmoid model was modified as shown in Equation (6) below:

$$F_x = \frac{L}{1 + e^{k(x - x_0)}} + y_0 \tag{6}$$

where L is the vertical range of model, k is the gradient or slope of the model, and x is the CCF coefficient data distribution. Furthermore, x_0 is determined as the transition frequency, and y_0 is the vertical correcting position of the model.

The sigmoid model provides a good example of a non-linear model, can predict the probability as an output, and makes computation easier. Therefore, the sigmoid model is reliable, which mimics the physiologically based data as in this study is from CCF coefficient data distribution [14]. This method utilizes curve fitting using a logistic function, which is optimized using the particle swamp optimization (PSO) technique. With this technique, the model with the lowest error was chosen as the best model providing the sigmoid model. In this paper, the lower frequency was referring to the frequencies below 0.5 Hz and high frequency was for frequencies above 0.5 Hz.

This model clearly can divide upper and lower distributions and the slope in between the distribution, as shown in Figure 3. In order to find the transition frequency for each subject, first we determined the initial slip and convergence of the model by the velocity of the sigmoid model data. The slope between the initial slip and convergence was defined as a transition phase, indicating that transition happens from the positive phase shift to the negative phase shift and vice versa. We calculated the coefficient of determination $\left(r^2\right)$, which is also defined as global goodness-of-fit of the sigmoid model, to clarify the fitness of the model towards CCF coefficient data. r^2 was calculated in order to define the fitness of the model since the model is a linear regression. The r^2 value of more than 0.039 with one variable was indicated as acceptable to the fitness of the model.

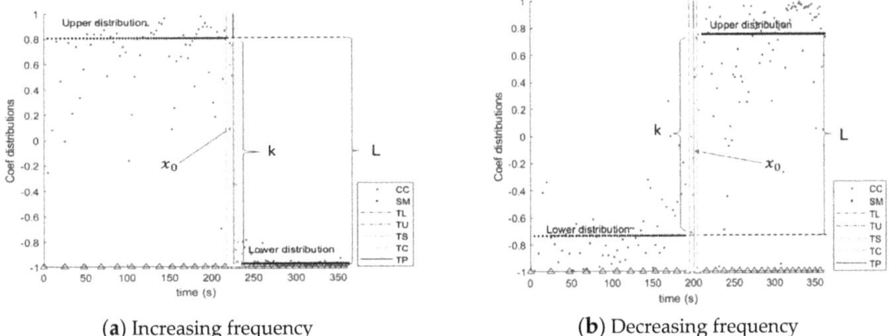

(**a**) Increasing frequency　　　　　　　(**b**) Decreasing frequency

Figure 3. Example of cross-correlation function (CCF) data distribution with the sigmoid model graph at (**a**) increasing and (**b**) decreasing frequency order conditions. The area between the upper distribution and lower distribution of each condition was determined as the transition phase (indicated as a light blue area) where the transition frequency was observed. CC: Cross-correlation function coefficient data; SM: sigmoid model; TL: transition lower; TU: transition upper; TS: transition slip; TC: transition convergence; TP: transition frequency point.

2.6. Statistical Analysis

The transition frequencies of all subjects are summarized in results section in terms of frequency. A comparison of the transition frequency between increasing and decreasing frequencies of COM–COP and ankle–hip angles was analyzed by using paired t-test analysis with a significance level of $p < 0.05$. The same statistical test was used to analyze a comparison between COM–COP and ankle–hip transition frequency in increased and decreased frequencies. Furthermore, two-way ANOVA analysis was used to observe the comparison between individual subjects. The Bonferroni post hoc test was used to determine the differences in all levels of comparison for the independent variables. All statistical analyses were completed using the MATLAB software.

3. Results

3.1. Displacement of Kinematics Parameters

All subjects were able to accomplish the given tasks perfectly. Figure 4 shows the waveforms of the representative subjects for COM, COP, ankle, and hip angle displacement during increasing and decreasing frequencies. Based on these waveforms, the COM and COP displacements decreased in frequency during increasing frequency. For ankle and hip angular displacement, the displacements were increased. Meanwhile, COM and COP displacements were increased, and ankle and hip angular displacements were decreased during decreasing frequency. With the presence of the transition between COM–COP and ankle–hip, at a certain time, the phase was changed from in-phase to anti-phase.

Figure 5 illustrates the average value of head peak-to-peak amplitude for each frequency during increasing and decreasing frequency. During increasing frequency perturbation, the peak-to-peak amplitude was decreasing. It was observed that the head peak-to-peak amplitude dropped about 50% between 0.42 Hz and 0.46 Hz. Besides, during decreasing frequency perturbation, the head peak-to-peak amplitude was gradually increasing. However, during low frequency, the head peak-to-peak amplitude during decreasing frequency was lower than increasing frequency.

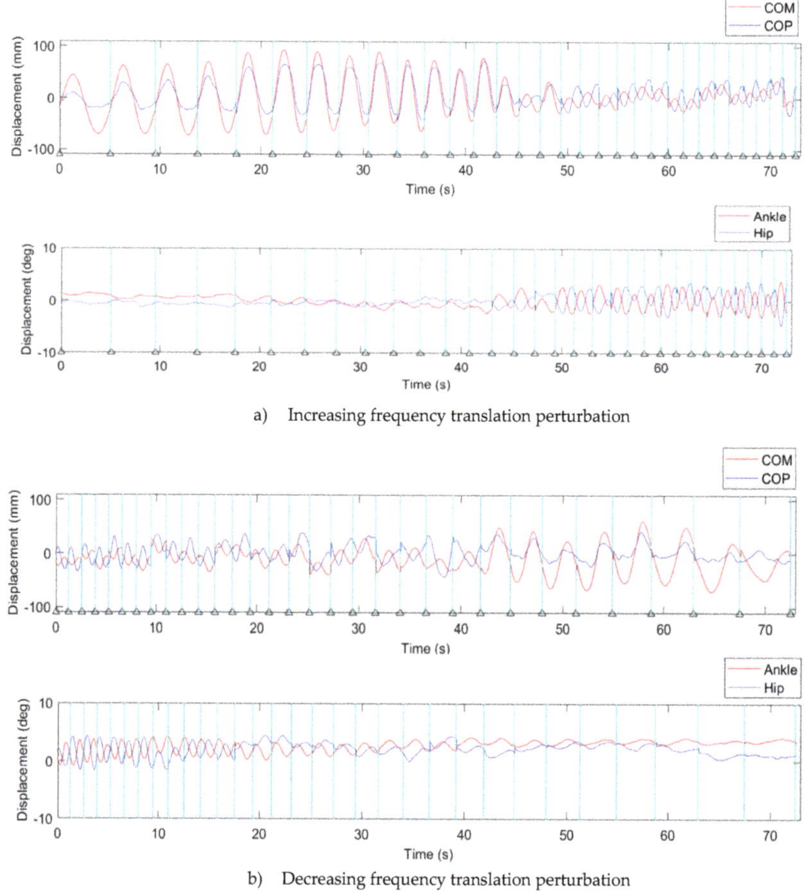

a)   Increasing frequency translation perturbation

b)   Decreasing frequency translation perturbation

Figure 4. The representative waveform of the center of mass (COM), center of pressure (COP), ankle joint, and hip joint displacement based on time during (**a**) increasing frequency and (**b**) decreasing frequency translation perturbation.

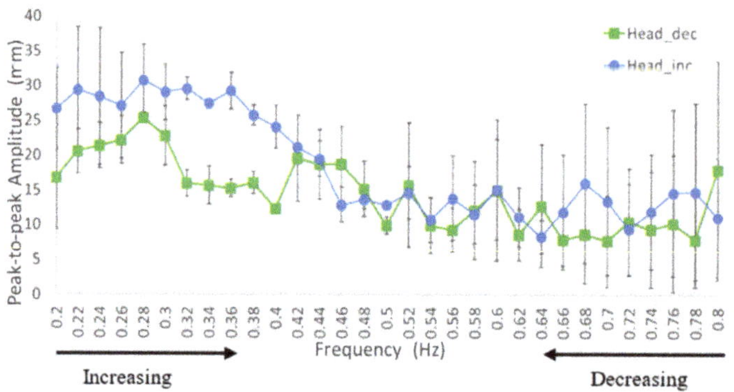

Figure 5. The average value of head peak-to-peak amplitude for each frequency during increasing and decreasing frequency.

3.2. Transition Frequency

The relation of COM–COP and ankle–hip angle was analyzed by using CCF coefficient data, and the transition frequency was determined with the sigmoid model. The COM–COP and ankle–hip sigmoid models of the representative subject are shown in Figure 6. All subjects showed the changes from upper to lower distribution and vice versa of CCF coefficient data, which can be explained as the degree of freedom of human balance being changed from one degree of freedom (DOF) to two DOF. The COM–COP and ankle–hip transition frequency for every subject and perturbation conditions are summarized and shown in Figure 7. Even though all subjects were free from neurological disease, they showed different transition frequencies at both COM–COP and ankle–hip transitions.

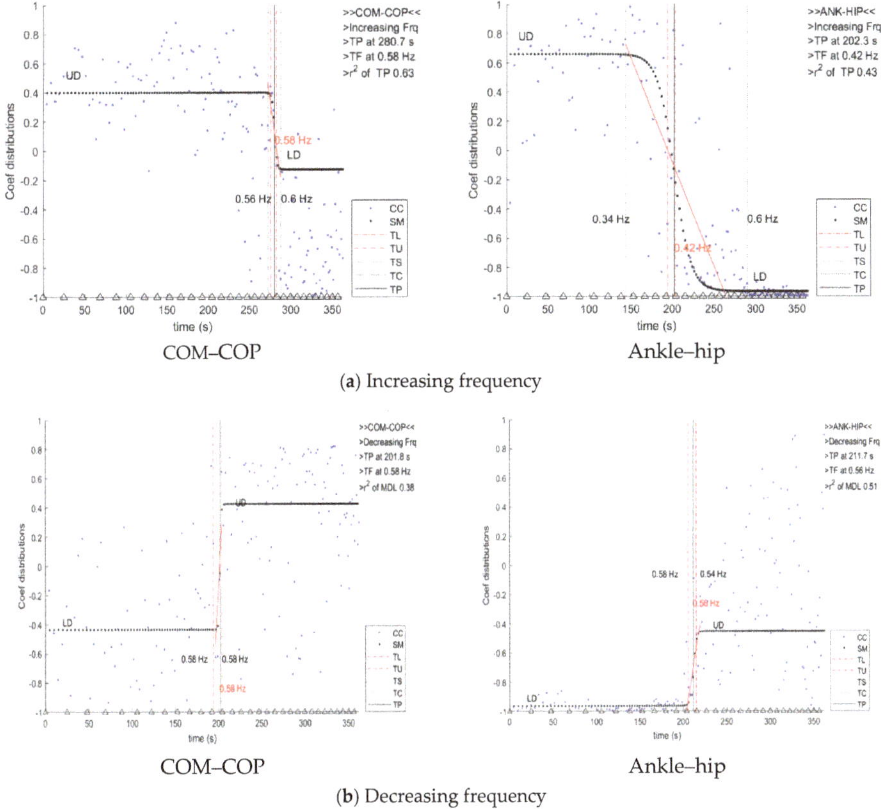

Figure 6. The representative of CCF coefficient data distribution with the sigmoid model during (**a**) increasing frequency and (**b**) decreasing frequency translation perturbation condition. Each blue point represents a CCF coefficient data for each frequency. The black line represents the sigmoid model. UD: Upper distribution; LD: Lower distribution.

For the COM–COP transition, the transition frequency varied between subjects. From the total of 40 transitions frequencies of COM–COP and ankle–hip for the 20 subjects, 26 transitions frequencies showed a transition of both COM–COP and ankle–hip transition at lower frequencies during increasing frequency perturbation conditions. For decreasing frequency, 16 transition frequencies showed the transition in the lower frequencies. The mean transition frequency of COM–COP was 0.44 Hz ($\pm$0.12) during increasing frequency, while it was 0.55 Hz ($\pm$0.12) during decreasing frequency. Furthermore, the mean transition frequency of ankle–hip during increasing frequency was 0.42 Hz ($\pm$0.12) and

0.56 Hz (±0.12) during decreasing frequency. From the mean value of transition frequency, it was observed that the COM–COP transition and ankle–hip transition occurred almost at the same frequency during both increasing and decreasing frequency. A significant difference was observed between increasing and decreasing frequency for COM–COP ($p = 0.003$) and ankle–hip transition frequencies ($p = 0.008$). However, no significant difference was observed between COM–COP and ankle–hip transitions frequency in both frequency conditions (increasing ($p = 0.56$, ns); decreasing ($p = 0.17$, ns)). No significant difference was observed between subjects during both frequency order conditions (increasing ($p = 0.08$, ns) and decreasing ($p = 0.34$, ns)). A significant difference was observed between increasing and decreasing frequency perturbation for head peak-to-peak amplitude ($p = 0.0361$). All subjects showed r^2 value more than 0.039, indicating that the model significantly fit with the CCF coefficient data distribution.

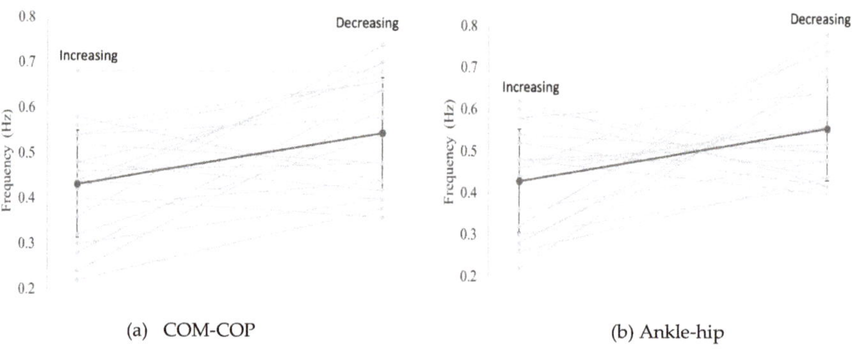

(a) COM-COP   (b) Ankle-hip

Figure 7. The transition frequency of (a) COM–COP and (b) ankle–hip during increasing and decreasing frequency translation perturbation. The dark line represents the mean value of transition frequency.

4. Discussion

The primary aim of this study was to observe and investigate the transition frequency of postural control strategies by using a proposed sigmoid model based on CCF coefficient data distribution. This paper is the first to use a sigmoid function model in order to observe the postural strategy transition frequency. The finding shows that the subjects' COM and COP displacements were decreased during increasing frequency and increased during decreasing frequency. In this paper, the transition frequency could be obtained from the proposed sigmoid model based on CCF coefficient data. All subjects showed difference transition frequencies at both COM–COP and ankle–hip. The transition frequency of ankle–hip occurred earlier than that of COM–COP during both increasing and decreasing frequencies.

4.1. Sigmoid Model Based on Cross-Correlation Coefficient Data for the Determination of Transition Frequency

Postural strategy transition can be defined as the changes of postural states between two attractors that result from the consequence of changes between programs operating at the level of the central nervous system (CNS) or of the self-organized nature of the postural system [9]. To date, several studies have investigated the transition between postural strategies by using a relative phase, either using point estimate relative phase (PRP) or continuous relative phase (CRP) [10,24–26]. However, in the present study, we proposed an analysis of the transition of postural strategy coordination by using a sigmoid model based on CCF coefficient data. The CCF coefficient is a statistical measure that calculates the strength of the relationship between the relative movements of two variables. Besides, since the CCF coefficient data distribution exhibits a similar pattern to the sigmoidal pattern, the sigmoid model was proposed to determine the transition frequency of the postural control strategies. This proposed model was shown to be suitable with any CCF coefficient data distribution even though the data were spread in a large distribution.

4.2. Kinematic Characteristic of Continuous Translation Perturbation Frequencies

The different frequencies, increasing and decreasing continuous translation perturbation platform, provided an experimental manipulation to investigate postural control strategies [10]. Based on kinematic results, COM and COP displacements were decreased towards increasing frequency and increased during decreasing frequency. This is in contrast with ankle and hip angle, where angular displacement increased during increasing frequency and decreased during decreasing frequency. Sensory information from visual, vestibular, and somatosensory systems is regulated by the central nervous system (CNS) to maintain balance and postural orientation. Different frequencies exposed to the body will change the coordinative patterns of the head, trunk, and legs to accommodate the different action on the body, such as the transition of postural strategy from ankle strategy to hip strategy [17]. Translating the body at different frequencies also moves the sensory systems within and outside of their optimal operating ranges. In this paper, at a low frequency, the subjects appeared to "ride" the platform depending on the movement of the platform itself. As suggested by Buchanan and Horak, the COM amplitude decreased during increasing frequency and the head motion became fixed in space, allowing vision scene oscillation produced by the translating support surface [6]. At the high frequency, the CNS tended to apply accurate control to reduce kinematic displacement in order to maintain balance in the desired position [27]. Besides, the CNS may suppress anticipatory postural adjustments (APAs) since APAs can reduce the effect of the forthcoming body perturbation [28]. Focusing on COP displacement, at a low frequency the displacement was smaller than COM and head displacement. In spite of this, the COP displacement was higher than COM at a high frequency, which agreed with the result of David A. Winter [2]. This indicated that the COP was greater in order to keep the COM within the COP line to maintain equilibrium. As shown in the results, at a low frequency the COM and COP were in-phase. As the frequency increased, COM and COP phases became anti-phase. An increasing transition frequency causes the COP amplitude to decrease and disturb the postural strategy of a subject as the COM amplitude can be out of bound, where a good prosecution of posture control strategy always requires inbound COM amplitude to avoid a stepping strategy. To avoid the involvement of a stepping strategy, the body increases joint stiffness and moment at the ankle joint [29]. At the lower frequency, the ankle strategy was used as a balance strategy. However, the physiological limits of performing the ankle strategy will be reached during increments in translation frequency; at this instance, it will try to compensate the balance strategy with contribution from hip movement [30,31]. This action of hip movement changes the COM acceleration, which in fact will affect the postural control strategy from an ankle to hip strategy. This changes the transition phase from in-phase to anti-phase while adapting to the perturbation translation's acceleration [32]. However, based on the result, the ankle–hip transition usually comes earlier than that of the COM–COP transition.

4.3. From a Single-Linked Model to Multi-Segmental Model

Quiet standing tends to approximate a single-link inverted pendulum. Recent studies have pointed out that hip joints play an important role in maintaining balance even in quiet standing [20,33–36], as well as in dynamic tasks, which show more ankle–hip joints characteristics [36] and which we can consider as a multi-segmental model. In this paper, the sigmoid model has shown that there was an upper and lower distribution of CCF coefficient data of COM–COP and ankle–hip, which revealed that there was a sudden transition from low to high frequency and vice versa at a certain frequency. These pattern changes are consistent with the results from Ko et al. [24]. The joint degree of freedom was regulated by a small amplitude as the motion from the moving platform was at a low frequency. However, as the motion was increased, the joint degree of freedom motion showed a large amplitude and forced the postural strategy to re-organize and dissipate the large force; i.e., the head fixed pattern and transition from ankle to hip strategy and vice versa.

The means of transition frequency for COM–COP and ankle–hip transitions were 0.44 Hz and 0.42 Hz for increasing frequency and 0.55 Hz and 0.56 Hz for decreasing frequency, respectively. This finding was parallel with Aviroop and Karl, who obtained a coordination change around of

0.4 Hz–0.6 Hz for the COM–COP transition frequency [11]. The transitions frequencies in both increasing and decreasing frequency were almost in the same frequency for both COM–COP and ankle–hip. Even though the transition occurred almost at the same frequency, a slightly difference was observed where the ankle–hip transition frequency was earlier than that of COM–COP. This finding was consistent with Ko et al., who stated that the COM–COP coordination is relatively robust and on a slower time scale of change compared to the coordination of joint components and their individual motions [24]. As the perturbation applied originated from the lower extremities, the perturbation moved to the sole then regulated muscle activation, which increased joint stiffness and moment at the ankle joint. Then, as the physiological limit was reached, the hip movement was involved. Within this process, the COM and COP positions were also moved from the origin.

The transition happened when the initial pattern was undergoing a state of instability. The differences of transition frequency were observed between increasing and decreasing frequency, where the mean transition frequency for increasing frequency condition occurred at the lower frequency than decreasing frequency condition. These different values for the transitions indicated that the coordination mode observed is dependent upon the direction of the changes in the control parameter (platform frequency) [37,38]. These results exhibit a hysteresis phenomenon when the frequency transition value is varied at different conditions of perturbation frequency. Hysteresis in this study means that the tendency for the postural control systems itself to remain in the current behavior state as the control parameter moves through the transition region, provides different frequency transition values depending on the direction of the control parameter [39]. Furthermore, the coordination of the postural control strategy was more stable when moving from anti-phase to in-phase (decreasing frequency) compared to in-phase to anti-phase (increasing frequency) [11]. In decreasing frequency, the instabilities decreased as the frequency perturbation decreased, which resulted in the earlier transition frequency compared to increasing frequency. Therefore, the mean value of decreasing transition frequency occurred earlier than that of increasing transition frequency, which indicates the subjects easily coordinated the postural strategies during high frequency perturbation tasks.

The differences and variability of the transition point frequency of COM–COP and ankle–hip among the subjects were also observed. The variation of transition frequency among the subjects might be related to the balance ability of each subject [40]. It was suggested that patients with low scores of the balance ability test, i.e., Functional Reach Test, have high joint stiffness [30]. However, in this study, no correlation was observed between balance ability and transition frequency among the subjects. This may have happened because there is no specified condition implemented that can normalize posture control strategies.

4.4. Study Limitations

Two limitations to the present study should be noted. The first is that only healthy and young subjects participated with high agility and flexibility, as the capability of each subject in terms of their motor sensory and reflex were considered. Without these attributes, the possibility of the subject being unable to produce a good posture control strategy during the manipulation of translation perturbation is high. In the future, we will include a contribution from subjects with impaired conditions such as having bad vision or vestibular disorder to determine this factor's influence in the identification of transition frequency. Besides for that, elderly people also need to be included. The second limitation is that both knees of the subjects were locked during perturbation to minimize the knee involvement in the postural control strategy, as our focus was to investigate the transition of ankle and hip strategy. This is because during an upright bipedal state, the major contribution of posture control strategies comes from ankle and hip movements, while the knee is involved only when the subject is out of balance [41].

5. Conclusions

In conclusion, a transition frequency identification of postural control from an ankle to hip strategy was presented. Two types of data were used: the first was COM–COP displacements and the second was ankle–hip angular displacements. These data were analyzed to obtain CCF coefficient data under increasing and decreasing continuous translation surface perturbation at different frequencies. The transition point of postural control strategies was identified based on the transition phase of CCF data (in-phase to anti-phase). This was achieved by implementing the logistic function of the sigmoid model to differentiate the upper and lower distribution of CCF coefficient data. Two tests were performed to determine the significant difference between frequencies and types of data; the first was a paired *t*-test and second was a two-way ANOVA. It was observed that ankle–hip angle opposed the COM–COP displacements during increasing and decreasing perturbation frequency, where the COM and COP displacements were decreased during increasing frequency perturbation while the ankle and hip angle displacement were increased and vice versa for decreasing frequency perturbation. Besides, the peak-to-peak amplitude of head displacement was decreased as the perturbation was increasing and vice versa. At some instant, there is a sudden change in the 'phase angle' of the data, where either the ankle strategy becomes passive or the hip becomes active. Based on the results, the mean transition frequency of COM–COP and ankle–hip for all subjects was determined. It showed the transition of COM–COP was evident at 0.44 Hz and the transition of ankle–hip was at 0.42 Hz during increasing perturbation. Meanwhile, a transition frequency of 0.55 Hz (COM–COP) and 0.56 Hz (ankle–hip) was observed during decreasing perturbation. Therefore, it was concluded that the transition frequency of postural control strategies for both the COM–COP and ankle–hip data of healthy subjects was apparent between 0.4 Hz and 0.6 Hz. The transition frequency was varied according to the direction of the platform frequency. Besides, the variation in the transition frequency of each subject occurs because there is no specified condition implemented that can normalize the posture control strategy such as a light touch and body-harness support. In addition, the implementation of a small step size of 0.02 Hz during increasing and decreasing continuous translation surface perturbation gives a more accurate reading of transition frequency as compared with previous studies, which indicated that the range of transition happened at 0.5 Hz and above. This study provides a new approach for analyzing the transition phase of postural control strategies.

Author Contributions: Conceptualization, M.A.M.D.; data curation, N.F.F.M.R.; formal analysis, N.F.F.M.R.; funding acquisition, S.-I.Y.; investigation, N.F.F.M.R.; methodology, N.F.F.M.R.; software, M.A.M.D.; supervision, S.-I.Y.; validation, N.F.F.M.R.; writing—original draft, N.F.F.M.R.; writing—review and editing, S.-I.Y.

Funding: This research received no external funding.

Conflicts of Interest: The authors declare no conflict of interest.

References

1. Lord, S.R.; Clark, R.D.; Webster, I.W. Postural stability and associated physiological factors in a population of aged persons. *J. Gerontol.* **1991**, *46*, 69–76. [CrossRef] [PubMed]
2. Winter, D.A. Human balance and posture control during standing and walking. *Gait Posture* **1995**, *3*, 193–214. [CrossRef]
3. Santos, M.J.; Kanekar, N.; Aruin, A.S. The role of anticipatory adjustment in compensatory control of posture: 2. Biomechanical analysis. *J. Electromyogr. Kinesiol.* **2013**, *20*, 398–405. [CrossRef] [PubMed]
4. Horak, F.B.; Nasher, L.M. Central programming of postural movement: Adaptation to altered support-surface configuration. *J. Neurophysiol.* **1986**, *55*, 1369–1381. [CrossRef]
5. Nasher, L.M.; McCollum, G. The organization of human postural movement: A formal basis and experimental synthesis. *Behav. Brain Sci.* **1985**, *8*, 135–172. [CrossRef]
6. Buchanan, J.J.; Horak, F.B. Emergence of postural patterns as a function of vision and translation frequency. *J. Neurophysiol.* **1999**, *81*, 2325–2339. [CrossRef]

7. Horak, F.B.; Macpherson, J.M. *Postural Orientation and Equilibrium*; Rowell, L.B., Shepard, J.T., Eds.; Handbook of Physiology: Section 12, Exercise Regulation and Integration of Multiple Systems; Oxford University Press: New York, NY, USA, 1996; pp. 255–292.

8. Bardy, B.G.; Oullier, O.; Bootsma, R.J.; Stoffregen, T.A. Dynamics of human postural transitions. *J. Exp. Psychol. Hum. Percept. Perform.* **2002**, *28*, 499–514. [CrossRef]

9. Kelso, S. Phase transitions and critical behavior in human bimanual coordination. *Am. J. Physiol.* **1984**, *246*, R1000–R1004. [CrossRef]

10. Hamill, J.; Bates, B.T.; Holt, K.G. Timing of lower extremity joint actions during treadmill running. *Med. Sci. Sports Exerc.* **1992**, *24*, 807–813. [CrossRef]

11. Dutt-Mazumder, A.; Newell, K. Transition of postural coordination as a function of frequency of the moving support platform. *Hum. Mov. Sci.* **2017**, *52*, 24–35. [CrossRef]

12. Sfar, A.R.; Challal, Y.; Moyal, P.; Natalizio, E. A Game Theoretic Approach for Privacy Preserving Model in IoT-Based Transportation. *IEEE Trans. Intell. Transp. Syst.* **2019**. [CrossRef]

13. Wu, G.; Say, B.; Sanner, S. Scalable Nonlinear Planning with Deep Neural Network Learned Transition Models. *arXiv* **2019**, arXiv:1904.02873.

14. Klimstra, M.; Zehr, E.P. A sigmoid function is the best fit for the ascending limb of the Hoffmann reflex recruitment curve. *Exp. Brain Res.* **2008**, *186*, 93–105. [CrossRef] [PubMed]

15. Gu, M.-J.; Schultz, A.B.; Shepard, N.T.; Alexander, N.B. Postural control in young and elderly adults when stance is perturbed: Dynamics. *J. Biomech.* **1996**, *29*, 319–329. [CrossRef]

16. Hughes, M.; Schenkman, M.; Chandler, J.; Studenski, S. Postural responses to platform perturbation: kinematics and electromyography. *Clin. Biomech.* **1995**, *10*, 318–322. [CrossRef]

17. Keshner, E.A.; Woollacott, M.H.; Debû, B. Neck, trunk and limb muscle responses during postural perturbations in humans. *Exp. Brain Res.* **1988**, *71*, 455–466. [CrossRef]

18. Woollacott, M.H.; Von Hosten, C.; Rösblad, B. Relation between muscle response onset and body segmental movements during postural perturbations in humans. *Exp. Brain Res.* **1988**, *72*, 593–604. [CrossRef]

19. Alexander, N.B.; Shepard, N.; Gu, M.J.; Schultz, A. Postural Control in Young and Elderly Adults When Stance Is Perturbed: Kinematics. *J. Gerontol.* **1992**, *47*, 79–87. [CrossRef]

20. Gage, W.H.; Winter, D.A.; Frank, J.S.; Adkin, A.L. Kinematic and Kinetic Validity of the Inverted Pendulum Model in Quiet Standing. *Gait Posture* **2004**, *19*, 124–132. [CrossRef]

21. Ko, Y.-G.; Challis, J.H.; Newell, K.M. Postural coordination patterns as a function of dynamics of the support surface. *Hum. Mov. Sci.* **2001**, *20*, 737–764. [CrossRef]

22. Winter, D.A. *Biomechanics and Motor Control of Human Movement*, 4th ed.; John Wiley & Sons, Inc.: Hoboken, NJ, USA, 2009; p. 86. ISBN 978-0-470-39818-0.

23. Nelson-Wong, E.; Howarth, S.; Winter, D.A.; Callaghan, J.P. Application of Autocorrelation and Cross-correlation Analyses in Human Movement and Rehabilitation Research. *J. Orthop. Sports Phys. Ther.* **2009**, *39*, 287–295. [CrossRef] [PubMed]

24. Ko, J.-H.; Challis, J.H.; Newell, K.M. Transition of COM–COP relative phase in a dynamic balance task. *Hum. Mov. Sci.* **2014**, *38*, 1–14. [CrossRef] [PubMed]

25. Bardy, B.G.; Marin, L.; Stoffregen, T.A.; Bootsma, R.J. Postural coordination modes considered as emergent phenomena. *J. Exp. Psychol. Hum. Percept. Perform.* **1999**, *25*, 1284–1301. [CrossRef] [PubMed]

26. Kato, T.; Yamamoto, S.I.; Miyoshi, T.; Nakazawa, K.; Masani, K.; Nozaki, D. Anti-phase action between the angular accelerations of trunk and leg is reduced in the elderly. *Gait Posture* **2014**, *40*, 107–112. [CrossRef]

27. Missenard, O.; Fernandez, L. Moving faster while preserving accuracy. *Neuroscience* **2011**, *197*, 233–241. [CrossRef]

28. Aruin, A.S.; Forrest, W.R.; Latash, M.L. Anticipatory postural adjustments in conditions of postural instability. *Electroencephalogr. Clin. Neurophysiol.* **1998**, *109*, 350–359. [CrossRef]

29. Azaman, A.; Yamamoto, S.I. Analysis of joint stiffness of human posture in response to balance ability and limited sensory input during dynamic perturbation. *Int. J. Exp. Comput. Biomech.* **2015**, *3*, 83–101. [CrossRef]

30. Martin, L.; Cometti, G.; Pousson, M.; Morlon, B. Effect of electrical stimulation training on the contractile characteristics of the triceps surae muscle. *Graefe's Arch. Clin. Exp. Ophthalmol.* **1993**, *67*, 457–461. [CrossRef]

31. Martin, L.; Cahouet, V.; Ferry, M.; Fouque, F. Optimization model predictions for postural coordination modes. *J. Biomech.* **2006**, *39*, 170–176. [CrossRef]

32. Runge, C.F.; Shupert, C.L.; Horak, F.B.; Zajac, F.E. Ankle and hip strategies defined by joint torques. *Gait Posture* **1999**, *10*, 161–170. [CrossRef]
33. Aramaki, Y.; Nozaki, D.; Masani, K.; Sato, T.; Nakazawa, K.; Yano, H. Reciprocal angular acceleration of the ankle and hip joints during quiet standing in humans. *Exp. Brain Res.* **2001**, *136*, 463–473. [CrossRef] [PubMed]
34. Zhang, Y.; Kiemel, T.; Jeka, J. The influence of sensory information on two component coordination during quiet stance. *Gait Posture* **2007**, *26*, 263–271. [CrossRef] [PubMed]
35. Sasagawa, S.; Ushiyama, J.; Kouzaki, M.; Kanehisa, H. Effect of the hip motion on the body kinematics in the sagittal plane during human quiet standing. *Neurosci. Lett.* **2009**, *450*, 27–31. [CrossRef] [PubMed]
36. Creath, R.; Kiemel, T.; Horak, F.; Peterka, R.; Jeka, J. A unified view of quiet and perturbed stance: simultaneous co-existing excitable modes. *Neurosci. Lett.* **2005**, *377*, 75–80. [CrossRef]
37. Kelso, J.S. *Dynamic Patterns*; MIT Press: Cambridge, MA, USA, 1995.
38. van Emmerik, R.E.; van Wegen, E.E. On Variability and Stability in Human Movement. *J. Appl. Biomech.* **2000**, *16*, 394–406. [CrossRef]
39. Bonnet, V.; Ramdani, S.; Fraisse, P.; Ramdani, N.; Lagarde, J.; Bardy, B.G. A structurally optimal control model for predicting and analyzing human postural coordination. *J. Biomech.* **2011**, *44*, 2123–2128. [CrossRef]
40. Duncan, P.W.; Weiner, D.K.; Chandler, J.; Studenski, S. Functional reach: A new clinical measure of balance. *J. Gerontol.* **1990**, *45*, M192–M197. [CrossRef]
41. Pollock, A.S.; Durward, B.R.; Rowe, P.J.; Paul, J.P. What is balance? *Clin. Rehabil.* **2000**, *14*, 402–406. [CrossRef]

Article

Center of Pressure Feedback Modulates the Entrainment of Voluntary Sway to the Motion of a Visual Target

Haralampos Sotirakis [1], Vassilia Hatzitaki [1,*], Victor Munoz-Martel [2,3], Lida Mademli [4] and Adamantios Arampatzis [2,3]

[1] Laboratory of Motor Behavior and Adapted Physical Activity, School of Physical Education and Sport Sciences, Aristotle University of Thessaloniki, 54024 Thessaloniki, Greece; sotiraki@phed.auth.gr

[2] Department of Training and Movement Sciences, Humboldt-Universität zu Berlin, 10115 Berlin, Germany; v.munozmartel@hu-berlin.de (V.M.-M.); a.arampatzis@hu-berlin.de (A.A.)

[3] Berlin School of Movement Science, Humboldt-Universität zu Berlin, 10115 Berlin, Germany

[4] Department of Physical Education and Sports Science at Serres, Aristotle University of Thessaloniki, 54024 Thessaloniki, Greece; lmademli@phed-sr.auth.gr

* Correspondence: vaso1@phed.auth.gr

Received: 13 June 2019; Accepted: 17 September 2019; Published: 20 September 2019

Abstract: Visually guided weight shifting is widely employed in balance rehabilitation, but the underlying visuo-motor integration process leading to balance improvement is still unclear. In this study, we investigated the role of center of pressure (CoP) feedback on the entrainment of active voluntary sway to a moving visual target and on sway's dynamic stability as a function of target predictability. Fifteen young and healthy adult volunteers (height 175 ± 7 cm, body mass 69 ± 12 kg, age 32 ± 5 years) tracked a vertically moving visual target by shifting their body weight antero-posteriorly under two target motion and feedback conditions, namely, predictable and less predictable target motion, with or without visual CoP feedback. Results revealed lower coherence, less gain, and longer phase lag when tracking the less predictable compared to the predictable target motion. Feedback did not affect CoP-target coherence, but feedback removal resulted in greater target overshooting and a shorter phase lag when tracking the less predictable target. These adaptations did not affect the dynamic stability of voluntary sway. It was concluded that CoP feedback improves spatial perception at the cost of time delays, particularly when tracking a less predictable moving target.

Keywords: posture; balance; weight shifting; target predictability; coherence; dynamic stability

---

1. Introduction

An integral component of balance rehabilitation exercises is real time visual feedback regarding the instantaneous center of pressure (CoP) or center of mass (CoM) trajectory in relation to a stationary or moving visual target position. Although this visualization increases the role of visual information in the multisensory integration process of controlling posture, its role in standing and dynamic balance control is not well understood.

Experimental evidence on the role of visual feedback in controlling standing balance when the goal of the task is to minimize sway around a reference position is conflicting. On one hand, real time CoP or CoM feedback decreased sway amplitude [1] and the instability induced by proprioceptive, somatosensory, and vestibular perturbations [2]. On the other hand, augmented feedback of the two-dimensional CoP or CoM position did not improve postural stability of a two-legged quiet standing posture [3], whereas, when standing participants were exposed to a complex visual room oscillation, provision of CoP feedback further destabilized standing sway [4]. Furthermore, the spatial

CoP variability of visually controlled whole-body leaning did not decrease when performance feedback was provided in addition to only target information [5]. A possible explanation could be the fact that visual feedback operates at a relatively low frequency scale and therefore it is not appropriate for correcting the high-frequency, small-amplitude CoP oscillations of standing sway [6]. On the other hand, feedback may have a more prominent impact on dynamic equilibrium control which involves greater, longer, and lower frequency CoP fluctuations (drifts) over the base of support.

Provision of augmented visual feedback during performance of dynamic balancing tasks revealed more systematic effects. Removing visual CoP feedback during lateral weight shifting decreased the speed and spatial accuracy of performance [7]. However, the same removal did not have an impact on performance when participants swayed voluntary while trying to produce different visually imposed ankle–hip coordination patterns represented in a complex Lissajous figure [8]. This suggests that more simple and direct forms of visual feedback are needed in order to improve performance. Indeed, the provision of two-dimensional instead of one-dimensional information resulted in faster lateral weight-shifting [9], while augmentation of the error feedback signal drove subjects to their steady-state performance faster than unaltered visual feedback [10]. Horizontally biased visual feedback induced horizontal compensatory postural adjustments, which increased CoP asymmetry even in quiet standing after practice [11]. This evidence suggests that augmented, real-time visual feedback regarding sway improves spatial accuracy and speed of performance when the goal of the task is to transfer the body toward a stationary visual target, such as in lateral weight shifting.

The role of visual feedback when actively tracking a moving visual target with the whole body is less studied. Work from our laboratory suggests that humans, regardless of their age, can couple their voluntary sway to a moving visual target regardless of whether this oscillates in a stereotypical (i.e., periodically) or in a more complex (i.e., chaotically) fashion [12]. However, the role of CoP feedback in visuo-postural entrainment and its dependence on target predictability is still not known. The extent to which the body's stability is challenged by the availability of feedback and target predictability during active target tracking is another unresolved issue. Local dynamic stability represents the ability of a system to maintain its movement pattern despite intrinsic and extrinsic perturbations [13,14]. Instantaneous visual feedback of the pelvis and trunk motion in the frontal plane during gait reduced the local dynamic stability of the aforementioned segmental movements, suggesting that visual feedback of the segment motions induces additional frontal plane instability during gait [15]. The reasons for this increase are still not well understood.

In order to address the above challenges, we examined the role of visual feedback on visuo-postural coupling and dynamic stability of sway when actively tracking either predictable (periodic) or less predictable (chaotic) target motion cues in the sagittal plane. We hypothesized that when actively tracking a visual target oscillating vertically with the body, the impact of feedback would depend on the predictability of the visual motion cues. Specifically, two predictions were tested: (a) That the removal of feedback would affect the strength of visuo-motor coupling and the dynamic stability of sway, and (b) that the impact of feedback would be greater when tracking the less predictable target motion.

2. Materials and Methods

2.1. Participants

Fifteen healthy adults (10 males, 5 females, height 175 ± 7 cm, body mass 69 ± 12 kg, age 32 ± 5 years, mean ± SD), recruited among the university staff participated in this study. None of the participants had a history of neuromuscular impairments or balance-related dysfunctions. No participants used orthotic insoles, and all had normal or corrected-to-normal vision. All participants were informed about the experimental protocol and gave their informed consent prior to their inclusion in the study. The experiment was performed with the approval of the institution's ethics committee (HU-KSBF-EK_2018_0013, Humboldt-Universität zu Berlin) in accordance with the Declaration of Helsinki.

2.2. Apparatus, Stimuli, and Task

Postural performance was recorded using a force platform (60×90 cm, Kistler, 1000 Hz, Winterthur, Switzerland). Visual stimuli were displayed on a large TV Screen (47 inch, HD LG), located 1.5 m in front of the participant at eye level (Figure 1A).

The target's motion was constructed in MATLAB (version R2014b, Math Works Inc, Natick, MA, USA) using two signals of different degrees of predictability. The *predictable signal* was a sinewave with a single frequency (f) set at 0.25 Hz that was generated using the sin function [$sine = \sin(2pi \times f \times t)$]. This particular frequency was selected because it was the dominant frequency of intuitive, self-paced voluntary sway based on prior studies [12,16] and pilot testing. The *less predictable signal* was derived from a Lorenz attractor according to the parameters: $\sigma = 10$, $\beta = 8/3$, and $r = 28$ and the initial conditions: $x0 = 0.1$, $y0 = 0.1$, and $z0 = 0.1$. The signal characteristics were h (time resolution) = 0.0040, steps (number of points) = 10,000 (we choose 6000 data points from the y-axis [y (4000: 10,000)], and noise flag = 0 [17,18]. The frequency range for the Lorenz signal was confined between 0 and 1 Hz, which was an ecologically valid spectrum of frequencies for voluntary sway [19], and its power spectrum revealed a dominant frequency around 0.25 Hz.

Prior to the experiment, participants were asked to step on the platform and adopt the testing position with their arms freely hanging by their sides. The distance between the internal malleoli was set at 10% of body height. The antero-posterior component of the base of support (foot length) was measured using the distance of the most anterior point (toe) to the calcaneus in order to normalize the amplitude of target motion. The maximum (peak to peak) amplitude of target motion was set to 60% of the foot length (Figure 1B).

During the experiment, participants tracked the motion of the visual target, which moved vertically on the screen, with their bodies by shifting their CoP in the antero-posterior direction. Two different-colored dots were shown on the screen, specifically, a red dot, which illustrated the target to be followed, and a yellow dot, which depicted the participant's antero-posterior CoP component serving as the instantaneous performance feedback signal. The only instruction given to participants was to follow the motion of the red dot by controlling the motion of the yellow one, which was achieved by shifting their weight antero-posteriorly. The two stimuli (target and feedback) were synchronously represented on the TV monitor using custom build software (developed in MATLAB) while their position was updated at a rate of 50 Hz, providing 120 s of continuous target stimulus motion and resulting in 6000 data points for each signal. Participants performed one 120 s long tracking trial in each of the following conditions: (a) Tracking of the sinusoidal target with (yellow dot was visible) and without (yellow dot was not visible) visual feedback, and (b) tracking of the chaotic (Lorenz) target with and without feedback. Task conditions were randomized to avoid possible order effects. Participants were provided ample time to familiarize themselves with the task prior to actual testing. Prior to each trial, the room lights were dimmed. Between each trial, participants rested for about 2 min. The experimental session, including subject preparation lasted no longer than 60 min.

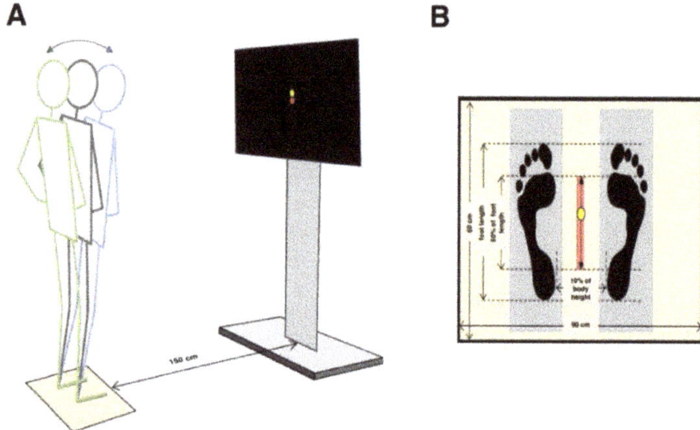

Figure 1. Graphical illustration of the apparatus and task. (**A**) Participants were asked to track the visual (red) target dot moving in the vertical direction with their body by shifting their center of pressure (CoP) (yellow dot) in the antero-posterior direction. An upward target motion was tracked by a forward CoP shift, while a downward target motion was tracked by a backward CoP shift. (**B**) The target's (red dot) motion amplitude was normalized to 60% of the participants foot length. Tracking lasted 120 s and was performed either with feedback (yellow dot was visible) or without feedback (yellow dot was not visible).

2.3. Data Analysis

Center of pressure (CoP) and target time series were processed and analyzed in MATLAB (R2014b). The original antero-posterior component of the CoP time series was down-sampled to 50 Hz in order to match the screen update rate of the target signal display prior to low pass filtering using a 4th order Butterworth low-pass filter, with a cutoff frequency at 5 Hz.

2.4. Spectral Coherence

The CoP-target motion coupling was assessed using spectral analysis in the frequency band between 0–1 Hz, based on the methods of Halliday et al. [20]. For this purpose, a customized version of a freely available software (NeuroSpec 2.0) was used. Both the CoP and target time series were interpolated at 64 Hz in order to achieve an appropriate frequency resolution for the spectral analysis, as we wanted to get a coherence, phase, and gain value of 0.25 Hz (frequency of the sinewave) and the neighborhood frequencies (for the Lorenz signal). Spectral analysis was performed in the interpolated time series of 7680 data points which were sampled at 64 Hz. Setting the segment length power at 2^{10} (1024 data points) gave a segment duration of 16 s. This resulted in a frequency resolution of 0.0625 Hz. The software output the means and confidence limits across the entire frequency band (0–1 Hz) according to three variables, i.e., spectral coherence, spectral phase, and spectral gain. The spectral coherence was used as a metric of correlation between the two signals (CoP-target position) in the frequency domain, illustrating their linear relationship. The spectral phase illustrated the temporal relationship between the two signals, expressed in degrees (°). The absolute synchronization between the two signals was illustrated by 0° phase lag, while positive and negative values indicated that the sway led or followed the target motion respectively. The spectral gain revealed information regarding the spatial (amplitude) coupling between the two signals (CoP-target), while a gain of 1 indicated an absolute spatial coupling at a given frequency point. Gain values less than or greater than 1 indicated target undershooting or overshooting (the CoP moved with smaller or greater amplitude than the target), respectively. For the sinewave, the coherence, phase, and gain values at the target's dominant frequency (0.25 Hz) were computed and registered for statistical analysis. Similarly, for the Lorenz target signal, the average of the coherence, phase, and gain values at 3 adjacent frequency bins (0.1875, 0.25, and 0.3125 Hz) was calculated.

2.5. Local Dynamic Stability

The local dynamic stability of the system in the current study was assessed using the maximum finite-time Lyapunov exponent (MLE), which quantifies the rate of divergence of nearby trajectories in state space [21,22]. The analysis followed the procedure used in a previous study [23]. The MLE was calculated on the norm of the anterio-posterior and mediolateral CoP. Initially, the original time-series were filtered using a 4th order Butterworth low-pass filter with a cut-off frequency of 20 Hz and consequently down-sampled to 20,000 data points. Due to the standardized overall trial duration (i.e., 120 s), no interpolation of the time-series was needed. To reconstruct the state space from the one-dimensional time series, we used delay-coordinate embedding [24] as follows:

$$S(t) = [z(t), z(t+\tau), \ldots, z(t+(m-1)\tau)], \tag{1}$$

where $S(t)$ is the m-dimensional reconstructed state vector, $z(t)$ is the input 1D coordinate series, τ is the time delay, and m is the embedding dimension. Time delays were selected based on the first minimum of the Average Mutual Information function [25]. For these data, $m = 3$ was sufficient to perform the reconstruction. Individually selected time delays were chosen by averaging the outcome delays deriving from both trials performed by the participants [26], with τ ranging from ~0.20 to ~0.27 of the overall cycle. Further, the average divergence of each point's trajectory to its closest neighbor was calculated using the Rosenstein algorithm [27]. The resulting MLE was calculated based on the delay of each participant, which ensured the standardization of the calculation for the MLE across all of the individuals. Commonly, the first peak in the resulting divergence curves corresponded to a delay of 0.5 (percentage of the average postural sway cycle). The final MLE value was calculated as the slope of the average divergence curves' linear fit which corresponded to the individuals' delay values at 0.25 of the average postural sway cycle (i.e., the most linear part of the curve).

2.6. Statistical Analysis

Prior to statistical analysis, the Shapiro–Wilk test was applied to test for violations of the normality assumption in each outcome measure. All measures were normally distributed and differences between the two target motions (periodic and chaotic) and feedback conditions (with and without visual feedback) on the CoP-target coherence, phase, and gain were compared by employing a 2 (target) × 2 (feedback) ANOVA model with repeated measures on both factors. Target by feedback interactions were further analyzed using pairwise comparisons between the feedback conditions, which were performed separately on each target motion. Differences in the resulting MLE values were examined using Student's *t*-test via pairwise comparisons between the two feedback conditions. Effect sizes were reported using h^2. Statistical analysis was performed using SPSS (v.24).

3. Results

Group-averaged CoP displacement and target signals are shown in Figure 2 for sinusoidal and chaotic target tracking, respectively, with and without feedback.

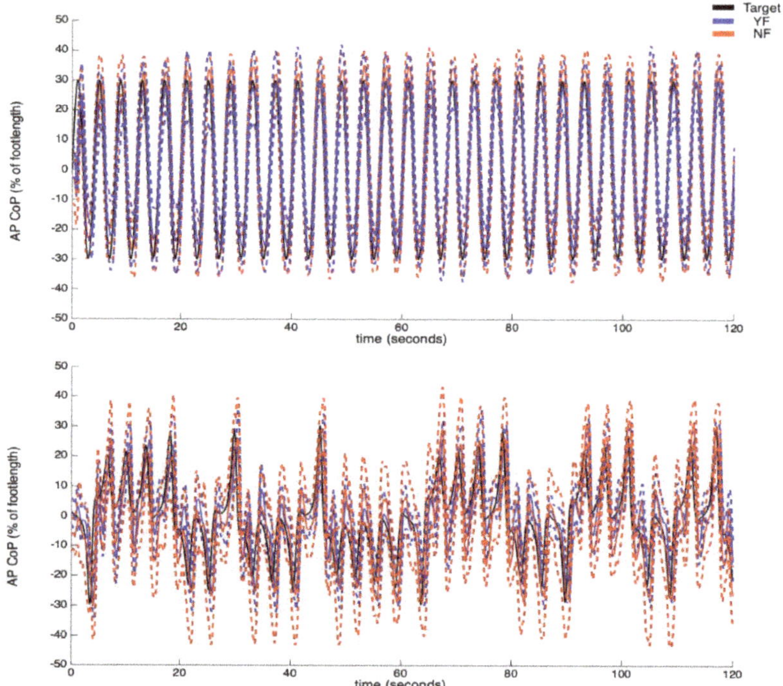

Figure 2. Group-averaged (*n* = 15) CoP and target (black) time series during performance of (**1st row**) the sinusoidal and (**2nd row**) the chaotic tracking with (blue) and without (red) feedback. Dotted lines indicate the group 95% confidence intervals.

3.1. Spectral Analysis

- Coherence

The CoP-target coherence values ranged between 0.75 and 0.99 for all participants under all tracking conditions (Figure 3A), suggesting that all participants were successful in tracking the target. However, CoP-target coherence decreased significantly with chaotic tracking relative to the periodic target (F(1,14) = 264, p = 0.000, h^2 = 0.950), although this was not different between the two feedback conditions.

- Phase

The CoP-target phase lag significantly increased when tracking the chaotic compared to the periodic target (F (1,14) = 67.19, p = 0.000, h^2 = 0.828). A negative phase lag value indicated that the CoP followed the target motion (Figure 3B). A significant feedback by target interaction effect on the phase lag (F (1,14) = 8.51, p = 0.011, h^2 = 0.378) suggested that the effect of feedback was dependent on the target motion. Post hoc analysis confirmed that the presence of visual feedback significantly increased the CoP-target phase lag compared to the no feedback condition only in chaotic target tracking (t (16) = 2.93, p = 0.010). On the other hand, the feedback signal had a non-significant effect on the CoP-target phase lag when tracking the periodic target motion.

- Gain

Removal of feedback resulted in a significant increase in the CoP-target gain (F (1,14) = 68.49, p = 0.000, h^2 = 0.830), suggesting that target overshooting occurred during both periodic and chaotic target tracking (Figure 3C). More target overshooting was observed when tracking the chaotic target compared to the periodic target (F (1,14) = 77.57, p = 0.000, h^2 = 0.847). The impact of visual feedback on CoP-target gain was significantly greater when tracking the chaotic compared to the periodic target, as confirmed by a significant feedback by target interaction effect (F (1,14) = 22.16, p = 0.000, h^2 = 0.613). Specifically, the increase in CoP-target gain induced by the removal of feedback was greater when tracking the chaotic compared to the sinusoidal target (t (16) = 7.81, p = 0.000).

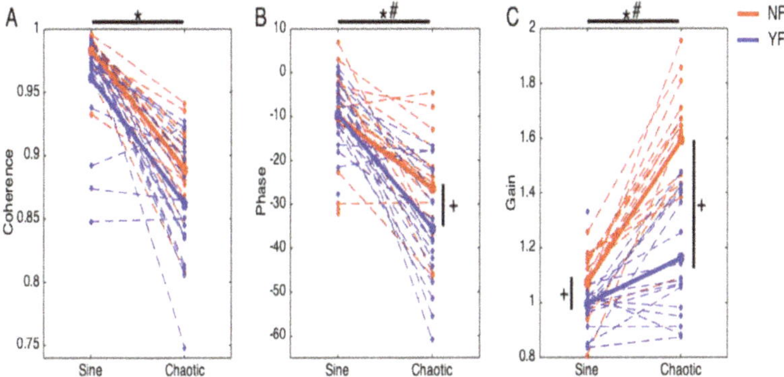

Figure 3. Center of pressure (CoP)-target: (**A**) coherence, (**B**) phase, and (**C**) gain during tracking of the sinusoidal and chaotic target motion with (YF, blue) and without (NF, red) CoP feedback. Dashed lines indicate individual cases and group means are shown with solid bold lines. +: Significant difference between feedback conditions at $p < 0.05$; *: Significant difference between target conditions at $p < 0.05$; #: Significant feedback by target interaction effect at $p < 0.05$.

3.2. Local Dynamic Stability

The group means and standard deviations for the MLE exponent are shown in Figure 4. The dynamic stability index of the normalized CoP displacement was not affected by the presence of feedback when tracking either the sinusoidal or the chaotic target. This was confirmed by the absence of a significant difference between the feedback conditions in both the periodic (p = 0.463) and chaotic (p = 0.276) target tracking.

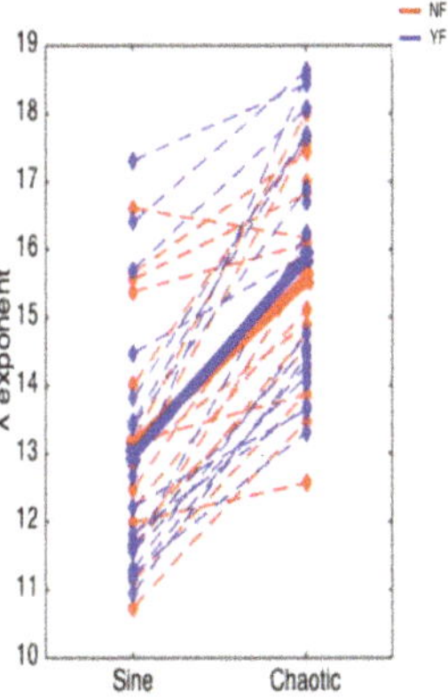

Figure 4. Maximum Lyapunov exponent (MLE) values of the CoP time series during tracking of the sinusoidal target motion and chaotic target motion with (YF, blue) and without (NF, red) CoP feedback. Dashed lines indicate individual cases and group means are shown with solid bold lines.

4. Discussion

In accordance with our hypotheses, the present results confirmed an effect of real-time CoP feedback on visuo-motor coupling during active postural tracking of a vertically moving visual target. The effect was stronger when tracking the more variable (i.e., chaotic) target motion. On the other hand, the presence/removal of feedback did not affect the dynamic stability of voluntary sway.

- Feedback improved spatial accuracy at the cost of time delays.

When real-time visual feedback of the CoP trajectory relative to the target motion was available during active target tracking, the CoP-target gain was close to one. This suggests that sway amplitude closely matched the target oscillation amplitude that was set to the individual stability boundaries in the antero-posterior direction (60% of foot length). We argue that real-time feedback regarding the instantaneous CoP position provided critical spatial information for actively controlling sway amplitude through an error correction process, which allowed performers to make corrective postural adjustments before horizontal CoP accelerations led to large deviations in postural sway. On the other hand, removing CoP feedback during tracking of the vertical target motion increased the CoP-target gain, suggesting target overshooting. Our results are concordant with those of a previous study showing that augmented CoP feedback improved the spatial accuracy of lateral weight shifting [7]. However, instead of translating the body toward a stationary target, in our protocol, the task goal required active and dynamic tracking of the moving target. Target overshooting in the absence of feedback in this case implies larger antero-posterior CoP acceleration and, consequently, increased difficulty and greater muscular effort in controlling the reversal in sway direction [28,29]. In this case, the control of active sway relied on an internal representation of the body schema in space, which tended to overestimate the sensory consequences of the actual tracking task relative to the boundaries of the target's motion, resulting in erroneously larger sway movements [30]. Internal representations of the body schema are based on prior knowledge and appear to be central in the planning and control of goal-directed actions [31,32]. The absence of error information in this case is related with exploratory behavior of the spatial task constraints and over-corrections [33,34]. Greater reliance on prediction is also associated with greater perceptual bias [30], which may explain the overestimation of the target boundaries and the target overshooting observed in the absence of feedback.

On the other hand, availability of feedback increased the phase lag between the target and the CoP motion when tracking the chaotic target, resulting in slower antero-posterior weight shifting. This finding contradicts previous studies showing that provision of visual feedback resulted in faster lateral weight shifting [10,35]. The discrepancy could be due to the stationary nature of the target used in the previous weight shifting paradigms as opposed to the moving target used in the present study. When the target was stationary, information about the target's position was picked up prior to the onset of movement, which allowed for faster lateral weight transfer. However, in the case of a moving target, the target's position was continuously updated during task performance, while real time feedback about the instantaneous CoP position imposed a greater need for online visuo-motor integration, resulting in time delays when making the appropriate postural adjustments. In this case, feedback may have acted as additional sensory noise that increased the time delay (phase lag) between the target and sway motion [36]. On the other hand, periodic target tracking did not increase the phase lag between postural sway and the target motion due to the predictable nature of the target motion. Once a particular visuo-motor transformation was learned after a couple of sway cycles, the participant was able to predict the target's motion, thereby reducing the load of online visual information processing during active target tracking [16].

Overall, our results suggest a trade-off in the control of the spatial and temporal properties of sway when tracking a vertically moving target which was modulated by the availability of visual CoP feedback. Feedback about the spatial error between the target and the CoP motion induced a prioritization of spatial over temporal coupling [37], while the latter decreased due to the speed accuracy trade-off [38].

- Feedback did not affect the dynamic stability of voluntary sway.

Removal of feedback did not affect the local dynamic stability of sway. This suggests that the additional visuo-motor processing load needed for matching the feedback (yellow dot) to the target signal (red dot) in the feedback condition did not challenge the overall dynamic stability of the voluntary sway task any more than the target tracking alone with no feedback. Along the same line of evidence, in an optic flow stimulation study, participants experienced less postural instability during the actual saccadic tracking of the directional stimuli compared to the preceding perceptual phase of the heading direction [39]. This suggests that when the visual motion of the target was congruent with the eye and/or body motion, the postural stability was not challenged. On the other hand, provision of real-time visual feedback during gait decreased the local dynamic stability and induced instability of the trunk and the pelvis in the frontal plane [15]. Several reasons could explain this disagreement. Actively tracking a moving target by shifting the body weight in the sagittal plane is a less automatic and slower task compared to gait. Provision of augmented feedback during gait could therefore lead to a disruption of automaticity and impose greater demand for visuo-motor processing. Weight-shifting, on the other hand, is a less automatized task that requires higher precision control, particularly when reaching close to the stability limits. The plane of motion could also be an important task constraint determining the impact of dynamic visual feedback cues on dynamic stability [40]. In the gait study, feedback was represented by a horizontal visual motion to inform about the instantaneous position of the pelvis and the trunk in the frontal plane, whereas in our paradigm, CoP feedback, provided by a vertically moving dot, did not seem to affect the stability of sway in the sagittal plane. It is known that non-gravity (horizontal)-related dynamic visual cues induce greater instability when compared to gravity-relevant visual cues [40]

- The contribution of feedback increased when tracking the less predictable target.

Tracking of the chaotic target resulted in lower coherence, greater target overshooting, and a longer phase lag between the target and the CoP motion when compared to periodic target tracking. These results confirm previous findings from our laboratory showing that voluntary, sagittal plane sway synchronizes better with predictable (i.e., periodic) target motion than less predictable (i.e., chaotic) target motion when actively tracking a vertically moving target [12,16]. It could be suggested that postural tracking of the chaotic target imposed more extensive visuo-motor processing due to the less predictable nature of the target's motion, which may have kept participants more actively engaged in the tracking task due to the continuous need to attend to the target motion. Tracking of the sinusoidal target motion, on the other hand, gradually decreased the need to attend to the target, as participants were able to predict it and adjust their sway, employing an open-loop type of control [32]. This was confirmed by the consistency of the target-CoP phase lag across the two feedback conditions when tracking the sinusoidal target motion. Moreover, synchronizing body sway to a chaotically oscillating target may also involve anticipatory control, which depends on the availability of visual feedback [41]. This process involves short-term prediction and correction of asynchronies between the target and the CoP motion [42]. Instead of prediction on a local time-scale, however, strong anticipation assumes coordination on a longer time-scale. Entraining to a chaotic target motion in this case would require the coordination of posture to the global temporal structure of the stimulus signal and not on the local temporal changes [5].

The greater target overshooting when tracking the less predictable target may also have been due to the different amplitude scaling of the chaotic signal. The chaotic target motion-imposed direction reversed at smaller sway amplitudes, as the target's motion did not consistently reach the stability limit, which was set at 60% of foot length in every sway cycle. The most important information for stabilizing coordination with an oscillating stimulus was available for the actor around the endpoint of its trajectory [43]. Thus, it may have been more difficult to control smaller oscillation amplitudes during active sway than target amplitudes which reached the stability limit [44].

5. Conclusions

Feedback may be more critical for spatial coupling when accuracy is a task requirement, while it imposes an extra processing constraint for synchronizing voluntary sway to the moving target stimulus. Models of postural control have highlighted possible roles for both feedback-based [45] and predictive, feed-forward control [46] in the control of voluntary sway during static and dynamic balance activities. Both control processes are compatible with these current behavioral results. The observed changes in postural sway control as a function of feedback availability may reflect an updating of internal models relating postural motor commands to their upcoming sensory consequences (i.e., forward models), facilitating accurate on-line detection and correction of postural deviations (feedback control).

Author Contributions: Conception and design of the work A.A., V.H., and L.M.; data acquisition H.S. and V.M.-M.; data analysis H.S. and V.M.-M.; data interpretation V.H., A.A., and L.M.; first draft H.S.; manuscript revision A.A., V.H., and L.M.; manuscript editing, H.S. and V.M.-M. All authors approved the submitted version to be published and agreed to be accountable for all aspects of the work.

Funding: This research was part of the project "FFP-AGE", which was funded by the German Academic Exchange Service (DAAD); grant number: 57339989.

Acknowledgments: Authors would like to thank Arno Schroll for the creation of the interface for data collection and the hardware setup, Antonis Ekizos for contributing to the data analysis, and Sebastian Bohm and Natasa Papavasileiou for their assistance in data collection and figure preparation.

References

1. Rougier, P. Visual feedback induces opposite effects on elementary centre of gravity and centre of pressure minus centre of gravity motions in undisturbed upright stance. *Clin. Biomech.* **2003**, *18*, 341–349. [CrossRef]
2. Cofré Lizama, L.E.; Pijnappels, M.; Reeves, N.P.; Verschueren, S.M.P.; van Dieën, J.H. Can explicit visual feedback of postural sway efface the effects of sensory manipulations on mediolateral balance performance? *J. Neurophysiol.* **2016**, *115*, 907–914. [CrossRef] [PubMed]
3. Kilby, M.C.; Slobounov, S.M.; Newell, K.M. Augmented feedback of COM and COP modulates the regulation of quiet human standing relative to the stability boundary. *Gait Posture* **2016**, *47*, 18–23. [CrossRef] [PubMed]
4. Li, R.; Peterson, N.; Walter, H.J.; Rath, R.; Curry, C.; Stoffregen, T.A. Real-time visual feedback about postural activity increases postural instability and visually induced motion sickness. *Gait Posture* **2018**, *65*, 251–255. [CrossRef] [PubMed]
5. Duarte, M.; Zatsiorsky, V.M. Effects of body lean and visual information on the equilibrium maintenance during stance. *Exp. Brain Res.* **2002**, *146*, 60–69. [CrossRef] [PubMed]
6. Ivanenko, Y.; Gurfinkel, V.S. Human postural control. *Front. Neurosci.* **2018**, *12*, 1–9. [CrossRef]
7. Dault, M.C.; De Haart, M.; Geurts, A.C.H.; Arts, I.M.P.; Nienhuis, B. Effects of visual center of pressure feedback on postural control in young and elderly healthy adults and in stroke patients. *Hum. Mov. Sci.* **2003**, *22*, 221–236. [CrossRef]
8. Faugloire, E.; Bardy, B.G.; Merhi, O.; Stoffregen, T.A. Exploring coordination dynamics of the postural system with real-time visual feedback. *Neurosci. Lett.* **2005**, *374*, 136–141. [CrossRef] [PubMed]
9. Kennedy, M.W.; Crowell, C.R.; Striegel, A.D.; Villano, M.; Schmiedeler, J.P. Relative efficacy of various strategies for visual feedback in standing balance activities. *Exp. Brain Res.* **2013**, *230*, 117–125. [CrossRef]
10. O'Brien, K.; Crowell, C.R.; Schmiedeler, J. Error augmentation feedback for lateral weight shifting. *Gait Posture* **2017**, *54*, 178–182. [CrossRef]
11. Shiller, D.M.; Veilleux, L.-N.; Marois, M.; Ballaz, L.; Lemay, M. Sensorimotor adaptation of whole-body postural control. *Neuroscience* **2017**, *356*, 217–228. [CrossRef] [PubMed]
12. Sotirakis, H.; Kyvelidou, A.; Stergiou, N.; Hatziitaki, V. Neuroscience Letters Posture and gaze tracking of a vertically moving target reveals age-related constraints in visuo-motor coupling. *Neurosci. Lett.* **2017**, *654*, 12–16. [CrossRef] [PubMed]
13. Lyapunov, A.M. The general problem of motion stability. *Int. J. Control* **1992**, *55*, 531–773. [CrossRef]

14. Ihlen, E.A.F.; van Schooten, K.S.; Bruijn, S.M.; Pijnappels, M.; van Dieën, J.H. Fractional stability of trunk acceleration dynamics of daily-life walking: Toward a unified concept of gait stability. *Front. Physiol.* **2017**, *8*, 516. [CrossRef] [PubMed]

15. Hamacher, D.; Hamacher, D.; Schega, L. Does visual augmented feedback reduce local dynamic stability while walking? *Gait Posture* **2015**, *42*, 415–418. [CrossRef]

16. Hatzitaki, V.; Stergiou, N.; Sofianidis, G.; Kyvelidou, A. Postural Sway and Gaze Can Track the Complex Motion of a Visual Target. *PLoS ONE* **2015**, *10*, e0119828. [CrossRef] [PubMed]

17. Sotirakis, H.; Kyvelidou, A.; Mademli, L.; Stergiou, N.; Hatzitaki, V. Aging affects postural tracking of complex visual motion cues. *Exp. Brain Res.* **2016**, *234*, 2529–2540. [CrossRef] [PubMed]

18. Suzuki, Y.; Nomura, T.; Casadio, M.; Morasso, P. Intermittent control with ankle, hip, and mixed strategies during quiet standing: A theoretical proposal based on a double inverted pendulum model. *J. Theor. Boil.* **2012**, *310*, 55–79. [CrossRef] [PubMed]

19. Cofré Lizama, L.E.; Pijnappels, M.; Reeves, N.P.; Verschueren, S.M.P.; van Dieën, J.H. Frequency domain mediolateral balance assessment using a center of pressure tracking task. *J. Biomech.* **2013**, *46*, 2831–2836. [CrossRef]

20. Halliday, D. A framework for the analysis of mixed time series/point process data—Theory and application to the study of physiological tremor, single motor unit discharges and electromyograms. *Prog. Biophys. Mol. Boil.* **1995**, *64*, 237–278. [CrossRef]

21. Dingwell, J.B.; Cusumano, J.P.; Cavanagh, P.R.; Sternad, D. Local dynamic stability versus kinematic variability of continuous overground and treadmill walking. *J. Biomech. Eng.* **2001**, *123*, 27–32. [CrossRef]

22. England, S.A.; Granata, K.P. The influence of gait speed on local dynamic stability of walking. *Gait Posture* **2007**, *25*, 172–178. [CrossRef] [PubMed]

23. Ekizos, A.; Santuz, A.; Arampatzis, A. Transition from shod to barefoot alters dynamic stability during running. *Gait Posture* **2017**, *56*, 31–36. [CrossRef] [PubMed]

24. Packard, N.H.; Crutchfield, J.P.; Farmer, J.D.; Shaw, R.S. Geometry from a time series. *Phys. Rev. Lett.* **1980**, *45*, 712. [CrossRef]

25. Fraser, A.M.; Swinney, H.L. Independent coordinates for strange attractors from mutual information. *Phys. Rev. A* **1986**, *33*, 1134–1140. [CrossRef] [PubMed]

26. Ekizos, A.; Santuz, A.; Schroll, A.; Arampatzis, A. The Maximum Lyapunov Exponent during Walking and Running: Reliability Assessment of Different Marker-Sets. *Front. Physiol.* **2018**, *9*, 1–11. [CrossRef] [PubMed]

27. Rosenstein, M.T.; Collins, J.J.; De Luca, C.J. A practical method for calculating largest Lyapunov exponents from small data sets. *Phys. D Nonlinear Phenom.* **1993**, *65*, 117–134. [CrossRef]

28. Varlet, M.; Coey, C.A.; Schmidt, R.C.; Marin, L.; Bardy, B.G.; Richardson, M.J. Influence of stimulus velocity profile on rhythmic visuomotor coordination. *J. Exp. Psychol. Hum. Percept. Perform.* **2014**, *40*, 1849–1860. [CrossRef]

29. Varlet, M.; Schmidt, R.C.; Richardson, M.J. Influence of stimulus velocity profile on unintentional visuomotor entrainment depends on eye movements. *Exp. Brain Res.* **2017**, *235*, 3279–3286. [CrossRef]

30. Wolpe, N.; Wolpert, D.M.; Rowe, J.B. Seeing what you want to see: Priors for one's own actions represent exaggerated expectations of success. *Front. Behav. Neurosci.* **2014**, *8*, 1–14. [CrossRef]

31. Bays, P.M.; Wolpert, D.M. Computational principles of sensorimotor control that minimize uncertainty and variability. *J. Physiol.* **2007**, *578*, 387–396. [CrossRef] [PubMed]

32. Franklin, D.W.; Wolpert, D.M. Computational Mechanisms of Sensorimotor Control. *Neuron* **2011**, *72*, 425–442. [CrossRef] [PubMed]

33. Wu, H.G.; Miyamoto, Y.R.; Nicolas, L.; Castro, G.; Smith, M.A.; Biology, E. Temporal structure of motor variability is dynamically regulated and predicts motor learning ability. *Nat. Neurosci.* **2015**, *17*, 312–321. [CrossRef] [PubMed]

34. Herzfeld, D.J.; Shadmehr, R. Motor variability is not noise, but grist for the learning mill. *Nat. Neurosci.* **2014**, *17*, 149–150. [CrossRef] [PubMed]

35. Kennedy, M.W.; Crowell, C.R.; Villano, M.; Schmiedeler, J.P. Effects of Filtering the Center of Pressure Feedback Provided in Visually Guided Mediolateral Weight Shifting. *PLoS ONE* **2016**, *11*, e0151393. [CrossRef] [PubMed]

36. Shadmehr, R.; Smith, M.A.; Krakauer, J.W. Error Correction, Sensory Prediction, and Adaptation in Motor Control. *Annu. Rev. Neurosci.* **2010**, *33*, 89–108. [CrossRef] [PubMed]

37. Posner, M.I.; Nissen, M.J.; Klein, R.M. Visual dominance: An information-processing account of its origins and significance. *Psychol. Rev.* **1976**, *83*, 157–171. [CrossRef] [PubMed]

38. Fitts, P.M. Cognitive aspects of information processing: III. Set for speed versus accuracy. *J. Exp. Psychol.* **1966**, *71*, 849–857. [CrossRef] [PubMed]

39. Piras, A.; Raffi, M.; Perazzolo, M.; Squatrito, S. Influence of heading perception in the control of posture. *J. Electromyogr. Kinesiol.* **2018**, *39*, 89–94. [CrossRef] [PubMed]

40. Balestrucci, P.; Daprati, E.; Lacquaniti, F.; Maffei, V. Effects of visual motion consistent or inconsistent with gravity on postural sway. *Exp. Brain Res.* **2017**, *235*, 1999–2010. [CrossRef] [PubMed]

41. Stepp, N. Anticipation in feedback-delayed manual tracking of a chaotic oscillator. *Exp. Brain Res.* **2009**, *198*, 521–525. [CrossRef] [PubMed]

42. Stephen, D.G.; Stepp, N.; Dixon, J.A.; Turvey, M. Strong anticipation: Sensitivity to long-range correlations in synchronization behavior. *Phys. A Stat. Mech. Its Appl.* **2008**, *387*, 5271–5278. [CrossRef]

43. Hajnal, A.; Richardson, M.J.; Harrison, S.J.; Schmidt, R.C. Location but not amount of stimulus occlusion influences the stability of visuo-motor coordination. *Exp. Brain Res.* **2009**, *199*, 89–93. [CrossRef] [PubMed]

44. Danion, F.; Duarte, M.; Grosjean, M. Fitts' law in human standing: The effect of scaling. *Neurosci. Lett.* **1999**, *277*, 131–133. [CrossRef]

45. Peterka, R.J. Sensorimotor Integration in Human Postural Control. *J. Neurophysiol.* **2002**, *88*, 1097–1118. [CrossRef] [PubMed]

46. Morasso, P.G.; Schieppati, M. Can Muscle Stiffness Alone Stabilize Upright Standing? *J. Neurophysiol.* **1999**, *82*, 1622–1626. [CrossRef]

Article

Influence of Pairing Startling Acoustic Stimuli with Postural Responses Induced by Light Touch Displacement

John E. Misiaszek [1,2,*], Sydney D. C. Chodan [1], Arden J. McMahon [1] and Keith K. Fenrich [1,2]

[1] Department of Occupational Therapy, Faculty of Rehabilitation Medicine, University of Alberta, Edmonton, AB T6G 2G4, Canada; schodan@ualberta.ca (S.D.C.C.); ajmcmaho@ualberta.ca (A.J.M.); fenrich@ualberta.ca (K.K.F.)

[2] Neuroscience and Mental Health Institute, University of Alberta, Edmonton, AB T6G 2E1, Canada

* Correspondence: john.misiaszek@ualberta.ca

Received: 5 December 2019; Accepted: 30 December 2019; Published: 4 January 2020

Abstract: The first exposure to an unexpected, rapid displacement of a light touch reference induces a balance reaction in naïve participants, whereas an arm-tracking behaviour emerges with subsequent exposures. The sudden behaviour change suggests the first trial balance reaction arises from the startling nature of the unexpected stimulus. We investigated how touch-induced balance reactions interact with startling acoustic stimuli. Responses to light touch displacements were tested in 48 participants across six distinct combinations of touch displacement (DISPLACEMENT), acoustic startle (STARTLE), or combined (COMBINED) stimuli. The effect of COMBINED depended, in part, on the history of the preceding stimuli. Participants who received 10 DISPLACEMENT initially, produced facilitated arm-tracking responses with subsequent COMBINED. Participants who received 10 COMBINED initially, produced facilitated balance reactions, with arm-tracking failing to emerge until the acoustic stimuli were discontinued. Participants who received five DISPLACEMENT, after initially habituating to 10 STARTLE, demonstrated re-emergence of the balance reaction with the subsequent COMBINED. Responses evoked by light touch displacements are influenced by the startling nature of the stimulus, suggesting that the selection of a balance reaction to a threatening stimulus is labile and dependent, in part, on the context and sensory state at the time of the disturbance.

Keywords: light touch; balance; startle; human; standing

1. Introduction

Unexpected disturbances to balance are often met with whole-body reactions to stabilize the body and mitigate the potentially catastrophic consequences of a fall [1]. Moreover, the opportunity for generating a functionally meaningful response to a balance disturbance is normally limited to the initial exposure to the disturbance; rarely is there a second chance to get it right. Therefore, accurate interpretation of the sensory feedback related to balance disturbances is critical to generating appropriate balance reactions. Recently, rapid, unexpected displacements of a light touch reference were shown to evoke reactions consistent with a balance correction when standing with eyes closed, despite the absence of a mechanical disturbance to balance per se [2,3]. However, this putative balance reaction was not consistently expressed across participants and was only observed following the first unexpected displacement of the touch reference. With subsequent touch displacements, participants tracked the motion of the touch reference with a simple arm movement. This suggests that the sensation at the fingertip during the first trial was misinterpreted as a sway of the body away from the touch reference but was correctly interpreted as a displacement of the touch reference on subsequent trials.

The sudden change in behaviour between the first and subsequent exposures to the light touch displacement raises the possibility that the first trial response reflects a startle response. Startle reflexes are commonly defined as involuntary motor reactions to unexpected sensory stimuli that habituate with repeated exposure to the stimulus. Although startle reflexes are often described in relation to sudden auditory stimuli, startles can be elicited from a variety of sensory modalities, including tactile stimuli (reviewed in [4]). Moreover, it is well documented that the first exposure to unexpected balance disturbances are of larger amplitude than subsequent exposures to the same perturbations [5–9]. Campbell et al. [6] demonstrated that a large first trial response to a balance disturbance likely arises due to the superimposition of the balance reaction with a startle response, which subsequently habituates with repeated exposure to the balance disturbance. A characteristic feature of the habituation of balance reactions and startle responses is the attenuation of muscle activity associated with the evoked response. In contrast, the putative balance reactions to light touch displacement we previously reported [2,3] do not appear to habituate with subsequent exposures, but instead are replaced by a different behaviour within a single trial making the contribution of a startle response to this first trial reaction ambiguous.

Evidence from both human and animal studies have demonstrated that startle responses summate when evoked from more than one stimulus source [4]. Moreover, the summated responses are larger when startling stimuli of different modalities (for example, tactile and acoustic) are combined [10,11]. Blouin et al. [12] exploited this property of summation of startle reflexes to restore the amplitude of habituated postural responses in neck muscles, providing strong evidence that startle contributes to the larger postural responses observed with the initial disturbance. In this study, we aimed to determine what impact introducing a startling acoustic tone would have on the responses evoked by a rapid displacement of a light touch reference during standing. We hypothesized that, similar to our previous studies, an unexpected displacement of the light touch reference would result in a balance response on the first trial followed by arm-tracking behaviour on subsequent trials; however, subsequently combining an acoustic startle with the light touch displacement would promote re-expression of the balance response. We further hypothesized that initially combining an acoustic startle with the light touch displacement would result in a larger first trial balance response than with light touch displacement alone, but that the arm-tracking behaviour would emerge with habituation of the startle response. Finally, we hypothesized that allowing participants to habituate to the acoustic startle before introducing an unexpected displacement of the light touch reference would result in touch-evoked responses comparable to those observed in the absence of an accompanying acoustic startle.

2. Materials and Methods

Forty-eight healthy volunteers (29 female; 44 right-handed; median age 21 years; range 18–29 years) provided written consent to participate in a protocol performed in accordance with the Declaration of Helsinki, and approved by the University of Alberta Research Ethics Board (Pro00070448). It was essential that participants were unaware that the touch reference would be displaced or that startling tones would occur. As such, participants were screened to verify they were unaware of the study's protocol. Full disclosure of the study's purpose and procedures was provided after the experimental session and participants were provided the opportunity to withdraw their consent. One participant reported with nonsyndromic autosomal recessive hearing loss. However, post hoc review of their responses to the acoustic startle showed that they were indistinguishable from other participants and therefore the data were retained within the set. Otherwise, none of the participants reported any neurological or musculoskeletal disorders.

2.1. Set-Up and Apparatus

For all conditions, participants stood in stocking feet, shoulder width apart, on an ethylene-vinyl acetate (EVA) foam pad placed atop a 6 axis force plate (AMTI OR6-7-1000, Advanced Mechanical Technology Inc, Watertown, MA, USA) (Figure 1). During the test condition, participants were asked to touch a 3D-printed plastic touch plate mounted to a steel rod that permitted the height of the touch plate to be adjusted to the participant's height. Participants were instructed to place the pad of the right index finger on a raised dimple (~0.5 mm) at the center of the touch plate, with the remaining fingers curled into the palm to avoid inadvertent contact with the plate. The use of the raised dimple was necessary as blind-folded participants will normally seek the edges of the touch plate. The height of the touch plate was adjusted so that participants could maintain contact of the finger pad with the wrist in a neutral position, the elbow flexed to approximately 90°, and a vertical alignment of the upper arm. During the No Touch conditions, the right arm was free to hang in a relaxed position at their side. The left arm was free to hang in a relaxed position at their side throughout the study. To produce a linear displacement of the touch plate, the plate was mounted on a square rail acme screw drive positioning stage (Lintech Positioning Systems 130 Series, Monrovia, CA, USA), driven by a computer-controlled two-phase stepper motor (Applied Motion Products 5023-124 2-phase hybrid stepper motor, Watsonville, CA, USA). The touch plate was displaced 12.5 mm, with a peak velocity of 124 mm/s. Stage position was measured using a linear displacement sensor (Penny & Giles SLS130, Penny & Giles Controls Limited, Christchurch, UK). The entire touch plate apparatus was on top of a 6 axis force plate (AMTI MC3A-100, Advanced Mechanical Technology, Inc., Watertown, MA, USA) to allow the vertical component of the touch force to be measured. The touch force was monitored online, and auditory feedback was provided if the force exceeded 1 N. Participants wore a pair of darkened goggles to block visual inputs.

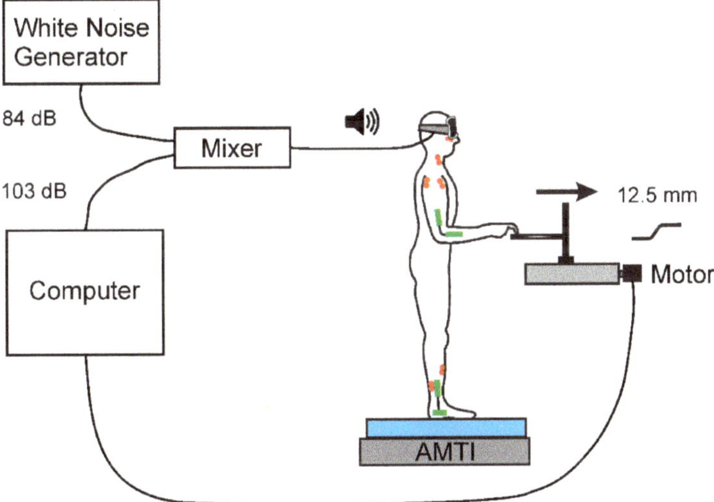

Figure 1. Schematic of the experimental set-up. Participants stood on foam atop a force plate, while lightly (<1N vertical load) touching a plastic touch plate. During DISPLACEMENT trials, the touch plate was unexpectedly translated away from the participant with the use of a computer controlled stepper motor. During STARTLE trials, the computer generated a startling tone delivered to the participant via a pair of indwelling earphones. During COMBINED trials, the touch displacement and startling tone were triggered simultaneously by the computer. In all conditions, the participant received white noise to mask the sound of the operating motor. Electromyographic (EMG) (red), goniometers (green), and center of pressure were used to characterize the evoked responses.

Auditory stimuli were delivered to the participants via a pair of Etymotic indwelling earphones (Etymotic ER-2, Etymotic Research Inc., Elk Grove Village, IL, USA). The startling acoustic stimulus (STARTLE) was generated from the PC sound card using a custom-written routine (LabView v8.2, National Instruments, Austin, TX, USA) and consisted of a 150 ms, 450 Hz tone delivered with a peak amplitude of 103 dB of sound pressure. Throughout the study, participants received white noise to mask the sound of the operating motor and other extraneous sounds. The white noise was calibrated at its maximum level and had a peak amplitude of 84 dB. Therefore, the minimum STARTLE signal to white noise was 19 dB of sound pressure. The STARTLE and white noise were calibrated using an Audioscan Verifit (Audioscan, Dorchester, ON, Canada) fitted with a 2-cc coupler. The STARTLE and white noise were mixed (Shure SCM268 Mixer, Shure Inc., Niles, IL, USA) prior to delivery to the earphones.

2.2. Protocol

Participants were randomly allocated to one of six cohorts. Each cohort represented one of the six possible sequences of the stimulus presentation orders for the three stimulus types: touch displacement alone (DISPLACEMENT), STARTLE, or simultaneous presentation of DISPLACEMENT and STARTLE (COMBINED). Pilot testing indicated that DISPLACEMENT and STARTLE evoked responses in tibialis anterior (TA) and anterior deltoid (AD) with comparable latency, but that STARTLE yielded shorter latency responses in sternocleidomastoid (SCM) and orbicularis oculi (OO) (also see Figure 8). Consequently, COMBINED stimuli were presented with a 0 ms lag (i.e., synchronized delivery via the computerized control program and verified post hoc from the linear displacement and acoustic tone recordings), given that the primary objective of this study was the interaction of STARTLE with DISPLACEMENT on responses related to activity in TA or AD. Each cohort completed four conditions: (1) standing eyes open, (2) standing eyes open with light touch, (3) standing eyes closed, and (4) the test condition. During the test condition participants were asked to stand with eyes closed with light touch. Approximately 10 s into the test condition the first stimulus trial was delivered. An additional 9 stimuli of that type were delivered, followed by 5 each of the other two stimuli types, for a total of 20 stimuli. Stimuli were separated by at least 8 s intervals. The test condition took up to 7 min to complete. Conditions 1 to 3 were 90 s each and were performed to create the expectation that the test condition would be uneventful. Participants rested for 2 min between conditions.

2.3. Recording and Data Acquisition

Electromyographic (EMG) activity was recorded from TA and soleus (SOL) of the right leg; AD and posterior deltoid (PD) of the right arm; and the right SCM and OO. EMG activity was recorded using pairs of Ag/AgCl electrodes (NeuroPlus A10040, Vermed, Bellows Falls, VT, USA) placed on the skin over the bellies of the intended muscles, with an inter-electrode distance of about 2 cm. Ground electrodes were placed over the right clavicle and the anterior tibia of the right leg. The skin at the limb EMG recording sites was shaved with a razor and cleaned with alcohol, while the skin below the eye and at the neck was only cleaned with alcohol. Electrode impedance was less than 20 kΩ at all recording sites (Grass F-EZM5 impedance meter, Astro-Med, Inc., West Warwick, Rhode Island, USA). The EMG signals were amplified and band-pass filtered (10 Hz-1 kHz with a 60 Hz notch filter, Grass P511 amplifiers, Astro-Med, Inc., West Warwick, RI, USA) prior to digitization. Electrogoniometers were placed across the right ankle (SG110A, Biometrics Ltd., Newport, UK) and elbow joints (SG110, Biometrics Ltd., Newport, UK). All analog signals were digitized at 2000 Hz (PCI-MIO-16E-4, National Instruments, Austin, TX, USA) and stored directly to hard drive using a custom-written LabView data acquisition routine.

2.4. Data Analysis

Post-processing of the signals was performed offline using custom-written LabView routines. The EMG signals were digitally full-wave rectified and then low-pass filtered (50 Hz, 4th order zero-lag Butterworth filter). The mechanical signals were low-pass filtered (20 Hz, 2nd order zero-lag Butterworth filter) and the position of the center of pressure was calculated from the force and moment signals from the force plate. For each stimulus, a 900 ms trace was extracted from the continuous data feed, aligned to the onset of the stimulus and included a 300 ms pre-stimulus period.

To determine whether the stimulus evoked an EMG response in the recorded muscles, a two standard deviations band around the mean EMG activity for the 100 ms prior to the perturbation onset was calculated. A response was considered to be present if following the onset of the stimulus the EMG trace exceeded this band for at least 20 continuous milliseconds. The onset latency of an evoked response was taken as the time following the stimulus onset that the EMG trace first exceeded the two standard deviations band. For consistency with our previous studies [2,3], only responses with onset latencies <200 ms were considered.

A key outcome measure of interest was the habituation of EMG response amplitudes with successive presentations of a stimulus. Therefore, it was important to estimate a response amplitude even in cases when the criteria for a response (described in the previous paragraph) were not met as it is possible that small amplitude responses were not identified. Consequently, EMG amplitude was calculated for each stimulus trial. To do so, analysis windows were established for each muscle by overlaying all 20 of the individual traces for that participant and manually placing cursors to capture the onset and offset of the apparent initial evoked response. This was preferred to using the average onset latency as average onset latencies will cleave the initial rise in a response in some of the trials due to the natural trial by trial variation or when the onset latency was different between stimulus types (for example, in OO and SCM, see Results and Figure 2). EMG response amplitudes were normalized to the maximum voluntary contraction obtained at the end of the experimental session.

As noted previously [3], participants sway considerably when standing on foam, making the two standard deviations method to identify stimulus-evoked events impractical for the mechanical signals. Consequently, the change in anterior–posterior position of the center of pressure (COP_{AP}) was calculated as the difference in position 300 ms following the stimulus onset, relative to the position at stimulus onset, for all trials. Changes in elbow and ankle angles were calculated using the same approach.

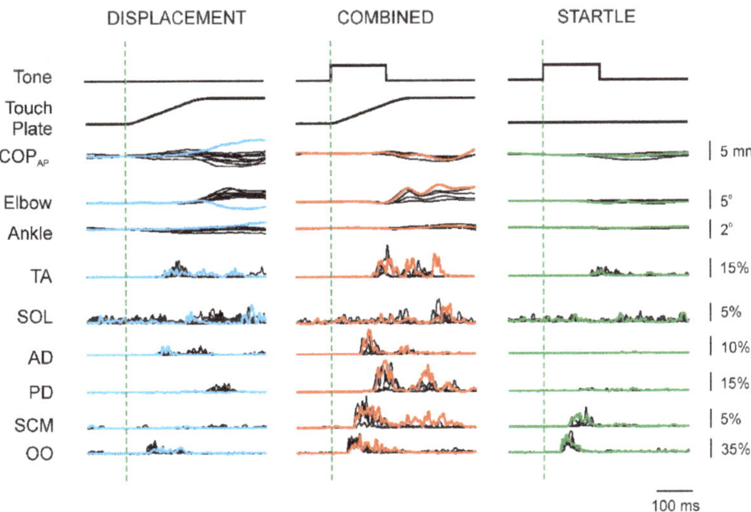

Figure 2. Sample data traces from a single participant who received the COMBINED stimuli after habituating to the DISPLACEMENT stimuli. Each cluster of traces represents the complete series of trials for that stimulus mode, with the coloured traces representing the first trial for that stimulus mode (Blue, DISPLACEMENT; Red, COMBINED; Green, STARTLE). The thin black lines of each cluster represent the subsequent trials. The vertical dashed line is aligned to the onset of the stimuli. Positive deflections in the COP_{AP} traces represent forward, elbow traces represent extension, and ankle traces represent plantarflexion. COP_{AP}, center of pressure anterior–posterior; TA, tibialis anterior; SOL, soleus; AD, anterior deltoid; PD, posterior deltoid; SCM, sternocleidomastoid; OO, orbicularis oculi.

2.5. Statistics

Amplitude of evoked EMG responses or mechanical events were compared across repeated trials within a protocol cohort using one-way repeated measures analyses of variance (rmANOVA). Six levels were used for each rmANOVA: Trial 1, the mean of Trials 8–10, Trial 11, Trial 15, Trial 16 and Trial 20. The mean of Trials 8–10 was used to represent the habituated response to the initial stimulus, whereas Trials 1, 11, 15, 16 and 20 represent the transition points between stimulus types. Significant main effects were then assessed using Tukey's Honest Significant Difference (HSD) tests, but only immediately adjacent points in the sequence were compared (e.g., Trial 1 vs. Trials 8–10, Trials 8–10 vs. Trial 11 etc.).

Note that for each EMG response or mechanical event, only participants with complete datasets for that outcome measure were included for analysis. Thus, if an EMG response was contaminated, such as by unexpected activity prior to the stimulus onset, then the data from that participant for that muscle were excluded. EMG data were also excluded on three occasions because of persistent noise in the recording that could not be eliminated. Due to a technical issue with the electrogoniometers, the elbow and ankle recordings were not included in the datasets of 10 and 5 participants, respectively. Additional participants were recruited to ensure a minimum of 5 complete datasets were available for each outcome measure for each cohort, while maintaining balanced numbers in all cohorts. Doing so brought the total number of participants in each cohort to 8. Consequently, the number of datasets for each outcome measure varied between 5 and 8 across the cohorts. This is reflected in the n reported in the Tables of statistics presented in the Results.

A comparison of EMG and mechanical response amplitudes was also performed between the DISPLACEMENT and COMBINED stimuli for those participants who received these stimuli during the initial 10 trial sequence of stimuli. Doing so allowed for up to 16 participants to be included in each group by combining the participants from the two cohorts receiving each initial stimulus type. For this analysis, a two-way ANOVA with one repeated factor was employed. Stimulus type (DISPLACEMENT vs. COMBINED) was the independent factor, while Trial (Trial 1 vs. mean of Trials 8–10) was the repeated factor. Tukey's HSD tests were used to assess the nature of significant interaction terms or main effects.

EMG response onset latencies were compared only for trials with a demonstrable evoked response. Therefore, the dataset of 20 trials for any given participant was rarely complete. Indeed, no observable responses were identified in one or more muscles of some participants. As the primary objective of the latency comparison was to assess the effect of combining the displacement with the acoustic startle, we took the average onset latency across all trials for a given stimulus type for each participant for each muscle. In this way, each participant contributed a single value for each stimulus type to be considered in the analysis. If a participant did not express at least one response for each stimulus type then that participant was excluded from the analysis for that muscle. A one-way rmANOVA was then employed, with 3 levels for stimulus type (DISPLACEMENT, STARTLE, COMBINED).

Statistical analyses were performed with the *Real Statistics Using Excel* Resource Pack software (Release 6.2). Copyright (2013–2019) Charles Zaiontz [13]. All comparisons were made with $\alpha = 0.05$. Descriptive statistics are presented as the mean ± standard error of the mean (SE). The complete dataset, along with participant characteristics, is available in the supplemental material (Spreadsheet S1).

3. Results

3.1. Superimposition of Acoustic Startle on Habituated Touch Displacement Responses

Unexpected touch displacements evoked responses comparable to what has been reported previously [2,3]. The leftmost column of traces in Figure 2 depicts data from one participant who displayed a forward sway reaction (COP_{AP}) coupled with an elbow flexion after the initial unexpected touch displacement (blue traces), but adapted to an arm-tracking behaviour (elbow extension) with minimal forward sway after the subsequent displacements (black traces). STARTLE was superimposed with the touch displacement for the subsequent five trials, displayed in the middle column of traces in Figure 2. For this participant, the initial COMBINED stimulus (red traces) resulted in the continued expression of the arm-tracking behaviour, but with larger amplitude EMG responses, or emergence of additional EMG responses, compared with the DISPLACEMENT responses. For the final five trials, STARTLE was presented alone (rightmost column), which yielded distinct bursts of EMG activity in TA, SCM and OO, but minimal impact on the COP_{AP} position or elbow angle, relative to the preceding DISPLACEMENT or COMBINED stimuli.

The graphs in Figure 3A display average data from all participants who received the sequence of stimuli shown in Figure 2. As shown in the COP_{AP} and elbow angle graphs, the initial unexpected touch displacement resulted in large forward sway (Trial 1 = 8.1 ± 3.4 mm), coupled with a distinct elbow flexion (Trial 1 = 3.2 ± 1.4°). On subsequent touch displacement trials, the forward sway was substantially reduced, and the elbow switched to an extension response. By the last three trials of DISPLACEMENT, the forward sway was minimal (Trials 8–10 = 0.5 ± 0.3 mm) and the elbow extension was consistent (Trials 8–10 = 2.6 ± 0.6°). With the superimposition of STARTLE on the touch displacement elbow extension was markedly increased (Trial 11 = 7.5 ± 1.9°), but COP_{AP} sway was not evident (Trial 11 = −0.2 ± 1.6 mm). Subsequently, the augmented elbow extension abated (Trial 15 = 3.6 ± 1.0°), comparable to the extension observed at the end of the DISPLACEMENT trials. The subsequent STARTLE alone trials did not evoke responses in these two metrics. EMG response amplitudes are also shown for TA, AD, SCM and OO. The initial DISPLACEMENT resulted in a burst of TA activity (Trial 1 = 7.6 ± 3.2% MVC), but not AD (Trial 1 = 0.6 ± 0.3% MVC), consistent with a forward sway response. With subsequent DISPLACEMENT trials the TA burst decreased (Trials 8–10 = 0.2 ± 0.2% MVC), while AD demonstrated a clear burst (Trials 8–10 = 3.9 ± 0.7% MVC), consistent with the occurrence of the arm-tracking behaviour. EMG activity in SCM or OO was not consistently expressed with DISPLACEMENT alone. Superimposition of STARTLE and DISPLACEMENT resulted in large bursts of EMG activity in all four muscles displayed. The first COMBINED trial resulted in a large burst in TA (Trial 11 = 21.9 ± 8.5% MVC) that abated with repeated trials (Trial 15 = 8.0 ± 3.1% MVC). Similar effects were observed in AD (Trial 11 = 18.4 ± 5.7; Trial 15 = 8.9 ± 3.2% MVC) and SCM (Trial 11 = 42.1 ± 15.0% MVC; Trial 15 = 15.0 ± 6.0% MVC), but not OO which displayed a burst amplitude unaffected by the repetition of the stimulus (Trial 11 = 21.7 ± 8.4% MVC; Trial 15 = 30.7 ± 5.7% MVC). Subsequently, with the STARTLE alone, bursts in TA and AD were markedly reduced (Trial 16 = 2.0 ± 1.1% MVC and 1.9 ± 1.5% MVC, respectively), whereas bursts in SCM continued to progressively decline with continued stimuli. In contrast, burst amplitudes in OO continued to be strongly expressed and did not appear to habituate with repeated STARTLE. A complete report of the descriptive statistics and the result of the associated ANOVAs is provided in Table 1A.

Figure 3B displays data from participants who were first habituated to DISPLACEMENT and then exposed to the STARTLE alone, followed by COMBINED stimuli. As can be seen, the DISPLACEMENT data for this cohort of participants largely replicates the results for the cohort shown in Figure 3A. Similarly, STARTLE resulted in minimal changes to COP_{AP} or elbow angle. STARTLE also evoked minimal activity in TA and AD. In contrast, STARTLE initially evoked a pronounced burst of activity in SCM (Trial 11 = 20.5 ± 6.1% MVC) that rapidly decreased with repeated STARTLE trials (Trial 15 = 6.1 ± 2.7% MVC), whereas the large initial burst in OO (Trial 11 = 13.7 ± 4.4% MVC) was not influenced by repeated STARTLE trials (Trial 15 = 11.4 ± 4.7% MVC). With the STARTLE superimposed on the DISPLACEMENT, the arm-tracking behaviour was expressed with an elbow extension (Trial 16 = 4.2 ± 0.7°), which remained unchanged with repeated COMBINED trials (Trial 20 = 3.2 ± 0.4°). COMBINED stimuli evoked inconsistent, small bursts of activity in TA, whereas bursts in AD (Trial 16 = 4.4 ± 1.8% MVC) were comparable to the bursts expressed with DISPLACEMENT and were relatively stable in amplitude with repeated trials (Trial 20 = 3.8 ± 0.9% MVC). Activity in SCM with the COMBINED stimuli was not different from the habituated STARTLE response. OO burst amplitude with the COMBINED stimuli remained similar to the STARTLE response and did not change with repeated trials. A complete report of the descriptive statistics and the result of the associated ANOVAs is provided in Table 1B.

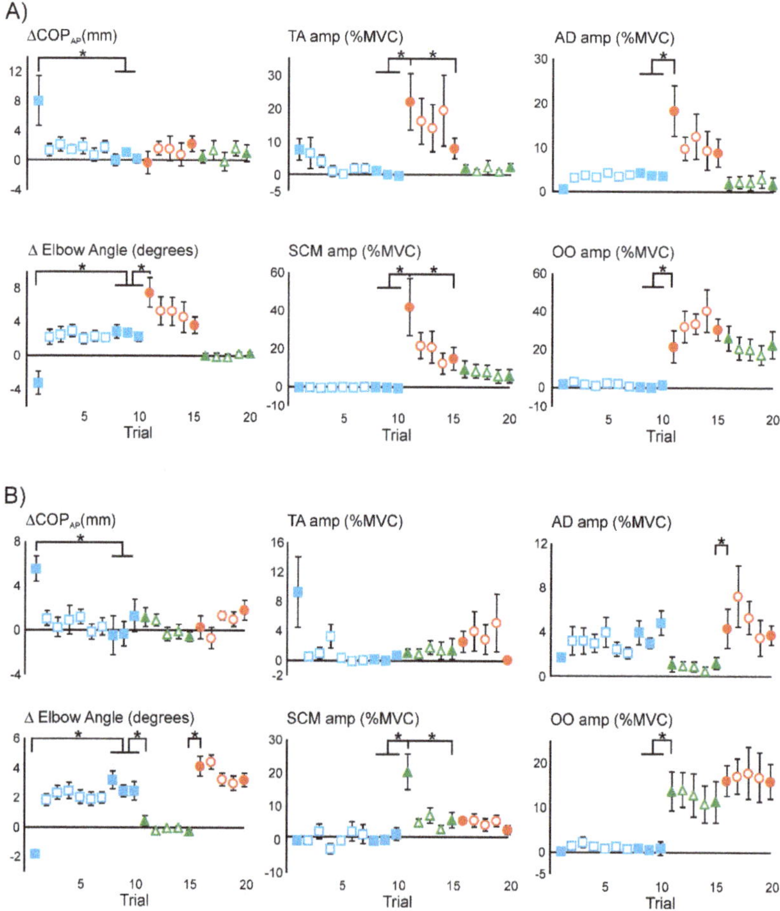

Figure 3. Trial-by-trial mean (SE) mechanical and muscle response amplitudes for participants who habituated to the DISPLACEMENT stimuli initially. (**A**) Participants subsequently received the COMBINED stimuli, followed by the STARTLE. (**B**) Participants subsequently received the STARTLE, followed by the COMBINED stimuli. Asterisks indicate differences identified by Tukey's HSD comparisons ($p < 0.05$). Blue squares, DISPLACEMENT; Red circles, COMBINED; Green triangles, STARTLE. Closed symbols indicate data points used in the statistical analysis. TA, tibialis anterior; AD, anterior deltoid; SCM, sternocleidomastoid; OO, orbicularis oculi; COP$_{AP}$, center of pressure anterior–posterior; MVC, maximum voluntary contraction.

3.2. Initial Superimposition of Acoustic Startle and Touch Displacement

Presenting naïve participants with the STARTLE superimposed on DISPLACEMENT in the first 10 trials evoked pronounced forward sway, coupled with elbow flexion in the initial trial (Figure 4, red traces). Subsequent exposures to the COMBINED stimuli continued to evoke a forward sway with elbow flexion (black traces), but of smaller amplitude. The participant displayed in Figure 4 continued to express a modest elbow flexion throughout the 10 COMBINED trials and never adopted an arm-tracking behaviour. Subsequent STARTLE trials did not evoke a discernible response in this participant, outside of a persistent blink reflex (OO). The DISPLACEMENT trials at the end of the sequence of stimuli resulted in a modest elbow extension on the first exposure (blue traces), that persisted thereafter (black traces) in this participant.

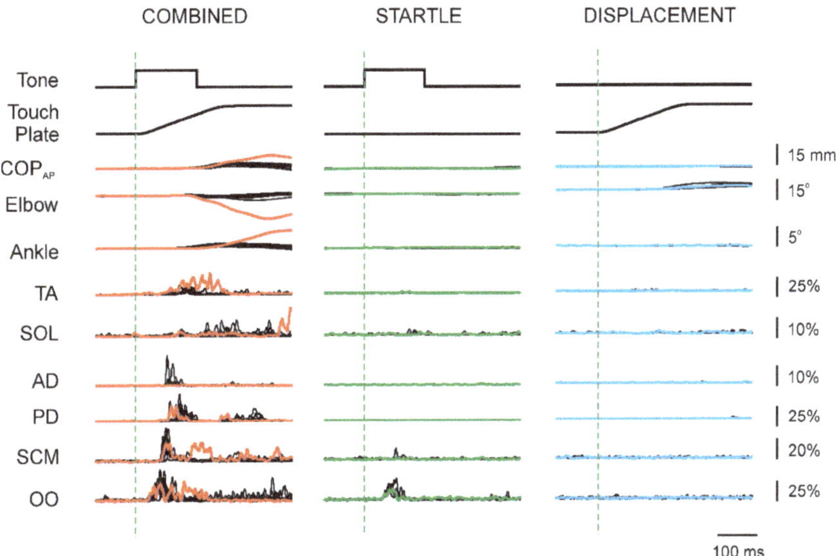

Figure 4. Sample data traces from a single participant who received the COMBINED stimuli initially. Each cluster of traces represents the complete series of trials for that stimulus mode, with the coloured traces representing the first trial for that stimulus mode (Red, COMBINED; Green, STARTLE; Blue, DISPLACEMENT). The thin black lines of each cluster represent the subsequent trials. The vertical dashed line is aligned to the onset of the stimuli. Positive deflections in the COP$_{AP}$ traces represent forward, elbow traces represent extension, and ankle traces represent plantarflexion. COP$_{AP}$, center of pressure anterior–posterior; TA, tibialis anterior; SOL, soleus; AD, anterior deltoid; PD, posterior deltoid; SCM, sternocleidomastoid; OO, orbicularis oculi.

Group averaged data for the cohorts that received the COMBINED stimulus initially are displayed in Figure 5. Figure 5A depicts the cohort that received the same sequence of stimuli as the participant in Figure 4. As can be seen, the initial pronounced forward sway (Trial 1 = 18.8 ± 6.0 mm) and elbow flexion (Trial 1 = 17.1 ± 3.3°) was a consistent outcome for all participants who received the COMBINED stimulus in the first trial. Subsequently, the forward sway persisted, but decreased (Trials 8–10 = 5.9 ± 2.7 mm), while the elbow flexion decreased towards no discernible response (Trials 8–10 = 0.8 ± 0.5° of flexion). Pronounced EMG bursts were evoked in each of the muscles depicted with the initial COMBINED stimuli. Where the response amplitudes in TA and SCM decreased by 2/3 of their first trial amplitude on average, AD burst amplitude decreased more modestly and OO burst amplitudes were unchanged with repeated exposure to the COMBINED stimuli. The subsequent STARTLE alone trials evoked a blink reflex in OO that was not different in amplitude from the response evoked in the preceding COMBINED trials, but little else of note in the other measures. Thereafter, the DISPLACEMENT trials resulted in modest forward sway (Trial 16 = 3.0 ± 2.1 mm) that decreased with repeated trials (Trial 20 = 0.8 ± 1.0 mm), and a minimal elbow extension (Trial 16 = 0.7 ± 0.4°) that persisted with repeated trials (Trial 20 = 1.0 ± 0.3°). Small, persistent bursts were observed in AD, while little activity in TA, SCM or OO was evoked with DISPLACEMENT trials. A complete report of the descriptive statistics and the result of the associated ANOVAs is provided in Table 2A.

Table 1. Change in COP_{AP} position and elbow angle, and normalized muscle response amplitudes.

A	Stimulus							Statistical Results	
	D_1	D_{8-10}	C_1	C_5	S_1	S_5	n	Main Effect	Tukey's HSD ($p < 0.05$)
Mechanical									
ΔCOP_{AP} (mm)	8.1 (3.4)	0.5 (0.3)	−0.2 (1.6)	2.3 (1.1)	0.6 (0.9)	1.1 (1.2)	8	3.38 ($p = 0.014$)	$D_1 > D_{8-10}$
ΔElbow (deg)	−3.2 (1.4)	2.6 (0.6)	7.5 (1.9)	3.6 (1.0)	−0.3 (0.3)	0.0 (0.2)	7	11.6 ($p = 0.000$)	$D_1 < D_{8-10}$, $D_{8-10} < C_1$
EMG (%MVC)									
TA	7.6 (3.3)	0.2 (0.2)	21.9 (8.5)	8.0 (3.1)	2.0 (1.1)	2.4 (1.1)	8	6.31 ($p = 0.000$)	$D_{8-10} < C_1$, $C_1 > C_5$
AD	0.7 (0.3)	3.9 (0.7)	18.4 (5.7)	8.9 (3.2)	1.9 (1.5)	1.8 (1.6)	8	7.24 ($p = 0.000$)	$D_{8-10} < C_1$
SCM	−0.2 (0.2)	−0.2 (0.1)	42.1 (15.0)	15.0 (6.0)	9.4 (4.4)	6.0 (3.5)	8	6.74 ($p = 0.000$)	$D_{8-10} < C_1$, $C_1 > C_5$
OO	1.8 (1.5)	0.7 (0.4)	21.7 (8.4)	30.7 (5.7)	26.2 (6.4)	22.7 (7.2)	8	9.71 ($p = 0.000$)	$D_{8-10} < C_1$

B	Stimulus							Statistical Results	
	D_1	D_{8-10}	S_1	S_5	C_1	C_5	n	Main Effect	Tukey's HSD ($p < 0.05$)
Mechanical									
ΔCOP_{AP} (mm)	5.5 (1.2)	0.2 (1.0)	2.8 (0.9)	−0.6 (0.4)	0.3 (1.0)	1.8 (0.9)	8	7.79 ($p = 0.000$)	$D_1 > D_{8-10}$
ΔElbow (deg)	−1.8 (0.3)	2.7 (0.3)	0.5 (0.3)	−0.2 (0.1)	4.2 (0.7)	3.2 (0.4)	5	29.4 ($p = 0.000$)	$D_1 < D_{8-10}$, $D_{8-10} > S_1$, $S_5 < C_1$
EMG (%MVC)									
TA	9.3 (4.8)	0.4 (0.3)	1.1 (0.5)	1.6 (1.4)	2.6 (1.4)	0.2 (0.2)	8	2.50 ($p = 0.049$)	
AD	1.7 (0.4)	4.0 (0.7)	1.1 (0.7)	1.2 (0.5)	4.4 (1.8)	3.8 (0.9)	8	4.12 ($p = 0.005$)	$S_5 < C_1$
SCM	−0.1 (0.2)	0.6 (0.8)	20.5 (6.1)	6.1 (2.7)	5.9 (0.6)	3.2 (1.6)	6	6.95 ($p = 0.000$)	$D_{8-10} < S_1$, $S_1 > S_5$
OO	0.1 (0.3)	0.8 (0.7)	13.7 (4.4)	11.4 (4.7)	16.2 (3.4)	16.0 (3.9)	8	12.1 ($p = 0.000$)	$D_{8-10} < S_1$

Values are means (SE), COP_{AP}, center of pressure anterior–posterior; TA, tibialis anterior; AD, anterior deltoid; SCM, sternocleidomastoid; OO, orbicularis oculi; D, touch displacement; S, startle; C, combined; MVC, maximum voluntary contraction. Statistical results are F statistics and p values. Tukey's HSD applied to adjacent points in sequence only.

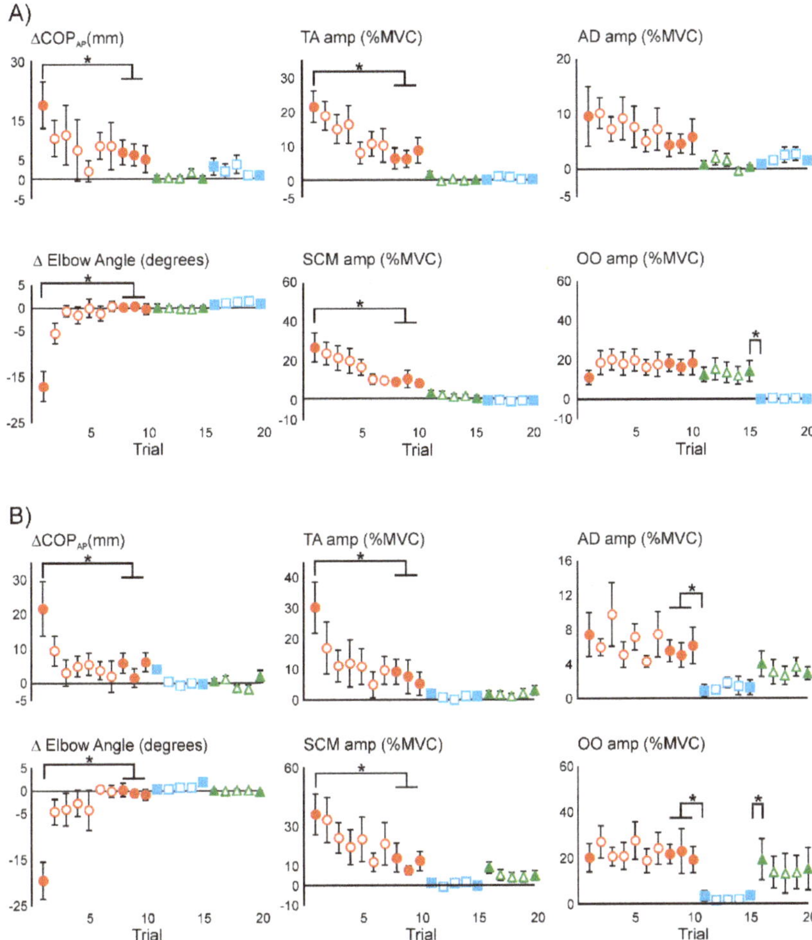

Figure 5. Trial-by-trial mean (SE) mechanical and muscle response amplitudes for participants who habituated to the COMBINED stimuli initially. (**A**) Participants subsequently received the STARTLE stimuli, followed by the DISPLACEMENT. (**B**) Participants subsequently received the DISPLACEMENT, followed by the STARTLE. Asterisks indicate differences identified by Tukey's HSD comparisons ($p < 0.05$). Blue squares, DISPLACEMENT; Red circles, COMBINED; Green triangles, STARTLE. Closed symbols indicate data points used in the statistical analysis. TA, tibialis anterior; AD, anterior deltoid; SCM, sternocleidomastoid; OO, orbicularis oculi; COP_{AP}, center of pressure anterior–posterior; MVC, maximum voluntary contraction.

Figure 5B displays the averaged data for the cohort of participants who received the COMBINED stimuli initially, followed by DISPLACEMENT and then STARTLE alone. The COMBINED stimuli data for this cohort are qualitatively similar to what is shown in Figure 5A. Of note, the COMBINED stimuli evoked a pronounced forward sway (Trial 1 = 21.6 ± 7.9 mm) that decreased with repeated trials (Trials 8–10 = 4.5 ± 2.4 mm), concomitantly with a pronounced initial elbow flexion (Trial 1 = 19.5 ± 4.0°) that progressively decreased until no response at the elbow was apparent (Trials 8–10 = 0.3 ± 0.9° of flexion on average). Subsequently, the data resulting from DISPLACEMENT and STARTLE are similar to the comparable stimuli of Figure 5A. A complete report of the descriptive statistics and the result of the associated ANOVAs is provided in Table 2B.

Table 2. Change in COP_{AP} position and elbow angle, and normalized muscle response amplitudes.

A			Stimulus					Statistical Results	
A	C_1	C_{8-10}	S_1	S_5	D_1	D_5	n	Main Effect	Tukey's HSD ($p < 0.05$)
Mechanical									
ΔCOP_{AP} (mm)	18.8 (6.0)	5.9 (2.7)	0.3 (1.0)	0.0 (0.7)	3.0 (2.1)	0.8 (1.0)	8	6.98 ($p = 0.000$)	$C_1 > C_{8-10}$
ΔElbow (deg)	−17.1 (3.3)	−0.8 (0.5)	0.0 (0.1)	0.0 (0.1)	0.7 (0.4)	1.0 (0.3)	8	27.1 ($p = 0.000$)	$C_1 > C_{8-10}$
EMG (%MVC)									
TA	21.3 (4.6)	7.0 (2.5)	1.7 (0.9)	0.1 (0.1)	0.1 (0.2)	0.2 (0.1)	8	17.7 ($p = 0.000$)	$C_1 > C_{8-10}$
AD	9.6 (5.5)	4.9 (2.1)	0.8 (0.7)	0.4 (0.3)	0.8 (0.6)	1.5 (0.8)	8	2.88 ($p = 0.028$)	
SCM	27.2 (7.8)	9.8 (2.0)	3.7 (1.5)	1.1 (1.4)	0.0 (0.1)	−0.1 (0.4)	7	11.5 ($p = 0.000$)	$C_1 > C_{8-10}$
OO	11.3 (3.5)	17.8 (4.4)	12.6 (3.6)	14.3 (5.3)	0.0 (0.5)	0.0 (0.3)	8	9.09 ($p = 0.000$)	$S_5 > D_1$

B			Stimulus					Statistical Results	
B	C_1	C_{8-10}	D_1	D_5	S_1	S_5	n	Main Effect	Tukey's HSD ($p < 0.05$)
Mechanical									
ΔCOP_{AP} (mm)	21.6 (7.9)	4.5 (2.4)	4.1 (1.3)	-0.2 (0.5)	0.6 (0.7)	1.8 (1.8)	8	5.99 ($p = 0.000$)	$C_1 > C_{8-10}$
ΔElbow (deg)	−19.5 (4.0)	−0.3 (0.8)	0.4 (0.2)	1.9 (0.7)	0.0 (0.3)	-0.2 (0.2)	8	24.1 ($p = 0.000$)	$C_1 > C_{8-10}$
EMG (%MVC)									
TA	30.3 (8.3)	7.3 (4.4)	2.1 (1.4)	1.3 (0.9)	1.8 (1.0)	3.1 (1.5)	8	10.1 ($p = 0.000$)	$C_1 > C_{8-10}$
AD	7.4 (2.6)	5.6 (1.3)	0.9 (0.7)	1.2 (0.9)	4.0 (1.5)	2.9 (0.8)	8	4.32 ($p = 0.004$)	$C_{8-10} > D_1$
SCM	36.0 (11.3)	11.2 (3.7)	1.3 (1.2)	−0.3 (0.2)	8.8 (3.1)	4.7 (2.6)	7	7.58 ($p = 0.000$)	$C_1 > C_{8-10}$
OO	20.2 (6.1)	21.3 (6.2)	3.7 (2.3)	4.0 (2.0)	19.3 (9.0)	15.2 (9.2)	8	4.10 ($p = 0.005$)	$C_{8-10} > D_1, D_5 < S_1$

Values are means (SE), COP_{AP}, center of pressure anterior–posterior; TA, tibialis anterior; AD, anterior deltoid; SCM, sternocleidomastoid; OO, orbicularis oculi; D, touch displacement; S, startle; C, combined; MVC, maximum voluntary contraction. Statistical results are F statistics and p values. Tukey's HSD applied to adjacent points in sequence only.

3.3. Comparing Initial COMBINED with Initial DISPLACEMENT

To compare the effect of superimposing STARTLE with DISPLACEMENT with the responses observed with DISPLACEMENT alone in the initial trials of naïve participants, the two cohorts of each stimulus type were collapsed. The group averaged data comparing the initial COMBINED stimuli with the initial DISPLACEMENT stimuli are presented in Figure 6. As seen in the COP_{AP} data, COMBINED resulted in a generally larger forward sway response than for DISPLACEMENT. The ANOVA indicated a significant Stimulus x Trial interaction (F_{130} = 4.44, p = 0.04), as the sway amplitude converged with the later trials. The first trial sway response was significantly larger for the COMBINED stimulus (Trial 1 = 20.2 ± 4.8 mm), compared with DISPLACEMENT (Trial 1 = 6.8 ± 1.8 mm). Responses at the elbow varied considerably between the COMBINED and DISPLACEMENT stimuli. In particular, with the DISPLACEMENT stimuli an initial elbow flexion was replaced with a persistent elbow extension by the second trial. In contrast, the large initial elbow flexion with the COMBINED stimuli progressively decreased in amplitude with repeated stimuli, but an extension of the arm was not observed. The ANOVA indicated a significant Stimulus x Trial interaction ($F_{1,22}$ = 46.85, p < 0.0001), as the amplitude of change in elbow angle converged with the later trials. The amount of elbow flexion with the first trial was significantly larger for the COMBINED stimulus (Trial 1 = 22.2 ± 2.4°), compared with DISPLACEMENT (Trial 1 = 2.6 ± 0.8°). The amplitude of the evoked burst of EMG activity in TA was generally larger with the COMBINED stimuli. The ANOVA did not identify a Stimulus x Trial interaction (F_{130} = 3.92, p = 0.06), whereas there were significant main effects of both Stimulus (F_{130} = 13.26, p = 0.001) and Trial (F_{130} = 25.24, p < 0.0001). The data for burst amplitude in AD demonstrate a clear Stimulus x Trial interaction (F_{130} = 6.80, p = 0.01) as the burst amplitudes with DISPLACEMENT are initially small (Trial 1 = 1.2 ± 0.3% MVC), but increase with later trials (Trials 8–10 = 3.9 ± 0.5% MVC). In contrast, the AD bursts evoked by the COMBINED stimuli show little adaptation with repeated trials, with an initial amplitude of 8.5 ± 2.9% MVC in Trial 1 then decreasing to 5.2 ± 1.2% MVC in Trials 8–10. Burst amplitude in SCM was larger with COMBINED stimuli than with DISPLACEMENT. A significant Stimulus x Trial interaction (F_{126} = 12.26, p = 0.002) was observed as the amplitude of the burst in SCM with the COMBINED stimuli decreased from an initial amplitude of 31.6 ± 6.7% MVC in Trial 1 to an amplitude of 10.5 ± 2.1% MVC in Trials 8–10, in contrast to what was observed with DISPLACEMENT, which did not change from near zero values for all trials. Bursts of activity in OO were seldom evoked with DISPLACEMENT, but were consistently evoked with the COMBINED stimulus. This yielded a significant main effect of Stimulus (F_{130} = 25.59, p < 0.0001), in the absence of a Stimulus x Trial interaction (F_{130} = 1.46, p = 0.2) or main effect of Trial (F_{130} = 1.17, p = 0.3).

3.4. Superimposition of Touch Displacement on Habituated Acoustic Startle Responses

Figure 7 shows data from one participant who was habituated to the STARTLE stimulus and then received DISPLACEMENT stimuli, before receiving the COMBINED stimuli. The first DISPLACEMENT trial (blue traces) resulted in a distinct forward sway observed in both the COP_{AP} and Ankle traces, which was then not expressed with subsequent trials, consistent with the forward sway described in the preceding Results and previous studies [2,3]. However, with the reintroduction of the STARTLE the forward sway was again apparent with the first COMBINED trial (red traces) in the COP_{AP} and Ankle, and with concomitant elbow flexion. Subsequent COMBINED trials reverted to the arm-tracking behaviour described earlier. This finding was not a unique occurrence and was consistently observed in all 8 participants who received this sequence of stimuli (Figure 8A). The EMG activity that accompanied these behaviours was qualitatively comparable to the pattern of adaptation observed when the STARTLE was superimposed on the habituated DISPLACEMENT stimuli (Figure 3A). That is, the first COMBINED stimulus resulted in a pronounced burst in TA (Trial 16 = 12.0 ± 3.0% MVC) and AD (Trial 16 = 18.5 ± 4.9% MVC), both of which decreased with repeated trials (Trial 20 = 4.6 ± 2.2% MVC and 7.9 ±% MVC, respectively). A complete report of the descriptive statistics and the result of the associated ANOVAs is provided in Table 3A.

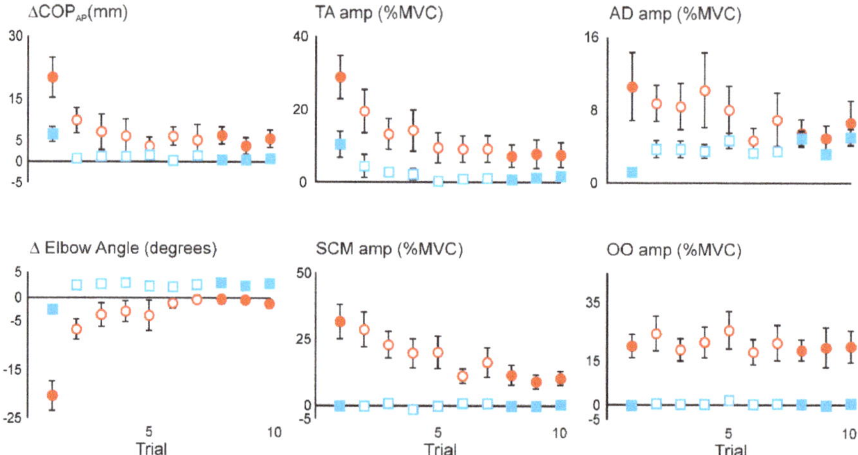

Figure 6. Trial-by-trial mean (SE) mechanical and muscle response amplitudes for all participants who received DISPLACEMENT (Blue squares) and COMBINED stimuli (Red circles) initially. Closed symbols indicate data points used in the statistical analysis. TA, tibialis anterior; AD, anterior deltoid; SCM, sternocleidomastoid; OO, orbicularis oculi; COP$_{AP}$, center of pressure anterior–posterior; MVC, maximum voluntary contraction.

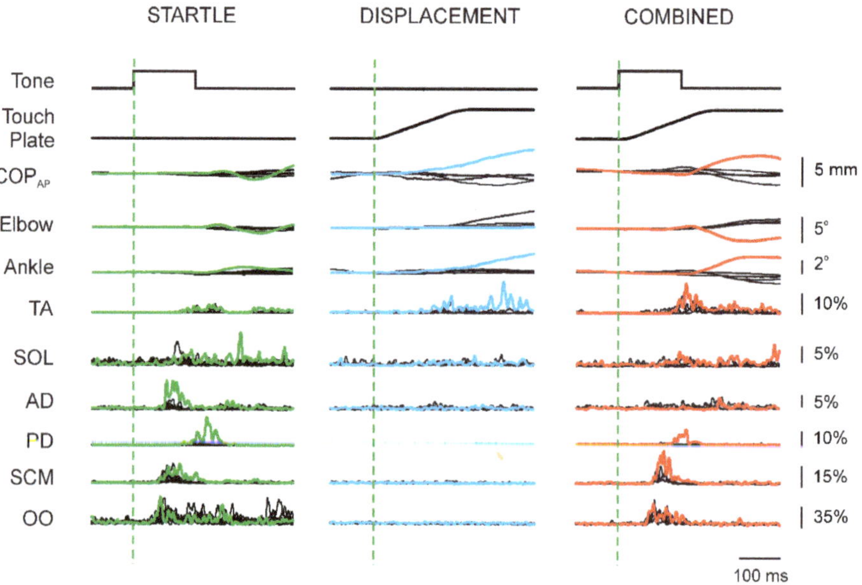

Figure 7. Sample data traces from a single participant who received the DISPLACEMENT stimuli after habituating to the STARTLE stimuli. Each cluster of traces represents the complete series of trials for that stimulus mode, with the coloured traces representing the first trial for that stimulus mode (Green, STARTLE; Blue, DISPLACEMENT; Red, COMBINED). The thin black lines of each cluster represent the subsequent trials. The vertical dashed line is aligned to the onset of the stimuli. Positive deflections in the COP$_{AP}$ traces represent forward, elbow traces represent extension, and ankle traces represent plantarflexion. COP$_{AP}$, center of pressure anterior–posterior; TA, tibialis anterior; SOL, soleus; AD, anterior deltoid; PD, posterior deltoid; SCM, sternocleidomastoid; OO, orbicularis oculi.

Table 3. Change in COP_{AP} position and elbow angle, and normalized muscle response amplitudes.

A	Stimulus							Statistical Results	
	S_1	S_{8-10}	D_1	D_5	C_1	C_5	n	Main Effect	Tukey's HSD ($p < 0.05$)
Mechanical									
ΔCOP_{AP} (mm)	1.0 (2.7)	0.8 (0.6)	5.2 (1.1)	1.0 (0.7)	6.7 (2.4)	3.4 (1.3)	8	2.07 ($p = 0.093$)	
ΔElbow (deg)	0.1 (1.2)	−0.2 (0.3)	−0.9 (0.3)	2.0 (0.4)	−3.9 (1.4)	3.3 (1.4)	5	5.65 ($p = 0.002$)	$D_5 > C_1, C_1 < C_5$
EMG (%MVC)									
TA	2.1 (0.8)	1.1 (0.4)	4.2 (1.5)	0.7 (0.5)	12.0 (3.0)	4.6 (2.2)	8	8.03 ($p = 0.000$)	$D_5 < C_1, C_1 > C_5$
AD	5.1 (3.1)	1.5 (0.6)	−0.2 (0.3)	2.7 (0.7)	18.5 (4.9)	7.9 (2.1)	7	8.78 ($p = 0.000$)	$D_5 < C_1, C_1 > C_5$
SCM	13.9 (4.4)	7.5 (3.6)	−0.3 (0.5)	0.5 (0.6)	19.1 (6.8)	11.1 (5.3)	7	4.87 ($p = 0.002$)	$D_5 < C_1$
OO	21.1 (11.0)	23.0 (9.8)	2.1 (1.0)	2.3 (1.1)	27.6 (14.4)	25.8 (10.6)	8	3.29 ($p = 0.015$)	$D_5 < C_1$

B	Stimulus							Statistical Results	
	S_1	S_{8-10}	C_1	C_5	D_1	D_5	n	Main Effect	Tukey's HSD ($p < 0.05$)
Mechanical									
ΔCOP_{AP} (mm)	0.8 (0.6)	−0.1 (0.8)	5.4 (0.7)	0.1 (1.2)	0.6 (1.0)	0.5 (1.6)	8	5.64 ($p = 0.001$)	$S_{8-10} < C_1, C_1 > C_5$
ΔElbow (deg)	−0.1 (0.4)	−0.2 (0.2)	-2.0 (0.6)	2.6 (1.0)	1.2 (0.6)	2.2 (1.0)	5	6.57 ($p = 0.000$)	$C_1 < C_5$
EMG (%MVC)									
TA	2.8 (1.2)	0.7 (0.4)	6.9 (1.9)	8.0 (2.6)	0.9 (0.7)	1.5 (1.3)	7	5.27 ($p = 0.001$)	$S_{8-10} < C_1, C_5 > D_1$
AD	1.1 (0.4)	0.9 (0.2)	2.3 (1.6)	3.3 (1.4)	1.5 (0.5)	1.3 (0.5)	7	1.12 ($p = 0.373$)	
SCM	8.4 (3.1)	2.9 (1.5)	5.0 (2.8)	6.3 (3.0)	−0.1 (0.3)	−0.4 (0.3)	7	2.74 ($p = 0.034$)	
OO	11.3 (3.7)	12.9 (5.0)	18.7 (6.8)	18.5 (5.7)	0.5 (0.4)	1.1 (0.8)	7	5.61 ($p = 0.001$)	$C_5 > D_1$

Values are means (SE), COP_{AP}, center of pressure anterior–posterior; TA, tibialis anterior; AD, anterior deltoid; SCM, sternocleidomastoid; OO, orbicularis oculi; D, touch displacement; S, startle; C, combined; MVC, maximum voluntary contraction. Statistical results are F statistics and p values. Tukey's HSD applied to adjacent points in sequence only.

In Figure 8B, group averaged data are shown for participants who received the STARTLE initially, followed by the COMBINED stimuli. STARTLE did not result in a consistent impact on the COP_{AP} and no demonstrable change in elbow angle, across participants. This was contrasted by bursts of EMG activity observed in each of the four muscles depicted. In TA, an initial burst (Trial 1 = 2.8 ± 1.2% MVC) decreased with repeated exposures (Trials 8–10 = 0.7 ± 0.4% MVC), whereas in AD a small initial burst (Trial 1 = 1.1 ± 0.4% MVC) showed little adaptation over repeated trials (Trials 8–10 = 0.9 ± 0.2 % MVC). A marked initial burst in SCM (Trial 1 = 8.4 ± 3.1% MVC) decreased with repeated trials (Trials 8–10 = 2.9 ± 1.5 % MVC). In contrast, pronounced initial bursts in OO (Trial 1 = 11.3 ± 3.7% MVC) were largely unchanged with repeated STARTLE exposure (Trials 8–10 = 12.9 ± 5.0% MVC). Subsequently, the DISPLACEMENT was superimposed with these habituated STARTLE responses. The initial COMBINED stimulus resulted in a forward sway (Trial 11 = 5.4 ± 0.7 mm), with an elbow flexion (Trial 11 = 2.0 ± 0.6°). With the subsequent COMBINED stimulus, the elbow flexion was replaced by an elbow extension, which remained largely unchanged (Trial 15 = 2.6 ± 1.0°). The forward sway in the COP_{AP} progressively decreased with repeated COMBINED stimuli (Trial 15 = 0.1 ± 1.2 mm). COMBINED stimuli evoked increased burst amplitudes in all four muscles depicted, with no apparent adaptation in amplitude with repeated trials. Thereafter, DISPLACEMENT trials continued to induce elbow extension (Trial 16 = 1.2 ± 0.6°; Trial 20 = 2.2 ± 1.0°), with a concomitant small burst in AD (Trial 16 = 1.5 ± 0.5% MVC; Trial 20 = 1.3 ± 0.5% MVC). DISPLACEMENT did not consistently evoke activity in TA, SCM or OO. A complete report of the descriptive statistics and the result of the associated ANOVAs is provided in Table 3B.

3.5. Response Latencies

Response latencies could only be estimated for trials that elicited a clear response (see Methods). Thus, not all trials yielded a value and therefore the average value from a participant for each stimulus type was used in the analysis. The subsequent group averaged response onset latencies for each stimulus type are presented in Figure 9, for TA, AD, SCM and OO. Responses latencies were generally longest in TA and shortest in OO, reflecting the conduction distances from the stimulus source to the target muscle. The latencies in TA were consistent across all three stimulus conditions with times of 127.5 ± 14.6 ms, 125.4 ± 5.8 ms, and 126.3 ± 12.7 ms for DISPLACEMENT, STARTLE, and COMBINED, respectively ($F_{2,15}$ = 0.05, p = 0.96). Responses in AD were also consistent across stimulus conditions with times of 104.9 ± 12.7 ms, 104.0 ± 5.3 ms, 99.1 ± 9.0 ms for DISPLACEMENT, STARTLE, and COMBINED, respectively ($F_{2,15}$ = 0.30, p = 0.74). In contrast, response latencies in SCM and OO were significantly slower with DISPLACEMENT alone, than with STARTLE or COMBINED stimuli, whereas the response times in STARTLE and COMBINED were not different. For SCM, the response latencies were 107.6 ± 14.3 ms, 82.8 ± 4.9 ms, and 74.1 ± 8.5 ms for DISPLACEMENT, STARTLE, and COMBINED, respectively ($F_{2,15}$ = 6.15, p = 0.01). For OO, the response latencies were 77.4 ± 17.1 ms, 47.5 ± 2.0 ms, and 48.4 ± 9.3 ms for DISPLACEMENT, STARTLE, and COMBINED, respectively ($F_{2,15}$ = 24.72, p < 0.001).

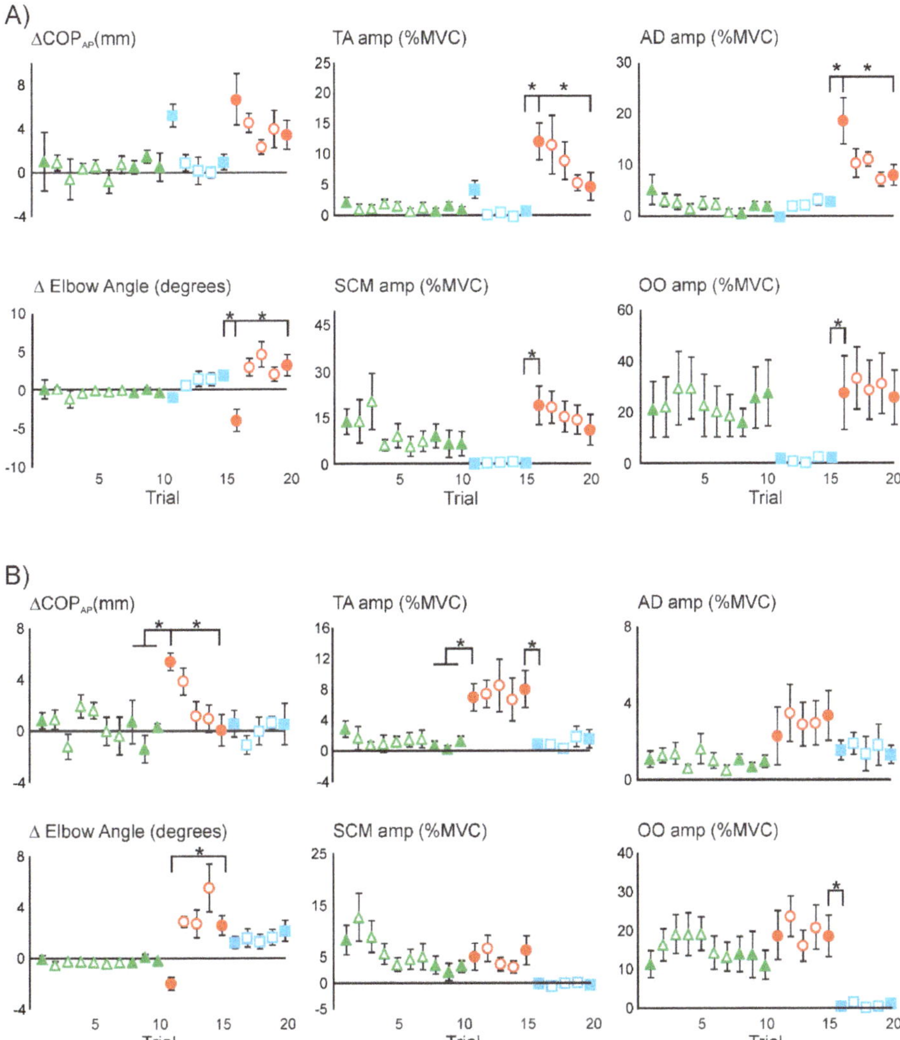

Figure 8. Trial-by-trial mean (SE) mechanical and muscle response amplitudes for participants who habituated to the STARTLE stimuli initially. (**A**) Participants subsequently received the DISPLACEMENT, followed by the COMBINED stimuli. (**B**) Participants subsequently received the COMBINED stimuli, followed by the DISPLACEMENT. Asterisks indicate differences identified by Tukey's HSD comparisons ($p < 0.05$). Blue squares, DISPLACEMENT; Red circles, COMBINED; Green triangles, STARTLE. Closed symbols indicate data points used in the statistical analysis. TA, tibialis anterior; AD, anterior deltoid; SCM, sternocleidomastoid; OO, orbicularis oculi; COP_{AP}, center of pressure anterior–posterior; MVC, maximum voluntary contraction.

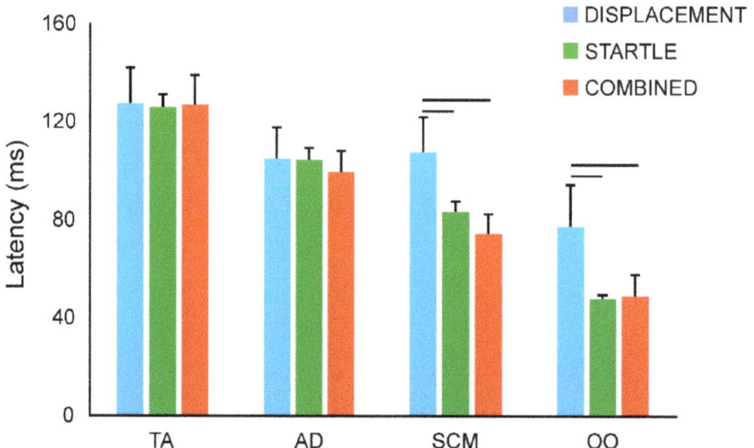

Figure 9. Mean (SE) onset latency for responses evoked in the respective muscle across all stimuli of the specific mode. Horizontal lines connect differences identified by Bonferroni adjusted, post hoc *t*-tests ($p < 0.05$). TA, tibialis anterior; AD, anterior deltoid; SCM, sternocleidomastoid; OO, orbicularis oculi.

4. Discussion

4.1. Interaction between Acoustic Startle and Habituated Touch Displacement Responses

Blouin et al. [12] demonstrated that introducing a STARTLE restored the amplitude of habituated neck postural responses in seated participants exposed to repeated accelerations of a sled. This finding strongly suggested that the initial response to the perturbation of the sled incorporated a postural response augmented by a superimposed startle response. Recently, we demonstrated that rapid displacement of a touch reference evoked a postural response on the first exposure, but an arm-tracking behaviour on subsequent exposures [2,3]. We speculated that the postural response evoked with the initial exposure might be related to the expression of a startle response and that the subsequent change in behaviour is related to the habituation of this startle. Contrary to our hypothesis, when we introduced a STARTLE after 10 trials of light touch displacement alone, the effect was to facilitate the arm-tracking behaviour that had been established, rather than restore the postural response (Figure 3A). However, when we introduced the STARTLE after only five trials of light touch displacement alone, the effect was to revert to the postural response initially (Figure 8A), supporting our hypothesis. These seemingly conflicting outcomes suggest that the interaction of the STARTLE with the reaction to the light touch displacement is dependent in part on the recent sensory history. In the first example, the facilitation of the arm-tracking behaviour suggests that the startle augmented a planned behavioural response, consistent with a StartReact-like effect. Whereas, in the second example, the restoration of the postural response suggests that the arm-tracking behaviour was not yet consolidated and that the startle biased the selection of the motor solution towards the balance corrective response or 'protective' option.

Startle is often observed to facilitate the execution of planned, voluntary movements [14]. The effect, referred to as the StartReact effect, has been observed with ballistic arm movements [15,16], ballistic ankle dorsiflexion [15], and stepping and obstacle avoidance [17], among other behaviours. Therefore, it was not entirely surprising that the arm-tracking behaviour in our study was facilitated when the STARTLE was co-presented with the touch plate displacement after 10 trials of touch plate displacement alone (Figure 3). The implication is that participants learned to preprogram the arm-tracking behaviour given the consistent direction and magnitude of the stimulus, which was then facilitated when the STARTLE was introduced. In stark contrast, when the STARTLE was co-presented with the touch plate displacement after only five trials, a postural response was evoked on the first COMBINED stimulus (Figure 8A, Trial 16). We note that the subsequent COMBINED stimuli (Trials 17–20) once again, immediately evoked the arm-tracking behaviour, but were larger than the preceding touch plate DISPLACEMENT trials, suggestive of a StartReact-like effect. This suggests that there was still uncertainty as to the appropriate motor solution in response to the stimulus at the finger after only five exposures, and that STARTLE facilitated whichever solution was expressed. Furthermore, it is important to note that the switch in motor response, from the postural to the arm-tracking response, occurred in the presence of STARTLE, indicating that facilitation of the responses did not specifically interfere with the process of selecting a motor solution. The sudden switches, within a single trial, from a postural response to the arm-tracking behaviour observed here are consistent with the affordance competition hypothesis, wherein it is argued that multiple motor solutions are encoded in parallel before selecting the one that is implemented [18,19].

4.2. Interaction of Acoustic Startle and Novel Touch Displacement Responses

Combining a startle with the touch plate displacement on the first instance in naïve participants resulted in a substantially larger postural response than in participants who received the touch displacement alone initially (Figure 6). This finding is consistent with the StartReact-like effects described above and suggests the startle facilitated the postural response. It is important to note, however, that combining the stimuli in this manner evoked first trial postural responses in all 16 participants we tested in this way. This is very different from our previous findings where we have observed first trial postural responses in only about 60% of participants who receive a touch plate displacement alone [2,3]. In the present study, of the 16 participants who initially received touch displacements alone, only eight exhibited the postural reaction on the first trial. This suggests that the startle facilitated the release of the postural response in those participants who would not have otherwise reacted to the light touch displacement. The implication is that in the presence of a concomitant startling stimulus, ambiguous balance-related sensory cues are more likely to generate a balance response, even if the stimulus is misinterpreted as a threat and the resulting balance correction is unnecessary.

A second key finding from combining the startle with the touch displacement initially in naïve participants is that the arm-tracking behaviour did not emerge with repeated exposures to the COMBINED stimuli (Figures 4 and 5). One possible explanation might be that the effect of the startle on the facilitation of the postural response had not habituated sufficiently to allow for the expression of the alternate behaviour. Indeed, as shown in Figures 5 and 6, a prominent response in SCM, a common marker for the presence of a startle response [14,20,21], was still apparent after 10 trials. However, the presence of a startle response in itself is not enough to prevent the switching of behaviour from a postural response to the arm-tracking behaviour, as Figures 7 and 8 demonstrate. We propose that when the COMBINED stimuli are presented in the first instance (Figure 5, Trial 1), the startling nature of the acoustic stimulus biases the motor solution selection process towards a protective behaviour, that is, the postural response observed in this case. When combined with our findings that STARTLE facilitates established arm-tracking behaviour (COMBINED data in Figure 3) and facilitates both the postural response and arm-tracking when the motor solution is labile (COMBINED data in Figure 8), we suggest that STARTLE likely facilitates all alternative motor solutions, but with a bias towards protective solutions, such as the postural response, when the sensory information is ambiguous.

4.3. Technical Considerations

We interpret our findings of the first trial response to a touch displacement as including a startle component. In large part, this is based on the consistent evidence in the literature that unexpected stimuli, including balance disturbances, evoke startle responses [4,9]. Moreover, it has been demonstrated that combining startle-inducing stimuli leads to larger startle responses [4], consistent with our findings. However, our recordings of SCM did not consistently display a response to the touch displacement alone, even with the first exposure in naïve participants (Figure 3, Trial 1). Responses in SCM are commonly used as a marker of startle, and clear responses, that habituated as expected, were apparent with the STARTLE in our study, which would support the use of SCM as a marker of startle. We suggest that in our study, the light touch displacement likely induced a weak startle, resulting in a small and inconsistent response in the SCM EMG recordings. This is corroborated in our study by the similarly inconsistent expression of the blink reflex, as indicated by the OO EMG recordings, in the DISPLACEMENT trials. The blink reflex is typically expressed as part of the startle response but is not generally considered a reliable marker of the startle response as it does not habituate [22]. Consistent with these previous findings, in our study, the STARTLE consistently evoked a blink reflex that did not habituate. The occasional, but inconsistent expression of the blink reflex and SCM response suggest that the light touch displacement was peri-threshold, or subthreshold, for the startle response. Previous work has also demonstrated that startle-like effects can be observed in the absence of SCM responses following a STARTLE [23–25]. Taken together, these findings suggest that startle responses need not be evident in SCM for unexpected stimuli to facilitate motor behaviours through a startle-like mechanism.

Whereas previous studies have shown interference between responses to stimuli delivered at different times (i.e., refractoriness, [26]), for our study, the purpose of the STARTLE was to provide a startling cue simultaneously with the onset of the touch displacement. We therefore used a 0 ms lag between the onset of the touch displacement and the STARTLE, which resulted in response latencies in TA and AD that did not differ across the three stimulus conditions (Figure 9). This was somewhat surprising as it has been demonstrated that the StartReact effect induces earlier onset reactions in voluntary reaction tasks (reviewed in [27]). The implication is that the postural responses evoked in TA and the arm-tracking responses evoked in AD might not be voluntary reactions per se, but perhaps are more automatic or reflexive, with less opportunity for reducing delays associated with the integrative processes within the neural chain.

This interpretation would be in conflict with the interpretation of Nonnekes et al. [23], who demonstrated that responses in TA to backward balance disturbances were quicker when accompanied by a STARTLE and suggested that the decrease in latency might be related to summative effects of medium latency postural responses and acoustic startle through a common relay in the reticular formation. Interestingly, Nonnekes et al. [23] also demonstrated that onset latencies of responses in gastrocnemius to forward balance perturbations were not influenced by the STARTLE, in contrast to the results in TA, which prompted the authors to suggest that the postural responses to forward and backward perturbations might be mediated via different neural circuits. In the present study, acoustic startle facilitated both the postural response (for example, Figure 5, Trial 1) and the arm-tracking response (for example, Figure 3 all COMBINED Trials) to light touch displacements. Moreover, the switching of responses, from postural to arm-tracking, was evident if the COMBINED stimulus occurred later in the protocol (for example, Figure 7), despite the apparent prevention of this behavioural switch if COMBINED was the initial stimulus (for example, Figure 6). Regardless, together these findings highlight the uncertainty surrounding the precise mechanism underlying the integration of startle with other sensorimotor circuitries and indicates that postural responses might arise, not from predetermined solutions to a sensory input, but from an ensemble of alternative motor solutions that are encoded simultaneously, and then expressed based upon the context and sensory state at the time of the disturbance [28,29].

4.4. Functional Implications

Unexpected sensory stimuli have been shown to generate stereotypical startle responses, regardless of the source or modality [4]. It has also been argued that the larger first trial responses to unexpected balance disturbances likely reflect the superimposition of a startle response on the underlying postural response [6,9,12]. Here, we demonstrate that overtly superimposing startle, by introducing an acoustic startle to the first occurrence of a light touch displacement, biases the evoked response towards a postural reaction, rather than the often observed arm-tracking response. This suggests that the functional outcome of the startling nature of unexpected stimuli might be to bias motor responses towards the most primitive, or protective interpretation of the stimulus. In this case, the slip detected at the fingertip could indicate the touch reference moved relative to the finger or that the finger moved relative to the touch reference. However, in the presence of the acoustic startle, naïve participants uniformly and robustly reacted as though they had moved relative to the touch reference, inducing a forward corrective sway. Unexpected balance disturbances are often isolated, unique events with only a single opportunity to select the correct motor solution. Our results suggest that when multiple, unexpected stimuli are presented simultaneously, as might happen in authentic balance disturbances in natural contexts, the cumulative effect of the startle-like facilitation might serve to bias the selection of the motor response. That is, startle may not simply facilitate preprogrammed motor behaviours, as has been demonstrated previously, but may also constrain or limit the motor solution options in threatening situations.

One interpretation of these findings is that provision of a supplemental startling stimulus may serve to facilitate balance reactions. Indeed, the results of Blouin et al. [12], wherein habituated neck responses were restored to larger representations, would suggest that such a tactic might have functionally relevant benefits. However, caution should be applied to this interpretation as startle in the present study facilitated, and in some circumstances restored, an inappropriate postural response to the light touch displacement. These augmented "false-positive" postural reactions may themselves be destabilizing and fall-inducing. Nevertheless, it remains possible that combining appropriate augmented stimuli with a supplementary startle could induce functionally relevant postural reactions, which might be a useful approach to mitigate balance impairments in those with compromised sensory systems or sensorimotor integration. For example, embedding an auditory tone within a mobility device may serve to enhance reactions to unexpected instabilities.

Supplementary Materials: The following are available online at http://www.mdpi.com/2076-3417/10/1/382/s1. Spreadsheet S1: Misiaszek et al. 2019.xlsx, includes the complete dataset reported in this paper, including data that were not described in detail in the Results. Participant characteristics are also provided.

Author Contributions: J.E.M.: Conceptualization, funding acquisition, methodology, resources, software, validation, data curation, and writing—original draft. S.D.C.C.: Formal analysis, investigation, and writing—review and editing. A.J.M.: Formal analysis, investigation, and writing—review and editing. K.K.F.: Conceptualization, methodology, project administration, supervision, and writing—review and editing. All authors have read and agreed to the published version of the manuscript.

Funding: This work was supported by a grant from the Natural Sciences and Engineering Research Council (NSERC) Canada (RGPIN-2017-04175 to JEM) and an NSERC studentship to SDC.

Acknowledgments: The authors thank William Hodgetts (audiologist, Department of Communication Sciences and Disorders, University of Alberta) for assistance with developing and calibrating the acoustic stimulus.

Conflicts of Interest: The authors declare no conflict of interest.

References

1. Marigold, D.S.; Misiaszek, J.M. Whole-body responses: Neural control and implications for rehabilitation and fall prevention. *Neuroscientist* **2009**, *15*, 36–46. [CrossRef] [PubMed]

2. Misiaszek, J.E.; Forero, J.; Hiob, E.; Urbanczyk, T. Automatic postural responses following rapid displacement of a light touch contact during standing. *Neuroscience* **2016**, *316*, 1–12. [CrossRef] [PubMed]

3. Misiaszek, J.E.; Vander Meulen, J. Balance reactions to light touch displacements when standing on foam. *Neurosci. Lett.* **2017**, *639*, 13–17. [CrossRef] [PubMed]

4. Yeomans, J.S.; Li, L.; Scott, B.W.; Frankland, P.W. Tactile, acoustic and vestibular systems sum to elicit the startle reflex. *Neurosci. Biobehav. Rev.* **2002**, *26*, 1–11. [CrossRef]

5. Allum, J.H.J.; Tang, K.S.; Carpenter, M.G.; Oude Nijhuis, L.B.; Bloem, B.R. Review of first trial responses in balance control: Influence of vestibular loss and Parkinson's disease. *Hum. Mov. Sci.* **2011**, *30*, 279–295. [CrossRef]

6. Campbell, A.D.; Squair, J.W.; Chua, R.; Inglis, J.T.; Carpenter, M.G. First trial and StartReact effects induced by balance perturbations to upright stance. *J. Neurophysiol.* **2013**, *110*, 2236–2245. [CrossRef]

7. Oude Nijhuis, L.B.; Allum, J.H.; Borm, G.F.; Honegger, F.; Overeem, S.; Bloem, B.R. Directional sensitivity of "first trial" reactions in human balance control. *J. Neurophysiol.* **2009**, *101*, 2802–2814. [CrossRef]

8. Oude Nijhuis, L.B.; Allum, J.H.J.; Valls-Solé, J.; Overeem, S.; Bloem, B.R. First trial postural reactions to unexpected balance disturbances: A comparison with the acoustic startle reaction. *J. Neurophysiol.* **2010**, *104*, 2704–2712. [CrossRef]

9. Sanders, O.P.; Savin, D.N.; Creath, R.A.; Rogers, M.W. Protective balance and startle responses to sudden freefall in standing humans. *Neurosci. Lett.* **2015**, *586*, 8–12. [CrossRef]

10. Li, L.; Yeomans, J.S. Summation between acoustic and trigeminal stimuli evoking startle. *Neuroscience* **1999**, *90*, 139–152. [CrossRef]

11. Li, L.; Steidl, S.; Yeomans, J.S. Contributions of the vestibular nucleus and vestibulospinal tract to the startle reflex. *Neuroscience* **2001**, *106*, 811–821. [CrossRef]

12. Blouin, J.S.; Siegmund, G.P.; Inglis, J.T. Interaction between acoustic startle and habituated neck postural responses in seated subjects. *J. App. Phsysiol.* **2007**, *102*, 1574–1586. [CrossRef] [PubMed]

13. Zaiontz, C. Real Statistics Using Excel. Available online: http://www.real-statistics.com/ (accessed on 22 March 2018).

14. Carlsen, A.N.; Maslovat, D.; Lam, M.Y.; Chua, R.; Franks, I.M. Considerations for the use of a startling acoustic stimulus in studies of motor preparation in humans. *Neurosci. Biobehav. Rev.* **2011**, *35*, 366–376. [CrossRef] [PubMed]

15. Valls-Solé, J.; Rothwell, J.C.; Goulart, F.; Cossu, G.; Muñoz, E. Patterned ballistic movements triggered by a startle in healthy humans. *J. Physiol.* **1999**, *516*, 931–938. [CrossRef]

16. Carlsen, A.N.; Chua, R.; Inglis, J.T.; Sanderson, D.J.; Franks, I.M. Prepared movements are elicited early by startle. *J. Mot. Behav.* **2004**, *36*, 253–264. [CrossRef]

17. Queralt, A.; Weerdesteyn, V.; van Duijnhoven, H.J.; Castellote, J.M.; Valls-Solé, J.; Duysens, J. The effects of an auditory startle on obstacle avoidance during walking. *J. Physiol.* **2008**, *586*, 4453–4463. [CrossRef]

18. Cisek, P. Cortical mechanisms of action selection: The affordance competition hypothesis. *Philos. Trans. R. Soc. B Biol. Sci.* **2007**, *362*, 1585–1599. [CrossRef]

19. Gallivan, J.P.; Logan, L.; Wolpert, D.M.; Flanagan, J.R. Parallel specification of competing sensorimotor control policies for alternative action options. *Nat. Neurosci.* **2016**, *19*, 320–326. [CrossRef]

20. Brown, P.; Rothwell, J.C.; Thompson, P.D.; Britton, T.C.; Day, B.L.; Marsden, C.D. New observations on the normal auditory startle reflex in man. *Brain* **1991**, *114*, 1891–1902. [CrossRef]

21. Bisdorff, A.R.; Bronstein, A.M.; Gresty, M.A. Responses in neck and facial muscles to sudden free fall and a startling auditory stimulus. *Electroencephalogr. Clin. Neurophysiol.* **1994**, *93*, 409–416. [CrossRef]

22. Wilkins, D.E.; Hallett, M.; Wess, M.M. Audiogenic startle reflex of man and its relationship to startle syndromes. A review. *Brain* **1986**, *109*, 561–573. [CrossRef]

23. Nonnekes, J.; Scotti, A.; Oude Nijhuis, L.B.; Smulders, K.; Queralt, A.; Geurts, A.C.H.; Bloem, B.R.; Weerdesteyn, V. Are postural responses to backward and forward perturbations processed by different neural circuits? *Neuroscience* **2013**, *245*, 109–120. [CrossRef]

24. Mackinnon, C.D.; Bissig, D.; Chiusano, J.; Miller, E.; Rudnick, L.; Jager, C.; Zhang, Y.; Mille, M.L.; Rogers, M.W. Preparation of anticipatory postural adjustments prior to stepping. *J. Neurophysiol.* **2007**, *97*, 4368–4379. [CrossRef]

25. Reynolds, R.F.; Day, B.L. Fast visuomotor processing made faster by sound. *J. Physiol.* **2007**, *583*, 1107–1115. [CrossRef]

26. Van de Kamp, C.; Gawthrop, P.J.; Gollee, H.; Loram, I.D. Refractoriness in sustained visuo-motor control: Is the refractory duration intrinsic or does it depend on external system properties? *PLoS Comput. Biol.* **2013**, *9*, e1002843. [CrossRef]

27. Nonnekes, J.; Carpenter, M.G.; Inglis, J.T.; Duysens, J.; Weerdesteyn, V. What startles tell us about control of posture and gait. *Neurosci. Biobehav. Rev.* **2015**, *53*, 131–138. [CrossRef]

28. Prochazka, A. The fuzzy logic of visuomotor control. *Can J. Physiol. Pharm.* **1996**, *74*, 456–462. [CrossRef]

29. Misiaszek, J.E. Neural control of walking balance: IF falling THEN react ELSE continue. *Exerc. Sport Sci. Rev.* **2006**, *34*, 128–134. [CrossRef]

 *applied sciences*

Article

Characteristics of Postural Muscle Activity in Response to A Motor-Motor Task in Elderly

Yun-Ju Lee [1],*, Jing Nong Liang [2] and Yu-Tang Wen [1]

[1] Department of Industrial Engineering and Engineering Management, National Tsing Hua University, Hsinchu 30013, Taiwan; pwiadr@gmail.com
[2] Department of Physical Therapy, University of Nevada Las Vegas, Las Vegas, NV 89154, USA; jingnong.liang@unlv.edu
* Correspondence: yunjulee@ie.nthu.edu.tw; Tel.: +886-3-5742943

Received: 9 September 2019; Accepted: 10 October 2019; Published: 14 October 2019

Featured Application: Prioritization of the postural component in a motor-motor task suggests future individualized exercise programs be developed with different postural control strategies in the elderly.

Abstract: The purpose of the current study was to evaluate postural muscle performance of older adults in response to a combination of two motor tasks perturbations. Fifteen older participants were instructed to perform a pushing task as an upper limb perturbation while standing on a fixed or sliding board as a lower limb perturbation. Postural responses were characterized by onsets and magnitudes of muscle activities as well as onsets of segment movements. The sliding board did not affect the onset timing and sequence of muscle initiations and segment movements. However, significant large muscle activities of tibialis anterior and erector spinae were observed in the sliding condition ($p < 0.05$). The co-contraction values of the trunk and shank segments were significantly larger in the sliding condition through the studied periods ($p < 0.05$). Lastly, heavy pushing weight did not change the timing, magnitude, sequence of all studied parameters. Older adults enhanced postural stability by increasing the segment stiffness then started to handle two perturbations. In conclusion, they were able to deal with a dual motor-motor task after having secured their balance but could not make corresponding adjustments to the level of the perturbation difficulty.

Keywords: postural control strategy; muscle activity; dual-task; older adult; translation perturbation

---

1. Introduction

Aging-related changes in postural control are associated with balance maintenance regardless of different types of perturbations. In response to perturbations, the central nervous system (CNS) reacts to fast, directional arm movement or reaching by employing feedforward as anticipatory [1–3] and feedback mechanisms as compensatory postural adjustments [4,5] to maintain and restore equilibrium. Older adults accommodate both anticipatory and compensatory postural adjustments related to the perturbations compared to young adults [6–11]. Specifically, for a perturbation from upper limb movement, such as pushing an object, older adults utilize a less efficient strategy such as muscle co-activation [12]. Alternatively, for a translation perturbation from lower limb induced by standing on an unstable or moving surface, older adults employ corresponding postural adjustments with increasing magnitudes of muscle activities and center of pressure displacement [13,14].

Effects of a single perturbation on postural control have been widely studied and well documented. Meanwhile, studies of aging effects on dual-tasking commonly focus on combinations of a cognitive task with a motor task. These studies have shown declined performance in the elderly during the combination of cognitive and motor tasks [15–17]. When dual-tasking involved walking with a visual cue, older adults showed more difficulties in making corrective step adjustments [18]. However, activities of daily living involve multiple perturbations to balance, such as walking while holding the cellphone, or pushing a shopping cart while walking around in the supermarket. Both activities engage upper and lower limb movements involved in two motor tasks.

Postural control in response to perturbations from upper and lower limb movements simultaneously, such as pushing an object while standing on the sliding board, has been studied in young adults [19]. Our previous findings indicated that when the surface was movable, the onset times of tibialis anterior and rectus femoris were delayed and the magnitudes of muscle activation were decreased. In addition, the ventral muscles (tibialis anterior, rectus femoris, and rectus abdominis) and dorsal muscles (medial gastrocnemius, biceps femoris, and erector spinae) initiated before and after the pushing movement respectively, which suggested that the reciprocal muscle activation pattern was utilized. It also revealed that the CNS of young adults handling these dual-motor tasks as prioritizing upper limb perturbation, pushing movement [19,20] or griping reactions [21], along with maintaining vertical posture. Subsequently, their postural control responded to lower limb perturbations, such as sliding translation [19–21].

For young adults, the CNS prioritizes motor tasks over postural maintenance because their balance is not in imminent danger during the task performance [22,23]. Contrarily, the CNS prioritizes postural maintenance over motor tasks when one or two tasks involve postural control [24] or threat to balance [25]. However, the organization of postural control is not well understood when older adults perform a dual motor-motor task rather than a cognitive-motor task. Thus, the objective of the present study was to investigate how older adults handle a combination of upper limb activity and a translational perturbation, and how that affect characteristics of muscular strategies used in balance maintenance and restoration. The experimental paradigm involved two body perturbations: the upper extremities performing the pushing of a cart, and the translation perturbation of standing on the sliding board. Hence, we hypothesized that the onset time of muscle activities would be affected by the presence of these perturbations. Additionally, the second hypothesis was that older adults would utilize the strategy of postural muscle co-contraction to maintain vertical posture when exposed to upper and lower limb perturbations.

2. Materials and Methods

2.1. Participants

Fifteen older adults (11 females, 4 males, age = 65.93 ± 3.59 years, height = 1.57 ± 0.06 m, mass = 63.11 ± 6.74 kg) participated in the study. All participants did not suffer any musculoskeletal disorder and neurologic disease that could affect performing the experimental tasks. Furthermore, their Mini-Mental State Examination was 28.13 ± 1.68 points and the Berg's Balance Score was 55.87 ± 0.35 points. The project was approved by the National Tsing Hua University Institutional Review Board, and all participants provided written informed consent before taking part in the experimental procedures.

2.2. Procedure and Instrumentation

Participants were instructed to stand on a sliding board wearing a safety harness with their feet shoulder-width apart and in parallel, in front of a pushing cart (length 0.74 m, width 0.48 m, and height 0.30 m from the wheels) and push on its horizontal handle. The sliding board (length 0.5 m, width 0.5 m, and height 0.21 m) was made of two layers and had a lock mechanism allowing the top layer to either be free to slide in the anterior-posterior direction or remain stationary. A lightweight (5% body mass) or heavyweight (30% body mass) was placed on the pushing cart. Participants stood with their upper limb in elbow flexion and wrist extension at 90 degrees, and palms were slightly contacting the handle. The height of the pendulum was adjustable to match the subject's hand position (Figure 1).

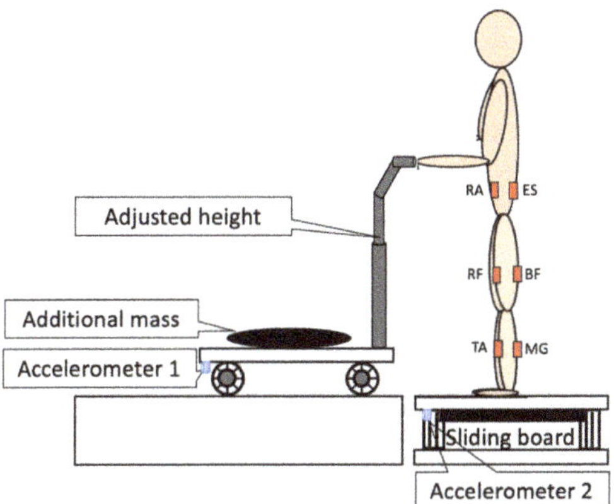

Figure 1. Schematic representation of the experimental setup. Additional mass: an additional 5% body weight or 30% body weight placed on the cart. Accelerometer 1: accelerometer attached to the cart, and Accelerometer 2: accelerometer attached to the sliding board. Red boxes represent the placements of electrodes. RA: rectus abdominis, ES: erector spinae, RF: rectus femoris, BF: biceps femoris, TA: tibialis anterior, and MG: medial gastrocnemius.

The instructions were that participants pushed the handle straight forward with both hands by using only trunk motion without wrist flexion and elbow extension as well as without taking a step or raising their heels from the surface of the board. The participants performed each trial in a self-paced manner after receiving the experimenter's command "push". Five trials were collected in each condition. Each participant was given two practice trials prior to data collection, to allow familiarization with the task. The condition of the secured sliding board would be referred to as the "fixed condition", while the condition of the free-moving sliding board would be referred to as the "sliding condition". The randomization of experimental conditions (two weights and two supporting-surface conditions) was applied.

Two Trigno IM Sensors were used as accelerometers in the experiment. The first accelerometer (Trigno IM Sensor, Delsys, INC., Natick, MA, USA) was attached to the pushing cart and was used to determine the moment of the cart pushed away (T_0). The second accelerometer (Trigno IM Sensor, Delsys, INC., Natick, MA, USA) was attached underneath the top layer of the board and was used to detect the moment of the board movement (bT_0).

The electrical activity of muscles (Electromyography, EMG) was recorded for the left side only due to the symmetric pushing task. EMG was obtained from the tibialis anterior (TA), medial gastrocnemius (MG), rectus femoris (RF), biceps femoris (BF), rectus abdominis (RA), and erector spinae (ES). Six Trigno IM Sensors (Delsys, INC., Natick, MA, USA) were used as EMG electrodes and attached to the muscle bellies after standard skin preparation procedures [26]. EMG signals were band-pass filtered (10–500 Hz) and amplified (gain 2000) by conducting in the Trigno^{TM} Wireless System (Delsys, INC., Natick, MA, USA). These Sensors were also included accelerometers and could represent movements of the trunk (from sensors on RA and ES), thigh (from sensors on RF and BF), and shank (from sensors on TA and MG) segments.

2.3. Data Processing

All data were processed offline using MATLAB software (MathWorks, Natick, MA, USA). The signals from the first and second accelerometers were used to determine the timing that the pendulum (T0) and the sliding board (bT0) started moving away. The onsets of the accelerometer signals were detected using the Teager-Kaiser onset time detection method [27,28]. The sensors on these segments also used the Teager-Kaiser method to identify the onset timing of segment movements.

All EMG data were high-pass filtered at 20 Hz, full-wave rectified, and low-pass filtered using linear envelope at 2 Hz (2nd order Butterworth) [26,29]. Subsequently, the Teager-Kaiser method was used to identify the onset of muscle activity for individual muscle (EMG_{onset}). The integrals of EMG activity of all studied muscles ($\int EMGs$) were calculated during the six epochs: (1) from -750 ms to -550 ms, (2) from -550 ms to -350 ms, (3) from -350 ms to -150 ms, (4) from -150 ms to $+50$ ms, (5) from $+50$ ms to $+250$ ms, and 6) from $+250$ ms to $+450$ ms in relation to T0. The 1–4 epochs were used to calculate components of feedforward postural adjustments and the 5–6 epochs for feedback postural adjustments [30,31]. Moreover, $\int EMGs$ of the baseline activity were obtained during a 200 ms time window at the beginning of the trial. Subtraction of $\int baseline$ was used to eliminate the effects of each muscle baseline activity. Values larger than zero ($\int EMG$-$\int baseline > 0$) were referred as activation of muscles and values smaller than zero ($\int EMG$-$\int baseline < 0$) as inhibition of muscles. Thus, the $\int EMG_{Epochs\ 1-6}$ were normalized by $\int EMGmax$ [26], which was the maximum value throughout all experimental trials for each muscle in each epoch and shown the example of $\int EMG_{Epoch\ 1}$ as:

$$\int baseline = \int_{0}^{200} EMG \tag{1}$$

$$\int EMG_{Epoch1} = \frac{\int_{T0-750}^{T0-550} EMG - \int baseline}{\int EMGmax} \tag{2}$$

Subsequently, the sums and differences between normalized $\int EMG$ values (Equation (2)) of RA and ES muscles for the trunk segment, RF and BF muscles for the thigh segment, and TA and MG muscles for the shank segment were calculated in Epochs 1–6, separately.

$$C = \int EMGventral + \int EMGdorsal \tag{3}$$

$$R = \int EMGventral - \int EMGdorsal \tag{4}$$

C indexes were calculated as the sum of $\int EMG$ of the antagonist-agonist muscle pairs to represent co-contraction and R indexes as the difference between $\int EMG$ in the muscle pairs to represent reciprocal activation [32]. Using the shank segment as an example, the C and R values in the Epoch 1 were calculated as:

$$C_{shank\ Epoch1} = \int TA_{Epoch1} + \int MG_{Epoch1} \tag{5}$$

$$R_{shank\ Epoch1} = \int TA_{Epoch1} - \int MG_{Epoch1} \tag{6}$$

The same C and R values were calculated for the thigh and trunk segments in the Epochs 2–6. All variables were calculated for each trial then averaged over five trials and presented with means and standard errors.

2.4. Statistics

Two-way repeated measures ANOVA were performed with two factors: board (2 levels: fixed and sliding) and weight (2 levels: 5% and 30% body mass) on EMG_{onset}, the onset timing of three segments movements, EMG integrals of $\int EMG_{Epochs\ 1-6}$ for individual muscles. Post hoc comparisons were performed using Tukey's Honestly Significant Difference test where statistically significant interactions were observed. For the analysis of C and R values, identification of either co-contraction (C) or reciprocal (R) activation pattern was done in the Epochs 1–6 for each segment. The EMG integrals of muscle coupling for the trunk, thigh, and shank segments were compared using C and R values by paired-sample *t*-tests. When ventral and dorsal muscles were activated (larger than the baseline), calculations from both positive values of coupling muscles revealed higher C value than R-value and vice versa. If R values were significantly larger than C values, this would indicate reciprocal activation [33]. Subsequently, if muscle coupling of the trunk, thigh, and shank segments had C value larger than R-value, two-way repeated measures ANOVA were performed with two factors: board (2 levels: fixed and sliding) and weight (2 levels: 5% and 30% body mass), to evaluate the C rather than R, and vice versa. Statistical significance was set at $p < 0.05$.

3. Results

Figure 2 illustrates the sequence of events in the conditions of standing on the sliding and fixed board while performing the pushing task. Both ventral and dorsal muscles were initiated before the T0 (0 ms) in both conditions. Two-way repeated measures ANOVA revealed that the onset of all postural muscles, except MG, was not affected by the factors of board and weight. All muscles were activated prior to the timing of pushing the cart away (T0). Thus, the RA onset was −577.20 ± 34.26 ms, the ES onset was −637.35 ± 56.21 ms, the RF onset was −698.71 ± 51.32, the BF onset was −684.96 ± 36.71 ms, the TA onset was −789.99 ± 56.22 ms, and the MG onset was −709.85 ± 35.93 ms, averaged across four conditions.

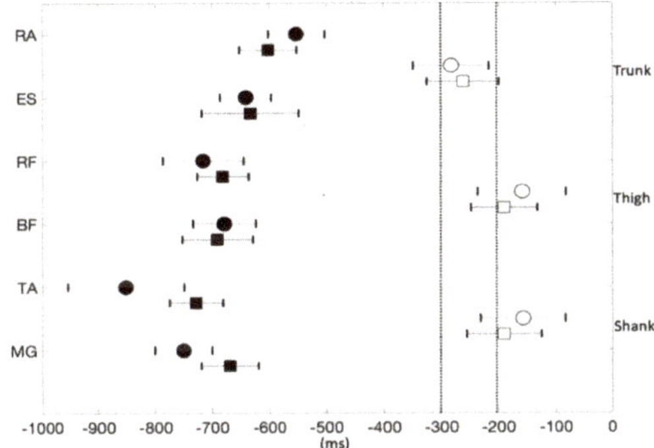

Figure 2. Onset times of muscle activation (black) for RA, ES, RF, BF, TA, and MG in the fixed condition (square) and in the sliding condition (circle). Onset times of the trunk, thigh, and shank segments movements (white) in the fixed condition (square) and in the sliding condition (circle). In addition, the dash vertical lines represent its mean ± standard error from −300.82 ms to −203.22 ms of the onset of the sliding board.

The onset timings of the trunk, thigh, and shank segment movements are shown in Figure 2. Three segments made movements before the T0 in both sliding and fixed conditions. Two-way repeated measures ANOVA revealed that the movement of segments was not affected by the factors of board and weight. In the sliding condition, the onset timing of the trunk segment movement was −282.85 ± 65.98 ms, similar to the timing of the board movement, followed by the thigh (−159.31 ± 77.21 ms) and shank (−157.36 ± 74.35 ms) segments.

Two-way repeated measures ANOVA revealed that the factor of the board only affected ES and TA in Epochs 1–6 (Table 1) and was higher in the sliding conditions than in the fixed conditions (Figure 3). In addition, ∫EMG was gradually decreased from the Epoch 1 to Epoch 6, particularly after the cart movement (Epoch 5 and Epoch 6). The C values were significantly larger than the R values and indicated co-contraction of muscles for the three segments in four conditions through Epoch 1 to Epoch 6 (Table 2). Subsequently, the C values of the trunk segment and the shank segment were significantly affected by the factor of board in Epochs 2–6 and in Epochs 1–5, respectively. In Epoch 1, the C value of the shank segment was significantly higher in the sliding condition (0.56 ± 0.05) than in the fixed condition (0.40 ± 0.06). In Epochs 2–5, the C values of the trunk segment were significantly higher in the sliding conditions (0.36 ± 0.04, 0.38 ± 0.06, 0.36 ± 0.05, 0.35 ± 0.05) than in the fixed conditions (0.27 ± 0.04, 0.25 ± 0.04, 0.24 ± 0.04, 0.24 ± 0.04). As well shown in Figure 4, the C values of the shank segment was significantly higher in the sliding conditions (0.54 ± 0.04, 0.54 ± 0.05, 0.50 ± 0.05, 0.46 ± 0.05) than in the fixed conditions (0.36 ± 0.06, 0.32 ± 0.05, 0.30 ± 0.05, 0.28 ± 0.04) in Epochs 2–5. In Epoch 6, the C value of the trunk segment was significantly higher in the sliding condition (0.33 ± 0.05) than in the fixed condition (0.21 ± 0.04). Finally, $\int EMG$ of all muscles were not significantly affected by the factor of weight in Epochs 1–6 (Table 1).

Table 1. Grand mean (mean of the combination of conditions) of normalized EMG integrals ($\% \int EMGmax$) for each epoch.

Muscle	Epoch 1	Epoch 2	Epoch 3	Epoch 4	Epoch 5	Epoch 6
RA	10.51 ± 2.51 [i]	12.68 ± 2.62	14.22 ± 2.62	13.77 ± 2.51	13.28 ± 3.14	13.15 ± 2.41
ES	20.18 ± 3.31 *	18.77 ± 3.27 *	16.97 ± 3.16 *	16.39 ± 3.22 *	16.23 ± 3.29 *	14.05 ± 3.21 *
RF	20.64 ± 4.07	21.10 ± 4.00	21.21 ± 4.17	19.02 ± 4.32	17.78 ± 4.50	16.82 ± 4.36
BF	24.45 ± 4.42	22.02 ± 4.06	19.82 ± 4.12	18.75 ± 4.10	18.44 ± 4.31	15.99 ± 4.30
TA	21.33 ± 3.52 *	21.02 ± 3.34 *	21.15 ± 3.11 *	19.82 ± 2.88 *	17.25 ± 2.56 *	15.22 ± 2.53 *
MG	26.55 ± 4.60	24.19 ± 4.34	21.60 ± 3.96	20.43 ± 3.68	19.00 ± 3.56	17.15 ± 3.52

* significant effect of the board, [i] significant effect of interaction between the factor of board and weight.

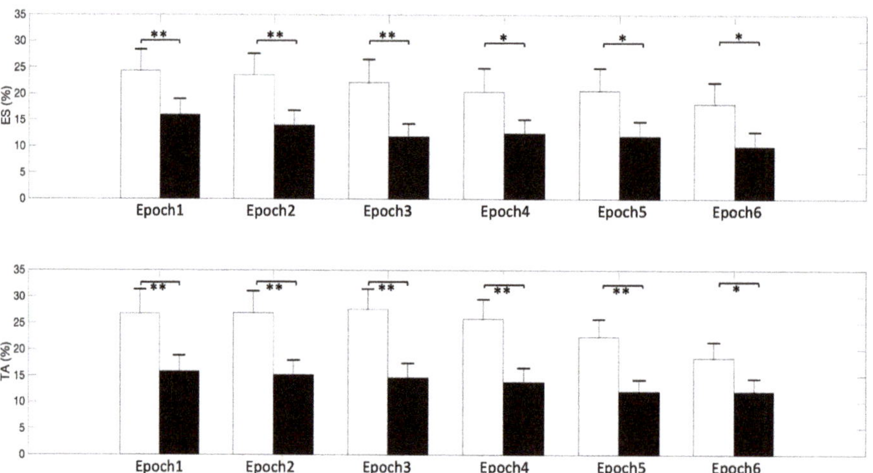

Figure 3. EMG integrals of ES and TA calculated for the sliding conditions (white) and fixed conditions (black) are shown for each epoch. * represents statistical significance ($p < 0.05$) of the factor of sliding board and ** for $p < 0.001$.

Table 2. The results of paired-samples *t*-test for the trunk, thigh, and shank segments in Epoch 1 to Epoch 6.

Segment	Epoch 1 *t* (59)	*p*	Epoch 2 *t* (59)	*p*	Epoch 3 *t* (59)	*p*	Epoch 4 *t* (59)	*p*	Epoch 5 *t* (59)	*p*	Epoch 6 *t* (59)	*p*
Trunk	−9.95	<0.001	−9.16	<0.001	−7.80	<0.001	−7.64	<0.001	−7.58	<0.001	−7.04	<0.001
Thigh	−8.66	<0.001	−8.74	<0.001	−7.80	<0.001	−7.78	<0.001	−7.46	<0.001	−6.71	<0.001
Shank	−9.76	<0.001	−9.15	<0.001	−8.22	<0.001	−8.12	<0.001	−8.19	<0.001	−7.94	<0.001

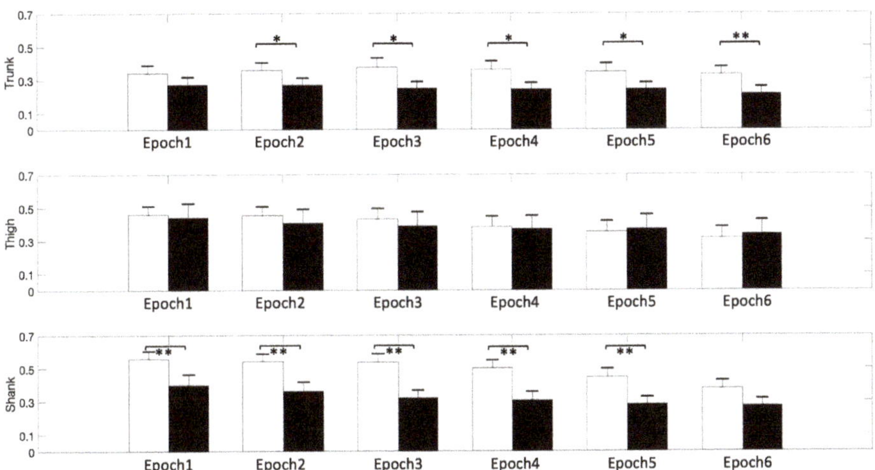

Figure 4. *C* values of the trunk, thigh, and shank segments for sliding conditions (white) and fixed conditions (black) are shown for each epoch. * represents statistical significance ($p < 0.05$) of the factor of sliding board and ** for $p < 0.001$.

4. Discussion

The current study investigated how older adults utilize postural control to push a cart when standing on a moveable sliding board (that induced translation perturbations) and the role of the weight of the cart is pushed. All these temporal events occurred prior to cart movement. The onset time of muscles and segment movements were not affected by standing on fixed or sliding board and pushing weight, which did not support our first hypothesis. The magnitudes of muscle activity in ES and TA were significantly higher in the sliding conditions compared to the fixed condition from −750 ms to +450 ms in relation to T0. Furthermore, muscle co-contraction of the three segments were observed through Epoch 1 to Epoch 6. The trunk and shank segments showed significantly higher *C* values in the sliding conditions than in the fixed conditions, which confirmed our second hypothesis.

The sequence of muscle activation onset was TA→MG→RFBF→ES→RA (Figure 2). The sequence of muscle activation observed in the elderly was very different from that observed in young adults (TA→RF→RA→ESRFMG) performing the same dual-task. When young adults were exposed to upper and lower limb perturbations, they initiated ventral muscles of all three segments, then the dorsal muscles [19]. This sequential activation of distal-to-proximal muscles is usually observed during translation perturbation from the lower limbs [34]. The earlier onset of either ventral or dorsal muscle activation depends on the direction of translation perturbation [20]. The function of this adjustment is to ensure that the body moves as an inverted pendulum to coordinate the reciprocal strategy [19,34], which was observed in young adults. Although older adults initiated the same distal-to-proximal sequence as young adults, the onsets of ventral and dorsal muscles of the three segments were very close and all occurred before the cart movement (Figure 2). Similar onset of ventral and dorsal muscle activation was associated with co-contraction strategy [35,36], which was also revealed by the *C–R* analysis in the current study.

Only ES and TA were significantly affected by the factor of board and larger magnitudes of muscle activities in the sliding conditions (Figure 3). All other muscles maintained similar magnitudes of muscle activities in all conditions. Hence, co-contraction patterns were observed in all three segments, and the corresponding *C* values were higher in the sliding conditions than in the fixed conditions on the trunk and shank segments (Figure 4). Older adults have shown to be utilizing less efficient postural control with the co-contraction strategy when performing the pushing task [12]. Together with the early onset time, co-contraction patterns indicated that older adults not only prepared for the pushing task but also enhanced the segment stiffness in preparation for potential translation perturbation [12,37,38]. Furthermore, the co-contraction strategy has been associated with increasing the available time to overcome the forthcoming perturbations [38,39] and for further motor responses [40]. Therefore, it might explain that older adults utilized co-contraction strategy much earlier than anticipatory postural adjustment attended to earn more time for postural stabilization. Additionally, the CNS increased significant muscle activation for the lower limb perturbation in the sliding condition.

In the current study, the trunk and board movements were observed with the comparable period in the sliding condition (Figure 2), which was different from young adults who coordinate shank segment for the moving surface while pushing [19]. Instead of the ankle strategy seen in young adults, older adults executed the trunk segment to counteract the board movement, followed by both shank and thigh segments. Young adults are capable of controlling the motor task and postural component when dealing with the dual-task perturbations. On the contrary, older adults initiated all postural muscles only to secure their posture first. The differences between young and older adults have also been reported in effects of visual movement during self-motion perception on postural control, in which young adults were able to adjust muscle activities, but older adults incorporated spatial conflicts, which further compromised their mobility performances [41–43]. Afterward, older adults could handle the upper (pushing) and significant extensive muscle activities for the lower limb (translational) perturbations. In addition, older adults did not adjust studied parameters for the different pushing weights, which was unlike the young adults [19,44]. It suggested that dealing with dual motor-motor tasking might still cost too much attention and could not allocate resources for additional postural adjustments [45]. Older adults exercised similar muscle activities regardless of the effects of pushing weight under the threat of dual motor-motor tasking. Perturbation-based balance training with single perturbation or dual cognitive-motor tasking has been conducted to reduce falls in elderly [46]. However, training effects may not be generalizable to handle functional activities with more than one motor task, which are commonly encountered in daily life. Specialized balance re-training paradigms involving dual motor-motor tasking could be an alternative approach, pending further investigation.

5. Conclusions

In older adults, when handling upper and lower limb perturbations, the chronological sequence of muscle onset time from distal-to-proximal segments is comparable to young adults, but older adults initiated ventral and dorsal muscles much earlier and almost simultaneously. Together with increased magnitudes of TA and ES muscle activity, older adults utilized co-contraction strategy on the trunk and shank segments when pushing while standing on the sliding board. In addition, older adults did not adjust corresponding muscle activities to different pushing weight. These results reveal that older adults increase the segment stiffness for the postural component to gain stabilization and prepare for the motor component. While it is capable of controlling the dual motor-motor task, it could not further adjust the magnitudes of muscle activities for the level of the primary motor task. For older adults, the co-contraction of postural muscles with similar magnitudes indicates that their balance was in imminent danger during the dual motor-motor task. The findings provide new insights for future studies focused on improving postural strategy while handling upper and lower extremity perturbations in the elderly. Furthermore, studying the combined effects of changes in the level of either motor task may contribute to establishing new rehabilitation approaches for improving motor control and resource allocation.

Author Contributions: Conceptualization, methodology, and writing-original draft preparation Y.-J.L.; formal analysis, Y.-T.W.; writing-review and editing, J.N.L.

Funding: This research was funded by the Young Scholar Fellowship Program by the Ministry of Science and Technology (MOST) in Taiwan, under Grant MOST-108-2636-E-007-002 and MOST-106-2314-B-007-005.

Acknowledgments: We thank the study participants for their exceptional cooperation.

Conflicts of Interest: The authors declare no conflict of interest. The funders had no role in the design of the study; in the collection, analyses, or interpretation of data; in the writing of the manuscript, or in the decision to publish the results.

References

1. Aruin, A.S.; Latash, M.L. The role of motor action in anticipatory postural adjustments studied with self-induced and externally triggered perturbations. *Exp. Brain Res.* **1995**, *106*, 291–300. [CrossRef] [PubMed]
2. Hodges, P.; Cresswell, A.; Thorstensson, A. Preparatory trunk motion accompanies rapid upper limb movement. *Exp. Brain Res.* **1999**, *124*, 69–79. [CrossRef] [PubMed]
3. Leonard, J.A.; Brown, R.H.; Stapley, P.J. Reaching to multiple targets when standing: The spatial organization of feedforward postural adjustments. *J. Neurophysiol.* **2009**, *101*, 2120–2133. [CrossRef]
4. Lowrey, C.R.; Nashed, J.Y.; Scott, S.H. Rapid and flexible whole body postural responses are evoked from perturbations to the upper limb during goal-directed reaching. *J. Neurophysiol.* **2017**, *117*, 1070–1083. [CrossRef]
5. Desmurget, M.; Grafton, S. Forward modeling allows feedback control for fast reaching movements. *Trends Cogn. Sci.* **2000**, *4*, 423–431. [CrossRef]
6. Carvalho, R.; Vasconcelos, O.; Goncalves, P.; Conceicao, F.; Vilas-Boas, J.P. The effects of physical activity in the anticipatory postural adjustments in elderly people. *Motor Control* **2010**, *14*, 371–379. [CrossRef]
7. Kubicki, A.; Bonnetblanc, F.; Petrement, G.; Ballay, Y.; Mourey, F. Delayed postural control during self-generated perturbations in the frail older adults. *Clin. Interv. Aging* **2012**, *7*, 65–75. [CrossRef]
8. Bleuse, S.; Cassim, F.; Blatt, J.L.; Labyt, E.; Derambure, P.; Guieu, J.D.; Defebvre, L. Effect of age on anticipatory postural adjustments in unilateral arm movement. *Gait Posture* **2006**, *24*, 203–210. [CrossRef]
9. Bugnariu, N.; Sveistrup, H. Age-related changes in postural responses to externally- and self-triggered continuous perturbations. *Arch. Gerontol. Geriatr.* **2006**, *42*, 73–89. [CrossRef]
10. Kanekar, N.; Aruin, A.S. The effect of aging on anticipatory postural control. *Exp. Brain Res.* **2014**, *232*, 1127–1136. [CrossRef]
11. Kanekar, N.; Aruin, A.S. Aging and balance control in response to external perturbations: Role of anticipatory and compensatory postural mechanisms. *Age* **2014**, *36*, 9621. [CrossRef] [PubMed]
12. Lee, Y.J.; Chen, B.; Aruin, A.S. Older adults utilize less efficient postural control when performing pushing task. *J. Electromyogr. Kinesiol.* **2015**, *25*, 966–972. [CrossRef]
13. Horak, F.B.; Nashner, L.M. Central programming of postural movements: Adaptation to altered support-surface configurations. *J. Neurophysiol.* **1986**, *55*, 1369–1381. [CrossRef]
14. Halicka, Z.; Lobotkova, J.; Bzduskova, D.; Hlavacka, F. Age-related changes in postural responses to backward platform translation. *Physiol. Res.* **2012**, *61*, 331–335.
15. Bergamin, M.; Gobbo, S.; Zanotto, T.; Sieverdes, J.C.; Alberton, C.L.; Zaccaria, M.; Ermolao, A. Influence of age on postural sway during different dual-task conditions. *Front. Aging Neurosci.* **2014**, *6*, 271. [CrossRef]
16. Boisgontier, M.P.; Beets, I.A.; Duysens, J.; Nieuwboer, A.; Krampe, R.T.; Swinnen, S.P. Age-related differences in attentional cost associated with postural dual tasks: Increased recruitment of generic cognitive resources in older adults. *Neurosci. Biobehav. Rev.* **2013**, *37*, 1824–1837. [CrossRef]
17. Li, K.Z.H.; Bherer, L.; Mirelman, A.; Maidan, I.; Hausdorff, J.M. Cognitive involvement in balance, gait and dual-tasking in aging: A focused review from a neuroscience of aging perspective. *Front. Neurol.* **2018**, *9*, 913. [CrossRef]
18. Mazaheri, M.; Hoogkamer, W.; Potocanac, Z.; Verschueren, S.; Roerdink, M.; Beek, P.J.; Peper, C.E.; Duysens, J. Effects of aging and dual tasking on step adjustments to perturbations in visually cued walking. *Exp. Brain Res.* **2015**, *233*, 3467–3474. [CrossRef]
19. Lee, Y.J.; Chen, B.; Liang, J.N.; Aruin, A.S. Control of vertical posture while standing on a sliding board and pushing an object. *Exp. Brain Res.* **2018**, *236*, 721–731. [CrossRef]
20. Dietz, V.; Kowalewski, R.; Nakazawa, K.; Colombo, G. Effects of changing stance conditions on anticipatory postural adjustment and reaction time to voluntary arm movement in humans. *J. Physiol.* **2000**, *524*, 617–627. [CrossRef]

21. Bateni, H.; Zecevic, A.; McIlroy, W.E.; Maki, B.E. Resolving conflicts in task demands during balance recovery: Does holding an object inhibit compensatory grasping? *Exp. Brain Res.* **2004**, *157*, 49–58. [CrossRef] [PubMed]

22. Chen, B.; Lee, Y.J.; Aruin, A.S. Control of grip force and vertical posture while holding an object and being perturbed. *Exp. Brain Res.* **2016**, *234*, 3193–3201. [CrossRef] [PubMed]

23. Mitra, S. Adaptive utilization of optical variables during postural and suprapostural dual-task performance: Comment on stoffregen, smart, bardy, and pagulayan (1999). *J. Exp. Psychol. Hum. Percept. Perform.* **2004**, *30*, 28–38. [CrossRef] [PubMed]

24. Muller, M.L.; Redfern, M.S.; Jennings, J.R. Postural prioritization defines the interaction between a reaction time task and postural perturbations. *Exp. Brain Res.* **2007**, *183*, 447–456. [CrossRef] [PubMed]

25. Shumway-Cook, A.; Woollacott, M.; Kerns, K.A.; Baldwin, M. The effects of two types of cognitive tasks on postural stability in older adults with and without a history of falls. *J. Gerontol. A Biol. Sci. Med. Sci.* **1997**, *52*, M232–M240. [CrossRef] [PubMed]

26. Konrad, P. *The ABC of EMG: A Practical Introduction to Kinesiological Electromyography*; Noraxon Inc.: Scottsdale, AZ, USA, 2005.

27. Li, X.; Zhou, P.; Aruin, A.S. Teager-kaiser energy operation of surface emg improves muscle activity onset detection. *Ann. Biomed. Eng.* **2007**, *35*, 1532–1538. [CrossRef]

28. Solnik, S.; Rider, P.; Steinweg, K.; DeVita, P.; Hortobagyi, T. Teager-kaiser energy operator signal conditioning improves emg onset detection. *Eur. J. Appl. Physiol.* **2010**, *110*, 489–498. [CrossRef]

29. Lee, Y.J.; Hoozemans, M.J.; van Dieen, J.H. Oblique abdominal muscle activity in response to external perturbations when pushing a cart. *J. Biomech.* **2010**, *43*, 1364–1372. [CrossRef]

30. Krishnan, V.; Latash, M.L.; Aruin, A.S. Early and late components of feed-forward postural adjustments to predictable perturbations. *Clin. Neurophysiol.* **2012**, *123*, 1016–1026. [CrossRef]

31. Lee, Y.J.; Aruin, A.S. Three components of postural control associated with pushing in symmetrical and asymmetrical stance. *Exp. Brain Res.* **2013**, *228*, 341–351. [CrossRef]

32. Slijper, H.; Latash, M.L. The effects of muscle vibration on anticipatory postural adjustments. *Brain Res.* **2004**, *1015*, 57–72. [CrossRef] [PubMed]

33. Chen, B.; Lee, Y.J.; Aruin, A.S. Anticipatory and compensatory postural adjustments in conditions of body asymmetry induced by holding an object. *Exp. Brain Res.* **2015**, *233*, 3087–3096. [CrossRef] [PubMed]

34. Vieira, T.M.; Farina, D.; Loram, I.D. Emg and posture in its narrowest sense. In *Surface Electromyography: Physiology, Engineering, and Allications*; Merletti, R., Farina, D., Eds.; Wiley: Hoboken, NJ, USA, 2016.

35. Krishnan, V.; Aruin, A.S.; Latash, M.L. Two stages and three components of the postural preparation to action. *Exp. Brain Res.* **2011**, *212*, 47–63. [CrossRef] [PubMed]

36. Latash, M.L.; Levin, M.F.; Scholz, J.P.; Schoner, G. Motor control theories and their applications. *Medicina* **2010**, *46*, 382–392. [CrossRef] [PubMed]

37. Chen, B.; Lee, Y.J.; Aruin, A.S. Role of point of application of perturbation in control of vertical posture. *Exp. Brain Res.* **2017**, *235*, 3449–3457. [CrossRef] [PubMed]

38. Kim, D.; Hwang, J.M. The center of pressure and ankle muscle co-contraction in response to anterior-posterior perturbations. *PLoS ONE* **2018**, *13*, e0207667. [CrossRef] [PubMed]

39. Finley, J.M.; Dhaher, Y.Y.; Perreault, E.J. Contributions of feed-forward and feedback strategies at the human ankle during control of unstable loads. *Exp. Brain Res.* **2012**, *217*, 53–66. [CrossRef]

40. Craig, C.E.; Goble, D.J.; Doumas, M. Proprioceptive acuity predicts muscle co-contraction of the tibialis anterior and gastrocnemius medialis in older adults' dynamic postural control. *Neuroscience* **2016**, *322*, 251–261. [CrossRef]

41. Piras, A.; Raffi, M.; Perazzolo, M.; Squatrito, S. Influence of heading perception in the control of posture. *J. Electromyogr. Kinesiol.* **2018**, *39*, 89–94. [CrossRef]

42. Raffi, M.; Piras, A.; Persiani, M.; Perazzolo, M.; Squatrito, S. Angle of gaze and optic flow direction modulate body sway. *J. Electromyogr. Kinesiol.* **2017**, *35*, 61–68. [CrossRef]

43. Ramkhalawansingh, R.; Butler, J.S.; Campos, J.L. Visual-vestibular integration during self-motion perception in younger and older adults. *Psychol. Aging* **2018**, *33*, 798–813. [CrossRef] [PubMed]

44. Argubi-Wollesen, A.; Wollesen, B.; Leitner, M.; Mattes, K. Human body mechanics of pushing and pulling: Analyzing the factors of task-related strain on the musculoskeletal system. *Saf. Health Work* **2017**, *8*, 11–18. [CrossRef] [PubMed]

45. Woollacott, M.; Shumway-Cook, A. Attention and the control of posture and gait: A review of an emerging area of research. *Gait Posture* **2002**, *16*, 1–14. [CrossRef]
46. Gerards, M.H.G.; McCrum, C.; Mansfield, A.; Meijer, K. Perturbation-based balance training for falls reduction among older adults: Current evidence and implications for clinical practice. *Geriatr. Gerontol. Int.* **2017**, *17*, 2294–2303. [CrossRef] [PubMed]

Article

Analysis of Three-Dimensional Circular Tracking Movements Based on Temporo-Spatial Parameters in Polar Coordinates

Woong Choi [1,*], Jongho Lee [2,*] and Liang Li [3]

[1] Department of Information and Computer Engineering, National Institute of Technology, Gunma College, Maebashi 371-8530, Japan
[2] Department of Clinical Engineering, Komatsu University, Komatsu 923-8511, Japan
[3] College of Information Science and Engineering, Ritsumeikan University, Kusatsu 525-8577, Japan; liliang@fc.ritsumei.ac.jp
[*] Correspondence: wchoi@ice.gunma-ct.ac.jp (W.C.); jongho.lee@komatsu-u.ac.jp (J.L.)

Received: 27 September 2019; Accepted: 10 January 2020; Published: 15 January 2020

Abstract: Motor control characteristics of the human visuomotor control system need to be analyzed in the three-dimensional (3D) space to study and imitate human movements. In this paper, we examined circular tracking movements on two planes in 3D space from a motor control perspective based on three temporospatial parameters in polar coordinates. Sixteen healthy human subjects participated in this study and performed circular target tracking movements rotating at 0.125, 0.25, 0.5, and 0.75 Hz in the frontal or sagittal planes in three-dimensional space. The results showed that two temporal parameter errors on each plane were proportional to the change in the target velocity. Furthermore, frontal plane circular tracking errors without depth for a spatial parameter were lower than those for sagittal plane circular tracking with depth. The experimental protocol and data analysis allowed us to analyze the motor control characteristics temporospatially for circular tracking movement with various depths and speeds in the 3D VR space.

Keywords: circular tracking movement; motor control; three-dimensional virtual space; polar coordinates; temporo-spatial parameters

---

1. Introduction

We studied and imitated human motion control mechanisms using a visually guided motor control system in the three-dimensional (3D) space. To understand the visually guided motor control of humans accurately, we need to analyze visually guided tracking movements in 3D space directly and analyze important 3D space motor control characteristics such as depth perception. This is vital for workspace distance determination and motion establishment. However, in previous target-tracking movement studies, no technology or system has provided an accurate visual target representation [1–10].

Recently, we developed a 3D visuomotor control evaluation system in a virtual reality (VR) environment [11]. This system enabled the analysis of VR space circular tracking movement to the millimeter level accuracy. We compared 3D circular tracking movement visuomotor control between monocular and binocular vision. As reported by Melmoth et al. [12], reaching and grasping movement accuracy in binocular vision had a 2.5 to 3.0 times advantage over monocular vision in the real environment. Our previous study obtained a similar result, where the circular tracking movement accuracy in binocular vision showed approximately 4.5 times the advantage over monocular vision on both frontal and sagittal planes in a 3D VR environment.

Depth estimation is an equally significant control metric for both motor control and depth perception in real and virtual 3D spaces. We analyzed two circular target-tracking movement types on the frontal and sagittal planes (relative to the subject). However, only the Cartesian coordinate spatial error was used to analyze the tracking movement accuracy in our previous study. The temporospatial relationship between the tracer and target should be investigated for a comprehensive study of the motor control characteristics of circular tracking movements. In this study, we redesigned the experiments to evaluate the impact of various target speeds.

Three polar coordinate kinematic parameters (ΔR, $\Delta\theta$, and $\Delta\omega$) are widely used in circular tracking movement analysis [13,14]. Our previous study analyzed the motor control and impact of speed and visual target feedback in 2D tracking movements based on these parameters [14]. In other words, ΔR allows us to examine spatial motor control characteristics in polar coordinates as a circular movement performance evaluation. However, $\Delta\theta$ and $\Delta\omega$ allow us to analyze motor control temporal characteristics in polar coordinates as a circular tracking movement evaluation of the positional and velocity-control precision. This inspired us to examine two planar circular tracking movements in 3D space from a motor control perspective based on temporospatial parameters.

Therefore, in this study, we analyzed the motor control characteristics of circular tracking movements in 3D space using three kinematic parameters in polar coordinates: difference in fixed pole distance, ΔR; position angle difference, $\Delta\theta$; and angular velocity difference, $\Delta\omega$. We investigated the parameter differences between circular tracking movements relative to the subject on the frontal and sagittal planes. We also examined parameter-based changes in motor control for four different target speeds.

2. Materials and Methods

2.1. Subjects and Experimental Setup

The subjects were 16 males, with a mean age of 20.1 ± 0.62 years (see Table 1). Three subjects were left-handed, and 13 subjects were right-handed. Fifteen subjects were right-eye dominant, and one subject was left-eye dominant. All had normal or corrected-to-normal vision. None had previously participated in similar studies. All subjects gave written informed consent before their participation. All experiments were conducted in accordance with relevant guidelines and regulations. The protocol was approved by the ethics committee of the National Institute of Technology, Gunma College.

Table 1. Characteristics of 16 participants.

ID	Age	Sex	Dominant Hand	Dominant Eye
1	20	M	L	R
2	20	M	R	R
3	20	M	R	R
4	20	M	R	L
5	20	M	R	R
6	20	M	R	R
7	19	M	R	R
8	21	M	R	R
9	21	M	R	R
10	19	M	R	R
11	21	M	R	R
12	20	M	R	R
13	20	M	R	R
14	21	M	R	R
15	20	M	L	R
16	20	M	L	R

The subjects were asked to perform a visually guided tracking task in a 3D VR environment [11], which involved tracking a target with a tracer (see Figure 1). The subjects held a hand-held controller of the HTC Vive HMD to move a tracer in the 3D VR space. The subjects used the tracer (visualized as a yellow ball) to track the target (visualized as a red ball) moving circularly in the clockwise direction. The green lines indicate the target path in the 3D VR space. The two graphs in the second trace show the target path as seen from the front (left) and side (center) from the subject's viewpoint. Insets in the second trace of (A) and (B) show how three outcome measures (ΔR, $\Delta \theta$, and $\Delta \omega$) were derived from the target (or the tracer) path data for each plane. The three lower graphs show a typical trial of the target path (green line) and the tracer path (black line) for each axis versus time. The target path was not displayed to the subjects during the experiment. The target was a virtual red ball with a radius of 1.5 cm. Instead of their own hands or HMD's controller, the subjects perceived a 20-cm-long virtual stick. The controller direction was synchronized with the virtual stick. Circular tracking was performed without the 3D hand and arm displayed in VR in this study; therefore, the virtual stick with the virtual tracer presents the hand position and direction information. The tracer was a virtual yellow ball of 1 cm radius, placed at the tip of the stick. The tracer position was synchronized with the subject's hand movements. In the experiment, the target moved continuously along an invisible circular orbit with a 15 cm radius. The rotational axis was set to two orientations based on experimental requirements.

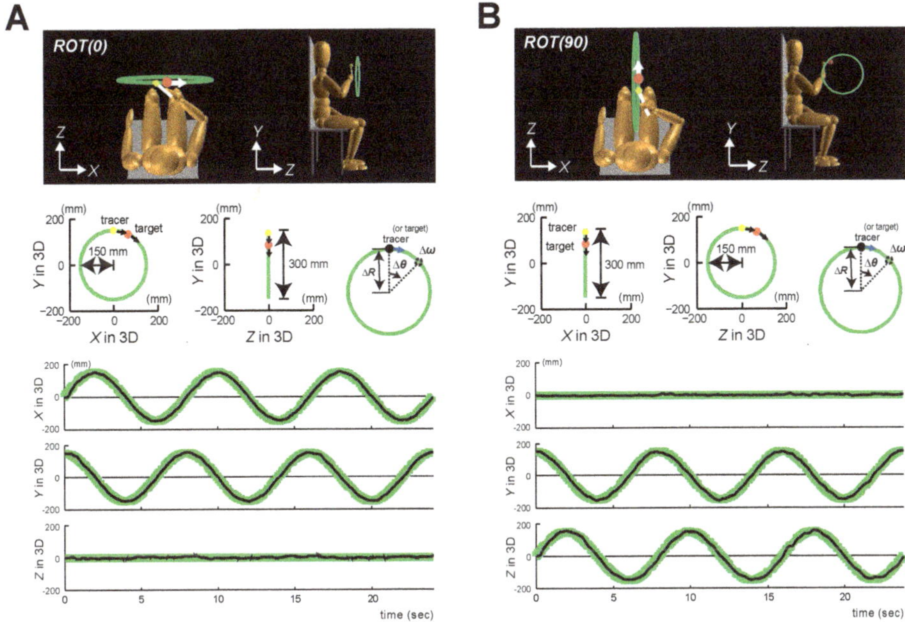

Figure 1. Experimental procedure. (**A**) The circular tracking experiment for the body's frontal plane (*ROT*(0)). (**B**) The circular tracking experiment for the body's sagittal plane (*ROT*(90)). For both (**A**) and (**B**), the top inset illustrates the circular tracking movement on each plane of the body in VR 3D space.

2.2. Movement Task

For this study, we performed an experiment to evaluate 3D visuomotor control quantitatively, using circular tracking movements for the frontal and sagittal planes relative to the subject in the VR space (see Figure 1). The subjects were seated on a chair built for the experiment and wore an HMD. Prior to the experiment, we orally confirmed that each subject could properly perceive stereoscopic vision. Subjects were asked to hold the physical controller in their dominant hand. We ran a calibration to locate the target's initial position optimized for the subject's arm length and height to minimize the different anthropometric parameter impacts on experimental results. The target rotated at 0.125,

0.25, 0.5, or 0.75 Hz along the orbit after a 3 s countdown with a sound effect. The subjects were asked to move the tracer to the target position during the countdown and then perform a circular tracking movement. The target stopped after three loops. At the end of each trial, the target stopped for 1 s and played a sound. Four trials each were performed with the target rotating in the frontal (*ROT*(0) in Figure 1A) and sagittal planes (*ROT*(90) in Figure 1B). The first trial for each plane was a practice run and was excluded from the analysis. Therefore, 32 experimental trials were performed (4 trials × 4 speeds × 2 planes) for each subject. To avoid a subject learning effect, the experiment was performed with random counterbalance.

2.3. Data Analysis

For data analysis, we transformed the Cartesian (X, Y, Z) data to radial displacement, angular displacement, and angular velocity on polar coordinates; R, θ, and ω, respectively.

$$\Delta R \ [\text{mm}] = \frac{\sum_{t=1}^{n} abs\left(R_{tracer}(t) - R_{target}(t)\right)}{n}, \tag{1}$$

where ΔR is defined as the radial position difference absolute value between the target and tracer from the origin.

$$\Delta \theta \ [\text{deg}] = \frac{\sum_{t=1}^{n} abs\left(\theta_{tracer}(t) - \theta_{target}(t)\right)}{n}, \tag{2}$$

where $\Delta \theta$ represents the angular displacement difference absolute value between the target and tracer.

$$\Delta \omega \left[\frac{\text{deg}}{\text{s}}\right] = \frac{\sum_{t=1}^{n} abs\left(\omega_{tracer}(t) - \omega_{target}(t)\right)}{n}, \tag{3}$$

where $\Delta \omega$ denotes the angular velocity difference absolute value between the target and tracer. These parameters were normalized with total time n of three trials for each target speed.

The units of ΔR, $\Delta \theta$, and $\Delta \omega$ are mm, deg, deg/s, respectively. We calculated the absolute difference averages of ΔR, $\Delta \theta$, and $\Delta \omega$ on three trials for each subject. Next, the mean (M) and standard deviation (SD) were calculated by using the ΔR, $\Delta \theta$, and $\Delta \omega$ averages for the 16 subjects.

Statistical analysis and data visualization were performed by SPSS Statistics V26, IBM and MATLAB, MathWorks. In general, Cronbach's α indicates the overall data reliability; it is understood that a value around 0.8 is respectable [15]. The reliability analysis for ΔR, $\Delta \theta$, and $\Delta \omega$ data was measured by Cronbach's α (Reliability Analysis function in SPSS Statistics, IBM), with values of 0.87, 0.82, and 0.84, respectively.

This study verified a relationship between the target speed and depth in 3D target-tracking movements. Therefore, we investigated the parameter differences of ΔR, $\Delta \theta$, and $\Delta \omega$ between frontal and sagittal plane circular tracking movements. In addition, we examined the visuomotor control differences for four different parameter-based target speeds. For the analysis of circular tracking movement differences based on ΔR, $\Delta \theta$, and $\Delta \omega$, we carried out a two-way repeated-measures analysis of variance (ANOVA), with a two-level plane factor (*ROT*(0), frontal plane; and *ROT*(90), sagittal plane) and a four-level speed factor (*V1*: 0.125 Hz, *V2*: 0.25 Hz, *V3*: 0.5 Hz, and *V4*: 0.75 Hz; each with $n = 16$). The total sample size was 128, calculated by a priori power analysis [16]. The main impacts and interaction of plane and speed factors in ΔR, $\Delta \theta$, and $\Delta \omega$ parameters were assessed by the Repeated Measures function in SPSS Statistics, IBM. We performed Mauchly's sphericity test to validate the ANOVA results. When sphericity was assumed ($p > 0.05$), the values corrected with Sphericity Assumed were used. When sphericity was not assumed ($p < 0.05$), the values corrected with Greenhouse–Geisser were used.

A posthoc test was conducted by pairwise comparison of the Bonferroni corrections. Except where noted, we describe data using M and SD; considering comparisons yielding $p < 0.05$ as statistically significant and those yielding $p < 0.01$ as highly significant differences. This analysis corresponds to the statistical tests shown in Tables S1–S3.

2.4. Power Analysis

We conducted the priori power analysis using G*Power software 3.1.9.4, determining the minimum required sample size in ANOVA: Repeated measures, within-between interaction [16]. With effect size = 0.25, alpha = 0.05, number of groups = 8, power = 0.80, number of measurements = 4, and nonsphericity correction = 1, the analysis showed that the sixteen participants were required to detect an actual power of 0.8.

Furthermore, to estimate the power of the two-way repeated-measures ANOVA used in this research, we conducted a post hoc power analysis using G*Power software 3.1.9.4. As shown in Item A in Tables S1–S3, the power values were approximately 1.0. We can be confident that the 16-participant sample size achieves enough power to detect the main effects and interactions.

3. Results

The following sections analyze motor control characteristics of circular tracking movements in 3D space based on each of the three polar coordinate parameters.

3.1. Differences in Circular Tracking Movement Based on ΔR in 3D VR Space

Figure 2 shows typical circular tracking movement examples at four target speeds (*V1*: 0.125 Hz, *V2*: 0.25 Hz, *V3*: 0.5 Hz, and *V4*: 0.75 Hz; each with $n = 16$). Figure 2A1,A2 show the circular tracking movement ΔR on the frontal and sagittal planes at each target speed. For the circular tracking movement on both the frontal and sagittal planes, the difference between target trajectories and the tracer tended to increase as the target speed increased. Further, there is a higher sagittal plane difference compared to the frontal plane at each target speed, as shown in Figure 2A1,A2. Hence, we examined the 3D space circular movement at the four target speeds using ΔR.

Two-way repeated-measures ANOVA on the ΔR performance differences revealed significant key impacts for the plane, $F(1,15) = 47.53$, $p = 0$, *partial* $\eta^2 = 0.76$, and speed, $F(1.43,21.46) = 76.15$, $p = 0$, *partial* $\eta^2 = 0.835$, as well as an interaction between the plane and speed factors, $F(3,45) = 7.4$, $p = 0$, *partial* $\eta^2 = 0.33$ (item A in Table S1). This indicates that the plane and speed factors affected the circular tracking movement ΔR. Further, the interaction between the target speed and depth in 3D target-tracking movements would affect the ΔR performance in estimating the spatial motor control characteristics in the task.

Therefore, we performed a pairwise comparison to analyze the speed factor between *ROT*(0) and *ROT*(90). As shown in Figure 3A, there was a statistically significant ΔR difference between *ROT*(0) and *ROT*(90) (item B in Table S1). The results show that the subjects found it particularly difficult to track the sagittal plane target radius ($M = 11.77$ mm, $SD = 5.2$ mm) than that on the frontal plane ($M = 8.93$ mm, $SD = 3.90$ mm) at target speeds of 0.25 Hz and higher ($p = 0$, $r = 0.656$; item C in Table S1).

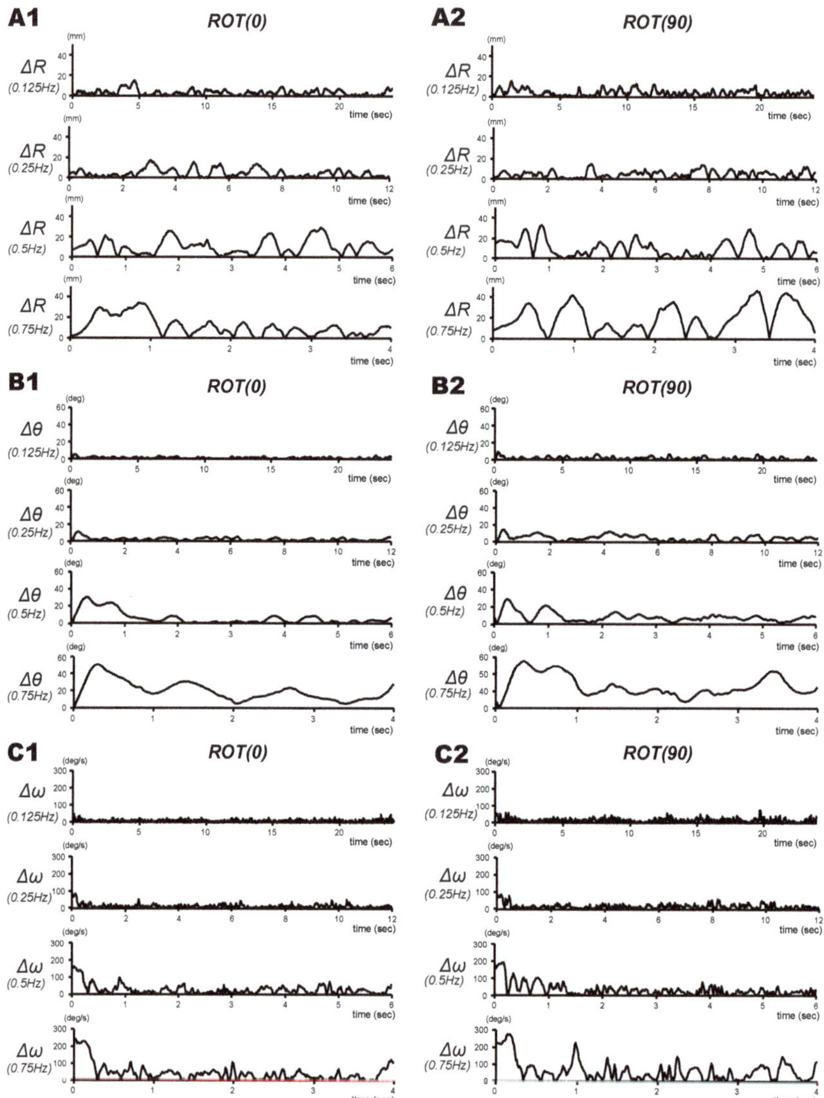

Figure 2. Typical examples of circular tracking movements for 0.125 Hz, 0.25 Hz, 0.5 Hz, and 0.75 Hz. (**A1**) Absolute values of ΔR for the frontal plane ($ROT(0)$) and (**A2**) the sagittal plane ($ROT(90)$). (**B1**) Absolute values of $\Delta\theta$ for $ROT(0)$ and (**B2**) $ROT(90)$. (**C1**) Absolute values of $\Delta\omega$ for $ROT(0)$ and (**C2**) $ROT(90)$.

Next, we examined the ΔR circular tracking movement on each plane at the four target speeds. Figure 3B shows that the pairwise comparison had a significant ΔR difference for $ROT(0)$ phase target speeds (item D in Table S1). The values of ΔR were 6.01 ± 3.53, 6.95 ± 3.24, 9.67 ± 1.72, and 13.11 ± 2.36 mm for 0.125, 0.25, 0.5, and 0.75 Hz, respectively. This indicates that ΔR increased significantly on the frontal plane ($ROT(0)$) as the target speed increased.

As shown in Figure 3C, there was a significant ΔR difference in target speeds for the $ROT(90)$ phase (item E in Table S1). The values of ΔR were 7.35 ± 3.23, 8.62 ± 2.54, 13.25 ± 2.45, and 17.86 ± 4.19 mm for 0.125, 0.25, 0.5, and 0.75 Hz, respectively. No significant ΔR difference between *V1* and *V2* was noted on the sagittal plane (*ROT*(90)) ($t(15) = 2.143$, $p = 0.29$, $r = 0.484$). This suggests that the subjects found it more difficult to track the sagittal plane (*ROT*(90)) target radius for speeds over 0.25 Hz.

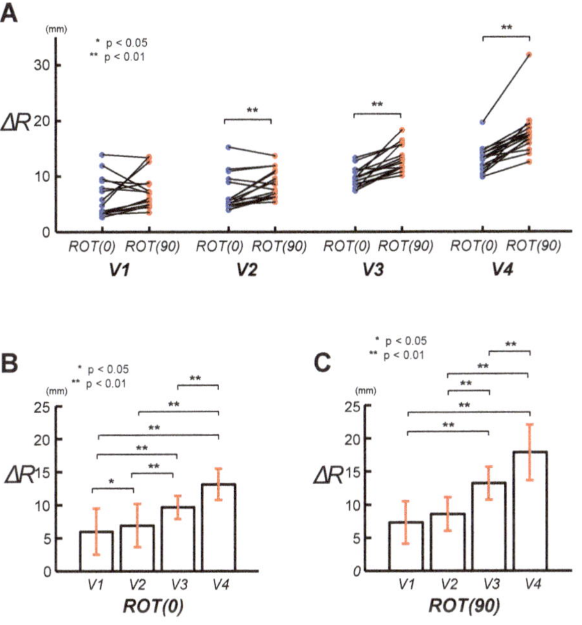

Figure 3. Evaluation of the circular tracking movement based on ΔR in 3D VR space. (**A**) The graphs indicate the pairwise comparison of ΔR analyzing the speed factor between *ROT*(0) and *ROT*(90). (**B**) The result of the pairwise comparison was indicated for ΔR, on the frontal plane (*ROT*(0)), at four target speeds. (**C**) The pairwise comparison was displayed for ΔR, on the sagittal plane (*ROT*(90)), at four target speeds.

We found that the circular tracking movement that maintains a constant distance from the circle center can be more accurately tracked on the frontal plane (*ROT*(0)) compared with that on the sagittal plane (*ROT*(90)).

3.2. Circular Tracking Movement Differences Based on $\Delta\theta$ in 3D VR Space

Figure 2B1,B2 show $\Delta\theta$ at the four target speeds on the frontal and sagittal planes, respectively. First, we compared $\Delta\theta$ between the frontal and sagittal planes at each target speed to investigate the impact of depth (distance from the subject) on $\Delta\theta$. Two-way repeated-measures ANOVA on $\Delta\theta$ performance-based differences revealed significant main impacts for the plane, $F(1,15) = 156.89$, $p = 0$, *partial* $\eta^2 = 0.913$, and speed, $F(1.07,16.12) = 64$, $p = 0$, *partial* $\eta^2 = 0.81$, as well as plane and speed interaction factors, $F(1.542,23.13) = 6.72$, $p = 0.008$, *partial* $\eta^2 = 0.309$ (item A in Table S2). This shows that the plane and speed factors affected $\Delta\theta$ in circular tracking movements. Also, the interaction between the frontal and sagittal planes at each target speed affected the $\Delta\theta$ performance in evaluating the precision of circular tracking movement position control.

Therefore, we performed a pairwise comparison analyzing the speed factors of *ROT*(0) and *ROT*(90). As shown in Figure 4A, there was a statistically significant difference in $\Delta\theta$ between *ROT*(0) and *ROT*(90) (item B in Table S2). Unlike ΔR, $\Delta\theta$ has a significant difference with respect to circular tracking accuracy on *ROT*(0) and *ROT*(90) over 0.125 Hz. This indicates that the subjects found it more difficult to synchronize the target and tracer positions on the sagittal plane ($M = 7.48°$, $SD = 5.81°$) than on the frontal plane ($M = 5.52°$, $SD = 4.44°$) at all target speeds ($p = 0$, $r = 0.845$; item C in Table S2).

Next, we examined the relationship between $\Delta\theta$ and target speed for each plane. Figure 4B shows that the pairwise comparison had a significant $\Delta\theta$ difference in target speeds for the $ROT(0)$ phase (item D in Table S2). The values of $\Delta\theta$ were $1.95 \pm 0.74°$, $2.79 \pm 0.85°$, $5.84 \pm 1.91°$, and $11.51 \pm 4.26°$ for 0.125, 0.25, 0.5, and 0.75 Hz, respectively. As is the case with ΔR, $\Delta\theta$ increased significantly on the frontal plane ($ROT(0)$) as the target speed increased. This indicates significant difficulty in circular tracking movement position control on the frontal plane ($ROT(0)$) as the speed increases.

Next, as shown in Figure 4C, there was a significant $\Delta\theta$ difference in target speeds for the $ROT(90)$ phase (item E in Table S2). The values of $\Delta\theta$ were $2.95 \pm 0.95°$, $4.12 \pm 1.2°$, $7.93 \pm 2.19°$, and $14.91 \pm 6.46°$ for 0.125, 0.25, 0.5, and 0.75 Hz, respectively. Also, with increased target speed, $\Delta\theta$ increased significantly on the sagittal plane ($ROT(90)$).

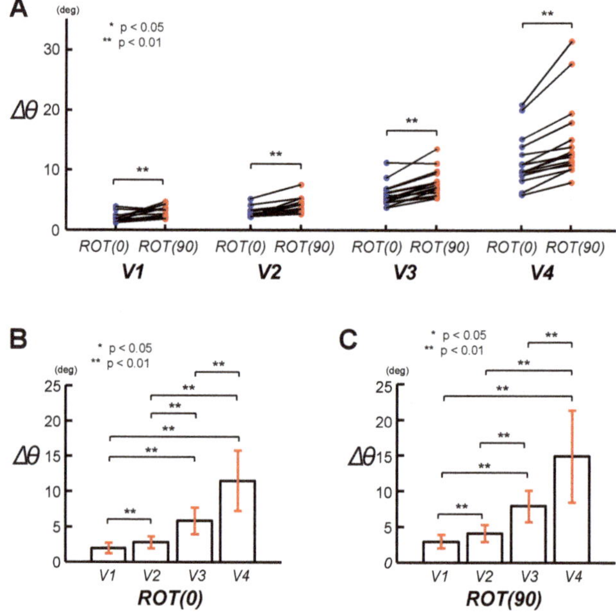

Figure 4. Evaluation of the circular tracking movement based on $\Delta\theta$ in 3D VR space. (**A**) The graphs indicate the pairwise comparison of $\Delta\theta$ analyzing the speed factor between $ROT(0)$ and $ROT(90)$. (**B**) The result of the pairwise comparison was indicated for $\Delta\theta$, on the frontal plane ($ROT(0)$), at four target speeds. (**C**) The pairwise comparison was displayed for $\Delta\theta$, on the sagittal plane ($ROT(90)$), at four target speeds.

The results indicate that as the speed increased, the subjects had substantially more difficulty with circular tracking movement position control on the sagittal plane ($ROT(90)$) compared with the frontal plane ($ROT(0)$).

3.3. Circular Tracking Movement Differences Based on $\Delta\omega$ in 3D VR Space

Figure 2C1,C2 show $\Delta\omega$ on the frontal and sagittal planes, respectively, at each of the four target speeds.

We compared $\Delta\omega$ between the frontal and sagittal planes at each target speed to investigate the impact of depth on velocity control accuracy. Two-way repeated-measures ANOVA on the performance differences based on $\Delta\omega$ revealed significant core impacts for the plane, $F(1,15) = 171.36$, $p = 0$, *partial*

$\eta^2 = 0.92$, and speed, $F(1.07,16.03) = 252.33$, $p = 0$, *partial* $\eta^2 = 0.944$, as well as an interaction between the plane and speed factors, $F(1.258,18.87) = 38.1$, $p = 0$, *partial* $\eta^2 = 0.717$ (item A in Table S3). This indicates that $\Delta\omega$ in the circular tracking was affected by the plane and speed factors. Further, the interaction between the target speed and depth in 3D target-tracking movements would affect the $\Delta\omega$ performance in evaluating the velocity-control precision in circular tracking movements.

As shown in Figure 5A, the pairwise comparison of the plane and speed factors revealed a statistically significant $\Delta\omega$ difference between $ROT(0)$ and $ROT(90)$) (item B in Table S3). This indicates that the subjects were more precise on the frontal plane ($M = 27.53 \degree \text{ s}^{-1}$, $SD = 18.27 \degree \text{ s}^{-1}$) than on the sagittal plane ($M = 37.17 \degree \text{ s}^{-1}$, $SD = 26.24 \degree \text{ s}^{-1}$) when synchronizing the target and tracer angular velocities ($p = 0$, $r = 0.855$; item C in Table S3).

Next, we examined the $\Delta\omega$ and target speed relationship for both planes using pairwise comparison. As shown in Figure 5B, there was a significant $\Delta\omega$ difference in target speeds for the $ROT(0)$ phase (item D in Table S3). The values of $\Delta\omega$ were 9.65 ± 1.61, 15.17 ± 2.05, 31.23 ± 4.8, and $54.07 \pm 10.13 \degree \text{ s}^{-1}$ for 0.125, 0.25, 0.5, and 0.75 Hz, respectively. $\Delta\omega$ increased significantly on the frontal plane ($ROT(0)$) as the target speed increased. In shows that it is increasingly difficult to control the frontal plane ($ROT(0)$) circular tracking movement velocity as the speed increases.

As shown in Figure 5C, there was a significant difference in target speeds $\Delta\omega$ for the $ROT(90)$ phase (item E in Table S3). The values of $\Delta\omega$ were 12.09 ± 1.5, 19.93 ± 3.01, 42.15 ± 5.35, and $74.54 \pm 18.66 \degree \text{ s}^{-1}$ for 0.125, 0.25, 0.5, and 0.75 Hz, respectively. Furthermore, $\Delta\omega$ increased significantly on the sagittal plane ($ROT(90)$) as the target speed increased.

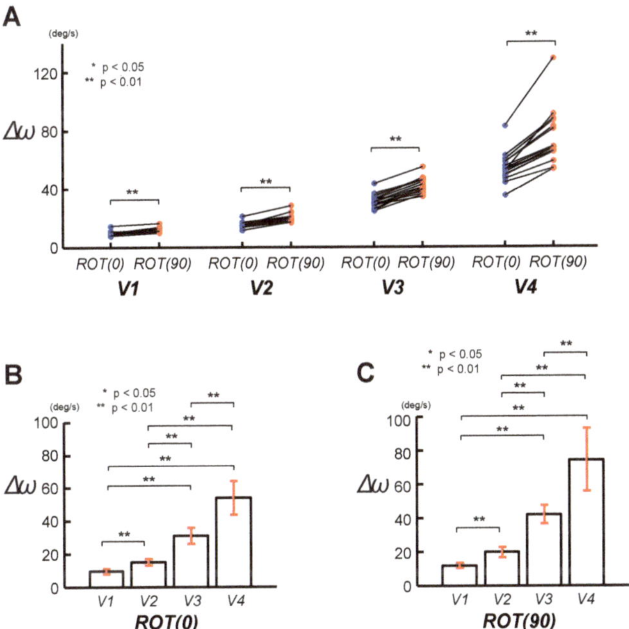

Figure 5. Evaluation of the circular tracking movement based on $\Delta\omega$ in 3D VR space. (**A**) The graphs indicate the pairwise comparison of $\Delta\omega$ analyzing the speed factor between $ROT(0)$ and $ROT(90)$. (**B**) The pairwise comparison was performed for $\Delta\omega$, on the frontal plane ($ROT(0)$), at four target speeds. (**C**) The pairwise comparison was performed for $\Delta\omega$, on the sagittal plane ($ROT(90)$) at four target speeds.

We found that as the speed increased, the subjects had more difficulty controlling the circular tracking movement position on the sagittal plane (*ROT(90)*) compared with that on the frontal plane (*ROT(0)*).

4. Discussion

In this study, we quantitatively evaluated motor control characteristics for circular tracking movements in 3D space. We analyzed the temporospatial relationship in 3D space between the circular tracking movements and target motion for various speeds and two different rotation axes. We measured three kinematic metrics (parameters) based on polar coordinates: differences in distance from the fixed pole, ΔR; position angle, $\Delta\theta$; and angular velocity, $\Delta\omega$.

We found that when the target speed increased, $\Delta\theta$ and $\Delta\omega$ (indicating temporal motor control characteristics in polar coordinates) increased for both the frontal and sagittal planes. This suggests that irrespective of the target rotation axis in 3D space, increasing the target speed makes it more difficult to synchronize the angle and angular velocities of the target and tracer temporally (Figure 4B,C and Figure 5B,C). Further, ΔR, which indicates spatial motor control characteristics in polar coordinates, increased for both the frontal and sagittal planes when the target speed increased. This suggests that, irrespective of the target rotational axis in 3D space, increasing the target speed over a specific velocity makes it more difficult to track the target spatially (Figure 3B,C).

To investigate the impact of the depth on motor control in 3D space, we compared the three parameters on both planes at each target speed. On the sagittal plane, $\Delta\theta$ and $\Delta\omega$ were significantly higher than on the frontal plane at all target speeds (Figures 4A and 5A). However, ΔR was significantly higher than that on the frontal plane over target speeds of 0.25 Hz (Figure 3A). Further, the differences between the two planes strongly increased along with the target speed (Figures 3A, 4A and 5A) for all parameters.

This result shows that the depth information is significant for target-tracking movements, particularly at high speeds. In the following, we discuss the impacts of the target speed and depth.

4.1. Impact of Target Speed on Circular Tracking Movement

Previous studies on motor control characteristics for circular tracking movements used Cartesian or polar coordinate parameters [13,14,17]. Three polar coordinate parameters are widely used: ΔR, $\Delta\theta$, and $\Delta\omega$ [13,14]. Based on these parameters, our previous study analyzed the motor control and impact of target speed and visual feedback about the target during 2D tracking movements [14].

In our previous study, Cartesian coordinate errors allowed us to confirm the circular tracking movement spatial accuracy (or spatial error) in terms of each axis in 3D space. However, $\Delta\theta$ and $\Delta\omega$ in polar coordinates of this study enable us to analyze motor control characteristics in 3D space in terms of the position and velocity control accuracy, respectively. Our results showed that a higher target speed increases position and velocity errors as well as the circular movement spatial error.

In this study, we confirmed these results for 3D space irrespective of the rotation axis (Figures 3–5). In our previous study, we mapped 3D joint movements onto a 2D plane. In contrast, for this study, we directly tracked motor control characteristics in 3D space. We found that for the circular tracking movement motor control at slow target speed in 3D space, $\Delta\theta$ and $\Delta\omega$ (which indicate the temporal characteristics) were more sensitive than ΔR (Figures 3–5). In other words, we confirmed that in 2D and 3D spaces, tracking errors (i.e., the three parameters) increased as the target speed increased. The observed movement delay is because the brain's neurotransmission, sensory processing, has a 200 ms delay in reaching the human visuomotor control system [18,19]. Furthermore, it is more difficult to acquire target motion visual information [20] when the target speed increases.

We also compared the frontal plane (*ROT(0)*) movements, similar to the 2D experiments, with those on the sagittal plane (*ROT(90)*), which are typical 3D movements. We quantitatively showed that circular tracking performance, position control accuracy, and velocity control accuracy in a 3D space is more difficult than in a 2D space (Figures 3A, 4A and 5A).

In future work, we will investigate the impact of the target visual feedback in terms of the control strategy characteristics in 3D circular tracking movements [14,21–26].

4.2. Impact of Depth on Circular Tracking Movement

Unlike 2D space, depth estimation is a significant control metric for depth perception and motor control in 3D space. In previous studies on target-tracking movements, particularly for circular tracking movements, no technology or system has been able to provide an accurate representation of the visual target. Therefore, the role of depth in 3D target-tracking movements, the relationship between target speed and depth in 3D target-tracking movements, and impact of depth on kinematic parameters, such as position, and velocity need to be explored further.

Previously, we developed a 3D visuomotor control evaluation system in a VR environment [11]. By evaluating the Cartesian coordinate spatial error, we analyzed circular tracking movements on the frontal and sagittal planes. However, the evaluation using the Cartesian coordinate spatial error in that study could not clearly show the difference between circular tracking movements on the two planes. Consequently, we did not explore the role of depth information in motor control of 3D target-tracking movements. Furthermore, we confirmed that, regarding the Cartesian coordinate spatial errors, a significant difference was observed between the two circular tracking movement types even when the target speed was increased to a higher level (see Figure 6). Accordingly, we needed to propose a new analysis method for various circular tracking movement types and speeds in a 3D environment and for exploring the role of depth information in 3D target-tracking movement motor control. In this study, based on the three polar coordinate parameters, we quantitatively showed that the depth information is significant for the performance of circular movements, position control accuracy, and velocity control accuracy in 3D movements. Furthermore, it was established that the depth information is even more significant when the target is moving faster (Figures 3–5).

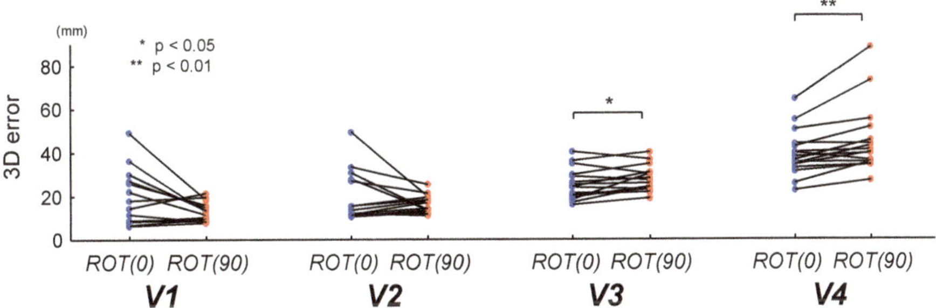

Figure 6. Evaluation of the circular tracking performance based on Δ3D *(3D error; spatial errors in Cartesian coordinates)*. Note that, there was a statistically significant Δ3D difference under *V3* and *V4* conditions between *ROT*(0) and *ROT*(90).

Target motion cognition in 3D space is significant for 3D circular tracking movement motor control. In our previous study, using a rubber hand illusion in a VR environment, we showed that depth perception is easier in left to right motion for the same depth than for different depths [27]. Also, we found that the limb position visual feedback is most accurate in the azimuth and least accurate in the depth direction [28]. Those previous studies helped us interpret our results in these studies. In other words, because target motion cognition is better on the frontal plane than on the sagittal plane, performance, position control, and velocity control are better on the frontal plane in circular tracking movements. The faster the target speed, the clearer the trend (Figures 3–5). In future studies, we will quantitatively analyze the hand-arm system sensitivity as well as the impact of the target cognition level for both planes in 3D space by modifying the target visual feedback [23]. Further, we will quantitatively analyze the difference in the weight of the 3D tracking movements by experimentally recording and analyzing muscle activity.

4.3. Control Strategy and Mechanism in 3D Tracking Movements

In general, the human visuomotor control system requires approximately 200 ms to perform visually guided error correction because of the brain's neurotransmission, sensory processing, and refractoriness in the human neuromotor control [18,19,29]. Previous studies have established theories for control strategies by considering the delay time during visuomotor processing. Intermittent feedback (FB) control and feedforward (FF) control using an internal model have been suggested as typical theories for the control strategy [21–24,30]. These studies demonstrated that control strategies (i.e., FB and FF controls) and major control parameters in kinematic features (i.e., position or velocity) for tracking movements vary irrespective of the target trajectory being random or consistent [13,17,21,22]. In other words, these studies reported that pseudorandom tracking movements are primarily controlled by intermittent FB control with position error (i.e., $\Delta\theta$) reduction. In addition, periodic sinusoidal tracking movements depend on more FF control with velocity error (i.e., $\Delta\omega$) reduction to predict future target motion. Our previous study examined the extent of the impact of the target visual information on the periodic circular tracking task control strategy as being the same in this study [14]. We also confirmed FF controls with velocity error (i.e., $\Delta\omega$) reduction in target-invisible regions for the periodic circular tracking task. In other words, the target visibility allows us to examine the impact of the target visual information on the control strategy by quantitatively comparing the three temporospatial parameters (ΔR, $\Delta\theta$, and $\Delta\omega$) relative changes during the target-visible phase responsible for stable FB control with the target-invisible phase for FF control. In our future studies, we will examine the same periodic circular tracking task with different speeds and visibility of the target based on the same ΔR, $\Delta\theta$, and $\Delta\omega$ parameters to observe control strategy changes and major kinematic feature control parameters during 3D circular tracking movement tasks.

For different movement types and target speeds, we should design different control strategies, considering cognition delay and visuomotor processing. Intermittent FB control and FF control, as an internal model, have been suggested as a suitable control strategy [13–15]. Based on previous studies in 2D space, FB control dominates when the target speed is slow; however, FF control dominates when the target speed is high [21–24]. In this study, similar results were found for 3D target-tracking movements. Despite adding a countdown sound to announce the start of the target movement, the initial errors were greater when the target speed was higher, indicating imprecise FF control, as observed for 2D movements (Figure 2) [25]. Our previous studies demonstrated that increasing target speed makes it more difficult to synchronize the angle and angular velocities of the target and tracer temporally. Considering our previous studies [31,32] showed the cerebellum plays an important role in the generation of motor commands for smooth velocity and position control, the method and analysis proposed in this study will apply to characterize the motor function of patients with cerebellar ataxia in the future.

Our recent study demonstrated that the smooth tracking movement of the wrist joint consists of two components with distinct motor commands for velocity and position controls [33]. In other words, the major component in a lower frequent range (named by the F1 component) of the tracking movement played the primary role to reproduce both the velocity and position of the target motion in a predictive manner. In contrast, the minor component in a higher frequency range (named by the F2 component) of the tracking movement encodes mostly the positions of small step-wise movements with intermittent FB control. In this study, the 3D tracking movement for higher target speeds and *ROT(0)* showed lower frequent components in two temporal parameters ($\Delta\theta$ and $\Delta\omega$). Therefore, we will quantitatively analyze the control strategy and mechanism of 3D tracking movements in the F1 and F2 components based on the predictive control and intermittent FB control in our future work.

4.4. Future Application and Limitation of Temporo-Spatial Parameters in Polar Coordinates

In this study, we adopted temporospatial parameters in an extrinsic coordinate frame to analyze motor control characteristics for two planes and various circular tracking movement speeds in a 3D environment. However, the temporospatial parameters in this study did not allow us to analyze different biomechanical constraints (i.e., arm posture of the kinematic chain, torque of each joint) in an intrinsic coordinate frame for the two planes. In other words, the different biomechanical constraints in the two planes could impact the control of the hand position (i.e., temporospatial parameters in polar coordinates) in an extrinsic coordinate frame. In the future, we will analyze the different biomechanics in an intrinsic coordinate frame between circular tracking movement on the two planes by experimentally recording and analyzing the arm posture and related muscle activities during circular tracking tasks [34–36].

The VR system used in this study allowed subjects to perform three degree-of-freedom visuomotor tracking movements in an immersive three-dimensional VR environment. In our future study, the proposed method and analysis parameters (i.e., temporospatial parameters in polar coordinates) can be used for evaluating the seriousness of illness and the effectiveness of rehabilitation for patients with hemiplegic upper limbs. The method can also be applied to perception studies of spatial neglect patients [37].

5. Conclusions

We examined two plane circular tracking movements in 3D space (from a motor control perspective) based on three temporo-spatial parameters. We found that errors of two temporal parameters on each plane were proportional to the change in the target velocity. Furthermore, for a spatial parameter, the errors in the circular tracking on the frontal plane ($ROT(0)$) without depth were lower than those in the circular tracking on the sagittal plane ($ROT(90)$) with depth. The experimental protocol and data analysis in this study allowed us to analyze the motor control characteristics temporospatially for circular tracking movement with various depths and speeds in 3D VR space.

Supplementary Materials: The following are available online at http://www.mdpi.com/2076-3417/10/2/621/s1, Table S1: A summary of statistical analysis of ΔR for the circular tracking movement; Table S2: A summary of statistical analysis of $\Delta\theta$ for the circular tracking movement; Table S3: A summary of statistical analysis of $\Delta\omega$ for the circular tracking movement.

Author Contributions: W.C. conceived, designed, and performed the experiments; analyzed the data; and wrote the paper. J.L. conceived, designed, and performed the experiments; analyzed the data; and wrote the paper. L.L. wrote the paper. All authors have read and agreed to the published version of the manuscript.

Funding: This work was supported by JSPS KAKENHI (Grant No. JP18K11594).

Conflicts of Interest: The authors declare that they have no competing interests.

Data Availability: Data are fully available through the corresponding author.

References

1. Georgopoulos, A.P.; Grillner, S. Visuomotor coordination in reaching and locomotion. *Science* **1989**, *245*, 1209–1210. [CrossRef] [PubMed]

2. Beggs, W.D.A.; Howarth, C.I. Movement control in a repetitive motor task. *Nature* **1970**, *225*, 752–753. [CrossRef] [PubMed]

3. Van den Dobbelsteen, J.J.; Brenner, E.; Smeets, J.B. Endpoints of arm movements to visual targets. *Exp. Brain Res.* **2001**, *138*, 279–287. [CrossRef] [PubMed]

4. Desmurget, M.; Prablanc, C.; Arzi, M.; Rossetti, Y.; Paulignan, Y.; Urquizar, C. Integrated control of hand transport and orientation during prehension movements. *Exp. Brain Res.* **1996**, *110*, 265–278. [CrossRef] [PubMed]

5. Jeannerod, M.; Arbib, M.A.; Rizzolatti, G.; Sakata, H. Grasping objects: The cortical mechanisms of visuomotor transformation. *Trends Neurosci.* **1995**, *18*, 314–320. [CrossRef]

6. Chieffi, S.; Gentilucci, M. Coordination between the transport and the grasp components during prehension movements. *Exp. Brain Res.* **1993**, *94*, 471–477. [CrossRef]

7. Paulignan, Y.; MacKenzie, C.; Marteniuk, R.; Jeannerod, M. Selective perturbation of visual input during prehension movements. 1. The effects of changing object position. *Exp. Brain Res.* **1991**, *83*, 502–512. [CrossRef]

8. Lacquaniti, F.; Caminiti, R. Visuo-motor transformations for arm reaching. *Eur. J. Neurosci.* **1998**, *10*, 195–203. [CrossRef]

9. Zelaznik, H.N.; Hawkins, B.; Kisselburgh, L. Rapid visual feedback processing in single aiming movements. *J. Mot. Behav.* **1983**, *15*, 217–236. [CrossRef]

10. Russell, D.M.; Sternad, D. Sinusoidal visuomotor tracking: Intermittent servo-control or coupled oscillations? *J. Mot. Behav.* **2001**, *33*, 329–349. [CrossRef]

11. Choi, W.; Lee, J.; Yanagihara, N.; Li, L.; Kim, J. Development of a quantitative evaluation system for visuo-motor control in three-dimensional virtual reality space. *Sci. Rep.* **2018**, *8*, 134–139. [CrossRef] [PubMed]

12. Melmoth, D.R.; Grant, S. Advantages of binocular vision for the control of reaching and grasping. *Exp. Brain Res.* **2006**, *171*, 371–388. [CrossRef] [PubMed]

13. Hayashi, Y.; Sawada, Y. Transition from an antiphase error-correction mode to a synchronization mode in interaction hand tracking. *Phys. Rev. E* **2013**, *88*, 022704. [CrossRef] [PubMed]

14. Kim, J.; Lee, J.; Kakei, S.; Kim, J. Motor control characteristics for circular tracking movements of human wrist. *Adv. Robot.* **2017**, *31*, 29–39. [CrossRef]

15. Field, A. *Discovering Statistics Using SPSS*, 3rd ed.; SAGE Publications Ltd.: Sauzendeaux, CA, USA, 2009; pp. 1–821.

16. Faul, F.; Erdfelder, E.; Lang, A.G.; Buchner, A. G*Power: A flexible statistical power analysis program for the social, behavioural, and biomedical sciences. *Behav. Res. Methods* **2007**, *39*, 75–91. [CrossRef]

17. Roitman, A.V.; Massaquoi, S.G.; Takahashi, K.; Ebner, T.J. Kinematic analysis of manual tracking in monkeys: Characterization of movement intermittencies during a circular tracking task. *J. Neurophysiol.* **2004**, *91*, 901–911. [CrossRef]

18. Keele, S.W.; Posner, M.I. Processing of visual feedback in rapid movements. *J. Exp. Psychol.* **1968**, *77*, 155–158. [CrossRef]

19. Van Roon, D.; Caeyenberghs, K.; Swinnen, S.P.; Smits-Engelsman, B.C.M. Development of feedforward control in a dynamic manual tracking task. *Child Dev.* **2008**, *79*, 852–865. [CrossRef]

20. Brown, B. Dynamic visual acuity, eye movements and peripheral acuity for moving targets. *Vision Res.* **1972**, *12*, 305–321. [CrossRef]

21. Miall, R.; Weir, D.; Stein, J. Manual tracking of visual targets by trained monkeys. *Behav. Brain Res.* **1986**, *20*, 185–201. [CrossRef]

22. Miall, R.; Weir, D.; Stein, J. Intermittency in human manual tracking tasks. *J. Mot. Behav.* **1993**, *27*, 53–63. [CrossRef] [PubMed]

23. Fine, J.M.; Ward, K.L.; Amazeen, E.L. Manual coordination with intermittent targets: Velocity information for prospective control. *Acta Psychol.* **2014**, *149*, 24–31. [CrossRef] [PubMed]

24. Miall, R.; Weir, D.; Stein, J. Planning of movement parameters in a visuo-motor tracking task. *Behav. Brain Res.* **1988**, *27*, 1–8. [CrossRef]

25. Beppu, H.; Nagaoka, M.; Tanaka, R. Analysis of cerebellar motor disorders by visually-guided elbow tracking movement. *Brain* **1984**, *107*, 787–809. [CrossRef] [PubMed]

26. Beppu, H.; Nagaoka, M.; Tanaka, R. Analysis of cerebellar motor disorders by visually-guided elbow tracking movement. 2. contribution of the visual cues on slow ramp pursuit. *Brain* **1987**, *110*, 1–18. [CrossRef] [PubMed]

27. Choi, W.; Li, L.; Satoh, S.; Hachimura, K. Multisensory integration in the virtual hand illusion with active movement. *BioMed Res. Int.* **2016**, *2016*, 8163098. [CrossRef] [PubMed]

28. Van Beers, R.J.; Wolpert, D.M.; Haggard, P. When feeling is more important than seeing in sensorimotor adaptation. *Curr. Biol.* **2002**, *12*, 834–837. [CrossRef]

29. Van de Kamp, C.; Gawthrop, P.J.; Gollee, H.; Loram, I.D. Refractoriness in sustained visuo-manual control: Is the refractory duration intrinsic or does it depend on external system properties? *PLoS Comput. Biol.* **2013**, *9*, e1002843. [CrossRef]

30. Sabes, P.N. The planning and control of reaching movements. *Curr. Opin. Neurobiol.* **2000**, *10*, 740–746. [CrossRef]
31. Lee, J.; Kagamihara, Y.; Tomatsu, S.; Kakei, S. The functional role of the cerebellum in visually guided tracking movement. *Cerebellum* **2012**, *11*, 426–433. [CrossRef]
32. Lee, J.; Kagamihara, Y.; Kakei, S. A new method for functional evaluation of motor commands in patients with cerebellar ataxia. *PLoS ONE* **2015**, *10*, e0132983. [CrossRef] [PubMed]
33. Kakei, S.; Lee, J.; Mitoma, H.; Tanaka, H.; Manto, M.; Hampe, C.S. Contribution of the cerebellum to predictive motor control and its evaluation in ataxic patients. *Front. Hum. Neurosci.* **2019**, *13*, 216. [CrossRef] [PubMed]
34. Kakei, S.; Hoffman, D.S.; Strick, P.L. Muscle and movement representations in the primary motor cortex. *Science* **1999**, *285*, 2136–2139. [CrossRef] [PubMed]
35. Kakei, S.; Hoffman, D.S.; Strick, P.L. Sensorimotor transformations in cortical motor areas. *Neurosci. Res.* **2003**, *46*, 1–10. [CrossRef]
36. Kambara, H.; Shin, D.; Koike, Y. A computational model for optimal muscle activity considering muscle viscoelasticity in wrist movements. *J. Neurophysiol.* **2013**, *109*, 2145–2160. [CrossRef]
37. Gross, D.P.; Zhang, J.; Steenstra, I.; Barnsley, S.; Haws, C.; Amell, T.; McIntosh, G.; Cooper, J.; Zaiane, O. Development of a computer-based clinical decision support tool for selecting appropriate rehabilitation interventions for injured workers. *J. Occup. Rehabil.* **2013**, *23*, 597–609. [CrossRef]

Article

Auditory Coding of Reaching Space

Ursula Fehse [1],*, Gerd Schmitz [1], Daniela Hartwig [1], Shashank Ghai [1,2], Heike Brock [3] and Alfred O. Effenberg [1],*

[1] Faculty of Humanities, Institute of Sports Science, Leibniz University Hannover, 30167 Hanover, Germany; gerd.schmitz@sportwiss.uni-hannover.de (G.S.); hartwigdaniela.dh@gmail.com (D.H.); shashank.ghai@mail.mcgill.ca (S.G.)
[2] School of Physical & Occupational Therapy (SPOT), McGill University, Montreal, QC H3G 1Y5, Canada
[3] Honda Research Institute Japan Co., Ltd., Wako-shi, Saitama 351-0188, Japan; h.brock@jp.honda-ri.com
* Correspondence: ursula.fehse@sportwiss.uni-hannover.de (U.F.); effenberg@sportwiss.uni-hannover.de (A.O.E.); Tel.: +49-511-762-5510 (A.O.E.)

Received: 19 November 2019; Accepted: 27 December 2019; Published: 7 January 2020

Abstract: Reaching movements are usually initiated by visual events and controlled visually and kinesthetically. Lately, studies have focused on the possible benefit of auditory information for localization tasks, and also for movement control. This explorative study aimed to investigate if it is possible to code reaching space purely by auditory information. Therefore, the precision of reaching movements to merely acoustically coded target positions was analyzed. We studied the efficacy of acoustically effect-based and of additional acoustically performance-based instruction and feedback and the role of visual movement control. Twenty-four participants executed reaching movements to merely acoustically presented, invisible target positions in three mutually perpendicular planes in front of them. Effector-endpoint trajectories were tracked using inertial sensors. Kinematic data regarding the three spatial dimensions and the movement velocity were sonified. Thus, acoustic instruction and real-time feedback of the movement trajectories and the target position of the hand were provided. The subjects were able to align their reaching movements to the merely acoustically instructed targets. Reaching space can be coded merely acoustically, additional visual movement control does not enhance reaching performance. On the basis of these results, a remarkable benefit of kinematic movement acoustics for the neuromotor rehabilitation of everyday motor skills can be assumed.

Keywords: reaching; motor control; sensory-motor integration; audition and movement; proprioception and movement; kinematic movement sonification; knowledge of results; knowledge of performance; visual-to-auditory substitution; neuromotor rehabilitation

---

1. Introduction

Reaching movements are everyday actions essential for coping with everyday life. Information about the position of the hand is transmitted via the visual and the proprioceptive sense [1]. In reaching movements, the movement of the arm is continuously adjusted [2]. Information about the target position and the location of the hand are continuously monitored [2]. Apparently, at least when the moving limb is not visible, but the target point is, proprioceptive information about the location of the arm is continuously compared to visual information about the target position and thus the ongoing movement is smoothly corrected [3].

Not only visual and proprioceptive, but also auditory information can be used to guide arm movements [4,5]. Boyer et al. [6] completed a reaching study with merely acoustically coded artificial targets in the transverse plane with blindfolded participants. Directional information was given by stereo headphones reconstructing the sound pressure of a sound source at a given position in the transverse plane. The average deviation from the target was lower when the target sound was presented for two seconds than when it was presented for 0.25 s. There was also a feedback condition in which a continuous auditory real-time feedback of the hand position, with the same systematic as for the target presentation, was given. As instruction, the target was also presented here for 0.25 s. Reaching performance in this condition did not differ from that of the two conditions without any feedback. When this feedback was shifted 18.5° left from the real hand position, reaching performance was worse than in the condition without any feedback and a target presentation of two seconds [6].

Sound and motion are closely linked. Sound is vibration and vibration is movement. There is no sound if there is no movement. The auditory perception has a high temporal resolution but can also convey spatial information and does not require focused attention. Most movements, however, do not provoke considerable acoustic effects. Additional artificial auditory movement information is needed to enhance the amount of auditory information of usually almost silent movements.

The idea to provide movement information via the auditory channel has been pursued for several decades [7]. Research has focused on the benefits of additional auditory movement information on motor perception, motor control, motor learning and cooperative motion [8–15]. As underlying neurophysiologic functions, multisensory integration [16–18] and intermodal sensorimotor movement representations [19–23] are discussed. Human ability to emerge supramodal action representations might enable the support of proprioception by additional acoustic information.

Acoustic movement information can be used to augment the perception of another modality, but movement acoustics alone also provides information about a movement pattern for the support of motor perception and motor performance. Research in this area has focused on natural [24–26] but also on artificial movement sounds [27–29]. Levy-Tzedek et al. [30] even developed a visual- to- auditory sensory substitution device to enable blind people to guide reaching movements.

Movement acoustics has been used in neurorehabilitation dedicated to sensory-motor deficits, for example in the rehabilitation of target-orientated arm movements [12,31–34]. Additional work has been realized with guiding- and feedback-acoustics. Thaut et al. succeeded in the rhythmic facilitation of gait for Parkinson's disease [35] and stroke patients [36], and also in the rhythmic cuing of reaching movements for stroke patients [37]. Ghai et al. reviewed further benefits of auditory cuing for gait impairments caused by aging [38], cerebral palsy [39], and Parkinson's disease [40]. Thaut provided an overview about therapeutic usable effects of music in neuromotor rehabilitation [41]. The findings of Schauer and Mauritz suggest that walking in time with music adjusted to the actual individual step frequency can improve not only step length and therefore also gait velocity but also gait symmetry in stroke patients more than traditional gait therapy [42]. Young et al. differentiated the benefit of different kinds of non-sonification auditory instruction for various gait parameters in Parkinson's disease [43].

There has been already some evidence that real-time movement acoustics can support neuromotor rehabilitation, albeit most of the studies use movement data to create auditory error feedback [44], target-orientated feedback [45] or more musical-oriented movement acoustics [46]. These kinds of acoustic information usually need processing on a conscious level of perception.

Initial approaches to the use of kinematic movement acoustics in neuromotor rehabilitation have also been developed [12,47]. Here, movement sonification is configured as continuous real-time kinematic auditory feedback intending to initiate audio-motor couplings and becoming integrated directly in multimodal perception as described above. Based on this "lower-level" efficiency, movement sonification should be appropriate for substituting reduced proprioception in stroke patients and for example so enhance the effectiveness of the rehabilitation process.

Such kind of sensory enhancements should increase the neuronal plasticity and support the functional reorganization of the sensory-motor system within the central nervous system. A first step is intended to focus on simple everyday actions of the upper extremities as goal-directed arm movements in three-dimensional action space. Preparatorily, an effective approach to represent this reaching space acoustically and the development of an appropriate sonification device is required.

This explorative study analyzes if it is possible, on the basis of our sonification device developed in the first steps, to perform reaching movements to merely acoustically instructed, invisible targets and to use acoustic feedback for the movement control. The aim was to study the implicit effectiveness of the available acoustic information about the target position and the movement. Therefore, the employed subjects were not experienced in the use of movement sonification and had no explicit knowledge of the used sound composition. Additionally, the impact of visual control of the movement on the precision of merely acoustically guided reaching movements in three planes in healthy subjects was analyzed. In this way, we expected to evoke an activation of audio-motor respectively audio-visuo-motor processes and to address sensory and motor control mechanisms.

Human movements are continuously exposed to alterations in direction, velocity, and forces. For this reason, a continuous mapping is required to adequately depict them. We decided to use continuous kinematic real-time auditory movement information to achieve a high level of auditory- proprioceptive congruence. Since the processing of error-related auditory information requires conscious cognitive processes as attention and we wanted to address implicit processes instead, we did not choose error-related, but kinematic movement sonification. These considerations are for example supported by Rosati et al. [34] who showed superiority of a task-related audio feedback compared to an error-related audio feedback in a visual tracking task.

Based on the considerations of Graziano [2], we decided to use data of the right metacarpophalangeal joints to generate acoustic information about the target position and the reaching movement. For the acoustic representation of the target position of the hand, three- dimensional spatial coordinates were sonified. For the continuous representation of the reaching movement, acoustic information about movement velocity was added. The systematics of the sonification was adopted from Vinken et al. [27], who found a high efficacy of such a four- dimensional sonification of a continuous kinematic trajectory in the auditory discrimination of arm movements. The coding of the vertical spatial component by pitch was also in accordance with Melara and O'Brien [48] and Scholz et al. [33,46]. The results of Küssner et al. [49], who studied the relations between musical parameters and intuitive gestures and found a positive correlation between pitch and height, also supported the choice of this mapping. The horizontal coding by stereo characteristics conformed to that applied by Boyer et al. [6]. The spatial depth was mapped on spectral composition. In this way, we intended to create an auditory space to enable the localization of reaching targets. Regarding the thoughts of Boyer et al. [6] and the findings of Vinken et al. [27], we added a velocity component to the spatial information for the movement sonification. In the light of the above-mentioned results of Boyer et al. [6] suggesting that the auditory position stimulus presentation of 0.25 s is too short for the generation of a precise and reliable target representation, our instruction sound was longer than 0.25 s in all conditions. In order to control the influence of the extent of given information, we worked with two different durations of target presentation.

In contrast to Boyer et al. [6] and to facilitate the task, we chose to design the instruction and feedback in the same manner. That means that the subjects were prompted to produce by their reaching movement the same sound as they had heard for instruction. To realize a clear temporal separation between instruction and feedback to avoid confusion, the participants' movement did not start before the instruction was finished. We therefore accepted a reduced reaching performance observed by other authors for the removal of target presentation prior to the completion of the reaching movement [3,6].

Since we think that it is possible to merely acoustically instruct target positions for reaching movements and that our kind of movement sonification is highly effective, we formulated the hypothesis that the blind as well as the sighted experimental group is able to hit the target points with a precision clearly above chance. In the light of the described important role of vision in reaching movements and the beneficial effects of audio-visual integration and audio-visuo-motor processing associated with movement sonification, we expect the sighted experimental group to be superior to the blind group in reaching precision. In spite of the merely acoustic target instruction, we suppose that the visibility of the moving arm enhances the precision of the reaching action. Furthermore, we hypothesize that additional continuous auditory instruction and real-time feedback about the movement is superior to a mere discrete sonification of target point. Considering the learning effect proved by Effenberg et al. [14] and the improvement of the performance over time observed by Vinken et al. [27] even without the availability of feedback, we assume an improvement of the reaching performance over time in the present study.

2. Materials and Methods

2.1. Participants

Twenty-four subjects (14 females, 10 males), aged 19 to 30 years (mean age: 22.2 ± 3.0), participated in this study. All participants had a normal or corrected-to-normal vision (standard vision test), a normal hearing (hearing test: HTTS, Version 2.10, 00115.04711, SAX GmbH, Berlin, Germany) and were right-handed (Edinburgh Inventory, 10-item version). To ensure musical sense, MBEA (Montreal Battery of Evaluation of Amusia) online version [50] was conducted. Following Peretz et al. [51], subjects who scored less than 70% were excluded. The investigations were conducted in accordance with the Declaration of Helsinki. All experimental procedures were approved by the "Central Ethics Committee" ZEK-LUH at the Leibniz University Hannover, Hanover, Germany. All participants gave their informed consent for inclusion prior to their participation. After the completion of the experiment, the subjects received a modest monetary compensation.

2.2. Experimental Setup

The participants were seated on a height-adjustable chair in front of an experimental apparatus resting on a table and wore circumaural headphones (Beyer Dynamic, DT 100, 30–20.000 Hz) as well as inertial sensors.

A digitizing tablet (WACOM Intuos4 XL PTK-1240, active surface 487.7 mm × 304.8 mm) was positioned successively in the *xy*-plane (with the long tablet side along the *y*-axis), *yz*-plane (with the long tablet side along the *z*-axis), or *xz*-plane (with the long tablet side along the *x*-axis) as shown in Figure 1.

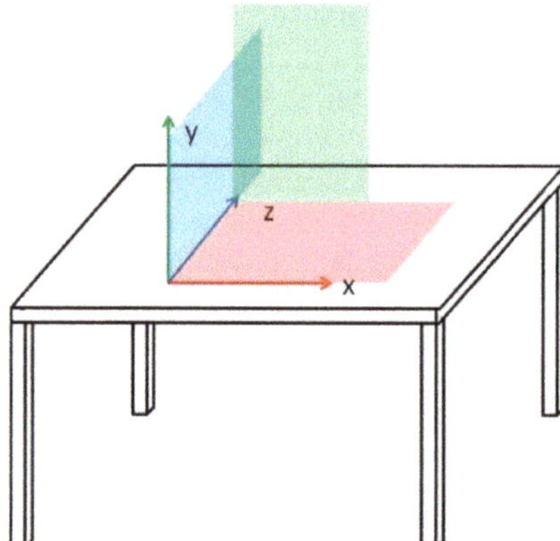

Figure 1. The tablet positions in the three used planes. *X*-axis is later called "transverse axis", *y*-axis is later called "longitudinal axis" and *z*-axis is later called "sagittal axis". The subject sits at the front edge of the table, looking in the direction of the *z*-axis.

The digitizing pen was positioned next to the tablet. A chin rest, positioning the chin at a height of 30 cm from table surface, in combination with a height-adjustable chair ensured a firm and standardized head position throughout the whole experiment. An arm rest provoking approximately a 90-degree abduction in the right shoulder joint was used in the calibration process of the inertial sensors. The subjects were instructed to execute reaching actions with a digitizing pen in their right hand towards invisible, merely acoustically coded target positions on the tablet (see Figure 2).

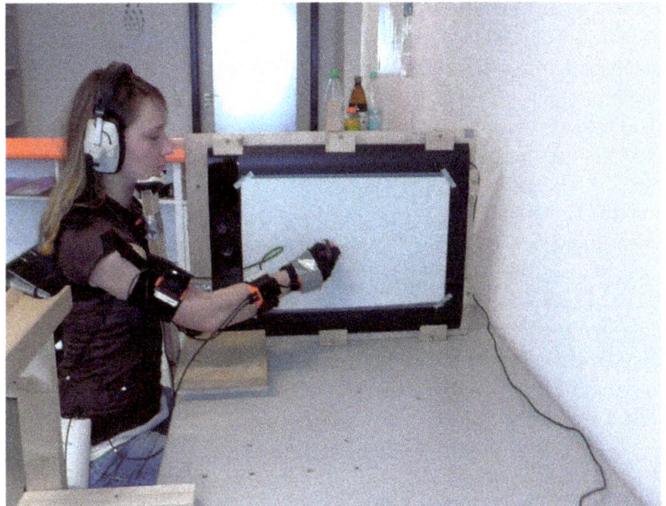

Figure 2. The experimental arrangement for the measurements in the yz-plane.

2.3. Stimulus Material

A human model was equipped with four inertial sensors (MTx miniature inertial 3DOF orientation tracker; Xsens Technologies BV, Enschede, The Netherlands) on the right arm. The sensors (size: 38 mm × 53 mm × 21 mm, weight: 30 g) were attached on the middle of the right shoulder, the right upper arm, the right lower arm, and the back of the right hand. The inertial sensors provided 3D acceleration data (up to eighteen times of the acceleration of gravity), 3D rate of turn (up to 1200°/s), three degrees of freedom orientation and a sampling rate up to 100 Hz depending on the number of sensors used [52,53]. Data of the inertial sensors were recorded while the model was executing reaching actions with the digitizing pen towards one of nine different target positions on the tablet (see Figure 3).

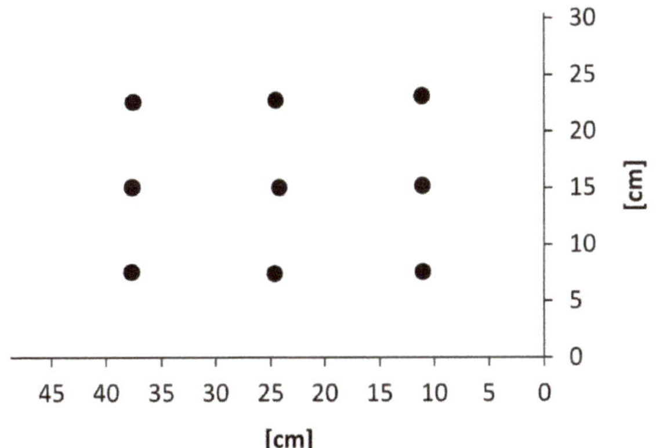

Figure 3. The originally invisible target point positions on the digitizing tablet. The lengths of the axes correspond to the dimension of the digitizing tablet.

In a next step, a forward kinematic model of the right arm was used to determine trajectory and velocity information of the right metacarpophalangeal joints [54]. Four specific kinematic parameters were chosen to be mapped on four different parameters of a sound (patch 100 "jupiter lead" from "Preset D Group", SonicCell, Roland) [54]: (a) The absolute velocity of the right metacarpophalangeal joints was mapped on the volume of the sound (the faster the movement, the louder the sound). (b) The position of the right metacarpophalangeal joints on the y-axis was mapped on the pitch (the higher the metacarpophalangeal joints in space, the higher the sound). The pitch was modulated by two octaves from G2 to G4. (c) The position of the right metacarpophalangeal joints on the z-axis was mapped on the spectral composition (the higher the value, the less overtones the sound has; standard MIDI Continuous Controller No. 74, controls the cutoff frequency of the low-pass filter for an overall brightness control). (d) The position of the right metacarpophalangeal joints on the x-axis was mapped on the stereo characteristics (the further the metacarpophalangeal joints were on the left side from the acting subject's perspective, the louder the sound was on the left audio channel and the quieter the sound was on the right audio channel and vice versa). In this way, sonification-recordings (1.0–1.65 s) of different reaching actions towards target positions were generated. Additionally, the position of the right metacarpophalangeal joints when having reached the target point with the digitizing pen was transformed into sound by using the three above mentioned positional mappings (b, c, and d). In this way, the instruction sounds were formed. The participants' real-time sonification feedback was created in the same way.

2.4. Experimental Conditions

The subjects, parallelized for sex, were randomized into two groups. While executing identical reaching tasks to invisible targets, 12 subjects were blindfolded and 12 subjects had their eyes open during the experiment. The experimental task was to hit the instructed target positions as exactly as possible. All participants received auditory instruction and feedback. There were three different conditions:

1. Continuous condition:

    - Instruction: Sonification of the model's reaching movement towards the target positions followed by the sonification of the model's final position for 1 s.
    - Feedback: Real-time sonification of the participant's reaching movement followed by the sonification of the participant's final position for 1 s.

2. Discrete short condition:

    - Instruction: Sonification of the model's final position for 1 s.
    - Feedback: Real-time sonification of the participant's final position for 1 s.

3. Discrete long condition:

    - Instruction: Sonification of the model's final position for the duration of the model's reaching movement and one more second (in total 2.0–2.65 s).
    - Feedback: Real-time sonification of the participant's final position for 1 s.

2.5. Procedure

After a standardized familiarization protocol, the test session began without any further practice trials. The experimental protocol took approximately 60 min (without including the familiarization). A total of 108 trials were given, consisting of nine trials (nine different target positions on the tablet, see Table S1) multiplied by four conditions (one discrete short block, one discrete long block, and two continuous blocks) multiplied by three tablet positions (xy-plane, yz-plane, xz-plane). The stimuli were displayed in blocks of nine trials (each target position was randomly presented once in the sequence). Four blocks (one set of blocks) were repeated in all three tablet positions, in the same block order (balanced and randomized among the subjects) but with a varied trial order. The order of the tablet positions was also balanced and randomized among the subjects. Between the three sets of blocks there were breaks of 5 min. The trials were presented to the participants in a standardized setting. Each trial consisted of (1) reaching for and grasping the pen, (2) the sonification sequence (see Instruction-Audios S1–81) presented once via headphones, (3) the execution of the reaching movement towards the digitizing tablet, (4) putting the pen down, and (5) putting the arm on the arm rest. The overall duration of a single trial was about 15 s.

2.6. Data Acquisition and Analysis

The two-dimensional Cartesian coordinates of the reaching point on the digitizing tablet were measured for each trial. The tablet sampled the x and y coordinates of the stylus hit with a spatial accuracy of 0.005 mm. The collected data were stored for later analyses. The reaching error was represented by the vector from the target to the reaching position. The absolute values of the single vector coordinates describe the sizes of the absolute deviations along the three directions in space (termed 'Absolute differences'). The magnitude of the vector represents the distance between the target and the reaching position (termed 'Absolute distance'). T-tests and analyses of variance with repeated measurements were used to compute the significance of the deviations to the target point. For the post hoc test, the Bonferroni test was used. A significance criterion of $\alpha = 5\%$ was established for all results reported.

3. Results

Figure 4 exemplifies the reaching positions of all participants for one target point. The endpoints of the reaching movements spread around the target point.

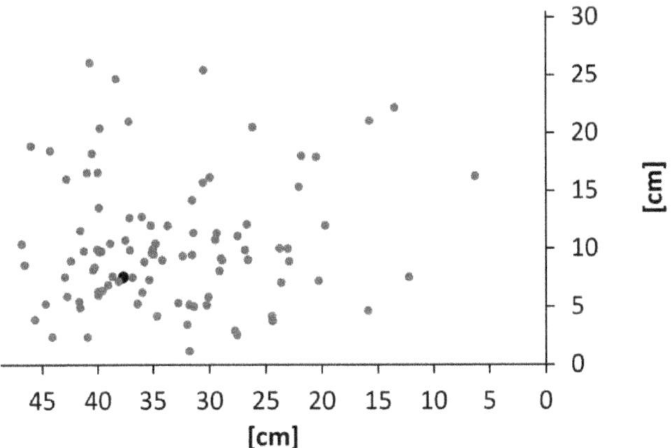

Figure 4. Reaching results for one target point in *xy*-plane. Target point: black, the participants' pointing results: grey.

The mean absolute distance from the target point was 11.68 cm (SD: 2.46 cm) over all subjects, planes and conditions and therefore significantly lower than 18.60 cm, the value for the absolute distance expected at random reaching ($t_{(23)} = -13.49$, $p < 0.001$). Accordingly, the averaged absolute differences between reaching and target point coordinates in the three directions in space over all subjects, axes and conditions was 7.25 cm (SD: 1.53 cm). This value is significantly lower than 11.75 cm, the value for the absolute axial differences expected at random reaching ($t_{(23)} = -14.13$, $p < 0.001$).

This error remained stable over time, repeated measures ANOVA (3 sequences × 4 blocks × 9 trials, sorted chronologically) revealed no significant main effect or interaction for absolute distance to target point. Figure 5 shows the time course of the average absolute distance.

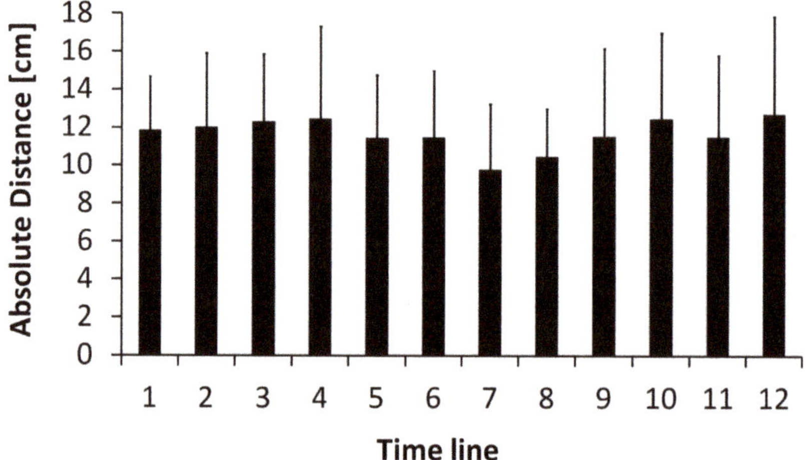

Figure 5. Time course of average absolute distance for the 12 blocks.

In the following, the absolute differences between reaching and target point coordinates in the three directions in space are considered.

For repeated measures ANOVA (3 axes × 2 tablet sides × 2 sonification classes × 2 blocks, sorted by condition), main effect "treatment" did not reach level of significance ($F_{(1, 22)} = 0.02$, $p = 0.887$, $\eta_p^2 < 0.01$). The blindfolded group (M = 7.30 ± 1.46 cm) did not reach significantly worse than the sighted group (M = 7.21 ± 1.59 cm). None of the interactions with the factor "treatment" reached level of significance.

There was a significant main effect "axis" ($F_{(2, 44)} = 8.92$, $p = 0.001$, $\eta_p^2 = 0.29$). The differences between the longitudinal axis and the transverse axis ($p = 0.022$) as well as between the longitudinal axis and the sagittal axis ($p < 0.001$) became significant, whereas the differences in the reaching precision between the transverse axis and the sagittal axis did not ($p = 0.580$) (see Figure 6). As the three axes were acoustically represented by three different sound parameters, this result cannot be interpreted purely regarding the spatial orientation.

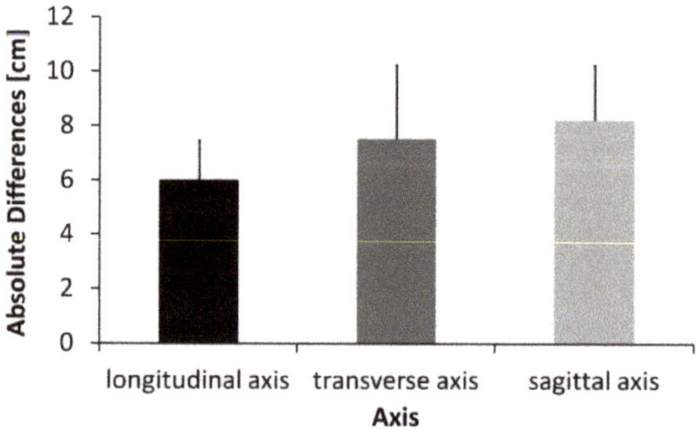

Figure 6. Means and standard deviations of absolute differences for the three axes. Longitudinal axis corresponds to *x*-axis, transverse axis to *y*-axis and sagittal axis to *z*-axis in Figure 1.

There was also a significant main effect "sonification class" ($F_{(1, 22)} = 6.81$, $p = 0.016$, $\eta_p^2 = 0.24$) with a better reaching performance in conditions with mere discrete instruction and feedback than in conditions with additional continuous instruction and feedback, as depicted in Figure 7.

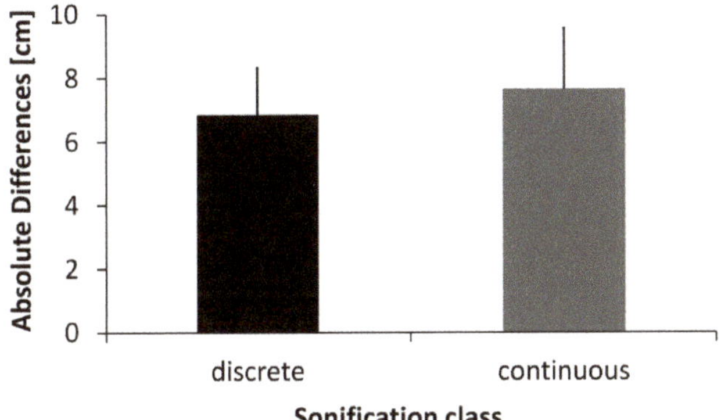

Figure 7. Means and standard deviations of absolute differences for the two sonification classes for instruction and feedback.

Looking for interactions with the factor "sonification class", a significant interaction "axis" × "sonification class" ($F_{(2, 44)} = 5.47$, $p = 0.008$, $\eta_p^2 = 0.20$) was revealed. The difference in the reaching precision between discrete and additional continuous auditory feedback was significant only for the longitudinal axis ($p < 0.001$). Figure 8 shows the interaction pattern.

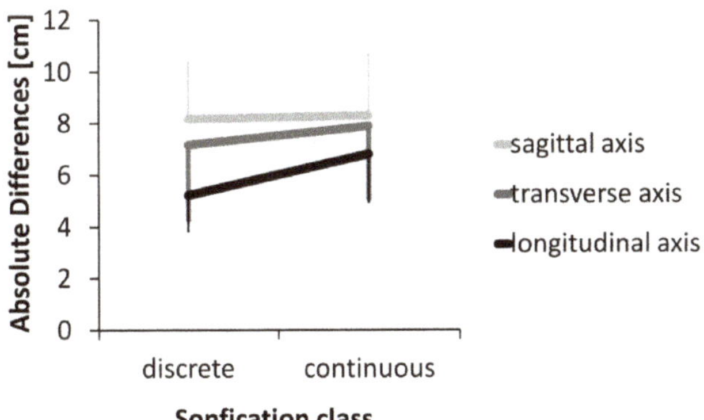

Figure 8. Means of absolute differences for the two sonification classes considered separately for the three axes.

Main effect "block" ($F_{(1, 22)} < 0.01$, $p = 0.955$, $\eta_p^2 < 0.01$) did not reach significance. The reaching performance in the discrete blocks in which both instruction and feedback took 1 s did not differ from that in the discrete blocks in which both instruction and feedback took as long as the continuous acoustic information did in the blocks with additional continuous information about the reaching movement.

4. Discussion

The present work investigated the efficacy of mere acoustic instruction and feedback of reaching movements of healthy subjects and the role of sight. We tested if the sonification device, developed by our work group is appropriate to represent reaching space merely acoustically. The participants were instructed to execute reaching movements to invisible, solely acoustically coded targets positioned in the frontal, the transverse and the sagittal plane in front of them. Differences in detail of the acoustic information shaped the experimental conditions. One of the two experimental groups was blindfolded.

Reaching is originally a visual action, but in this study, it was transformed into an auditory action. The mean absolute distance from the target point over all participants and measuring points was significantly lower than expected at random reaching. It can be concluded that it is possible to instruct action areas merely acoustically. Although the participants were not informed about how the sound was composed, they were able to orientate themselves towards the sonification. The developed sonification device enables the mere acoustic representation of reaching space. Regarding the invisibility of the target, with an average absolute distance of 11.68 cm from the target point in an active surface of 48.77 cm × 30.48 cm, the reaching performance was surprisingly high. Kinematic movement sonification can provide information for perception and action for humans, being classified as musical. This result corresponds to other studies with movement sonification [8,14,27,55].

The participants did not improve over time. No temporal effect became evident, even if the blocks with continuous and discrete instruction and feedback were regarded separately from those with only discrete sounds. The feedback did not seem to have any effect on the subsequent reaching performance. The reasons for the absence of a temporal effect need further investigation.

There was no difference between the two treatment groups with vs. without the possibility for visual movement control, indicating that visual perception is not required for the control of solely acoustically instructed reaching movements. Studies emphasizing the importance of sight for the precision of reaching movements work with visual targets [1,3,56,57]. In this case, vision is used for the adjustment of the movement to the target position. However, if the target cannot be seen, vision apparently cannot be applied to improve the reaching accuracy. This result indicates that we achieved a close proximity between the proprioceptive and auditory perception of the reaching position with the chosen sound design. For an application of kinematic movement acoustics in neuromotor rehabilitation, this result could mean that the sonification device is effective by itself and that additional visual focusing is redundant.

The continuous acoustic presentation of the hand position in the condition with discrete and continuous instruction and real-time feedback was supposed to function as auditory counterpart to the prevented continuous visual adjustments of the reaching movement, but it did not. The reaching accuracy in the experimental condition with additional continuous sonification as instruction and feedback was even impaired. A related effect was reported by Rosati et al. [34] who observed a worse performance in a visual tracking task with two kinds of visual feedback than in the task with only one visual feedback. The authors explain this effect with a saturation of the visual channel, so that the additional feedback acts as distraction, rather than providing useful information [34]. Boyer et al. [6] observed a similar effect of a continuous auditory real-time feedback of the hand position in reaching movements. In their above-described study, the addition of feedback in form of movement acoustics to an otherwise feedback-free acoustically instructed target did not improve the reaching accuracy. The authors discuss whether auditory feedback about the limb position provides benefitting information about a motor action. They consider that the participants might have been confused because the auditory modality overflowed with both the sonification of the target position and the positional movement sonification [6].

Possibly, complex acoustic sound sequences are not suitable to mark movements to discrete target points. For reaching movements, for which the target points are the focus, the movement itself might lose relevance. Another possibility is that the combination of a discrete and a continuous sonification component was responsible for the negative effect of the movement sonification. Either simply the additional cognitive load or the difference in the characteristic of the two sonification sounds might have led to an informational overload. Possibly, the combination of two different sonification characteristics (discrete vs. continuous or "knowledge of results" vs. "knowledge of performance") prevented the subjects from shaping an audio-proprioceptive action representation. In this way, the performance-based sonification might have functioned as an interference factor that disturbed the memory of the effect-/result-based sonification. Boyer et al. [6] assume that in this case the subjects do not use the acoustic feedback information but rely on the familiar and reliable proprioceptive information. Because the difference in the absolute differences in reaching precision between the two conditions dealt with here was 0.81 cm, the negative impact is not that serious.

The reported superiority of mere discrete instruction and feedback compared to the combination of a discrete and a continuous component for instruction and feedback becomes significant only for the longitudinal axis, but not for the transverse or the sagittal axis. Since the effect of alignment of axis and sound parameter are mixed here, it cannot be decided which factor is responsible for the differences in the reaching precision. We think it is more plausible that it is the coding by the pitch, rather than the longitudinal alignment of the axis, that is associated with the negative impact of the movement sonification on the reaching precision. Coincidently, the reaching precision on the longitudinal axis was significantly better than on the transverse and sagittal axes. The fact that the pitch (coding the position on the longitudinal axis) apparently transmits more information than the other two sound parameters might be the reason for the higher susceptibility to a disturbance by an additional movement sonification for target positions coded by pitch.

The acoustic presentation of the invisible target positions provided appropriate information for the alignment of the reaching movements. Acoustical instruction and feedback about the course of the reaching movement was not useful. However, for an application of continuous movement acoustics to patients with sensorimotor disabilities, the prerequisites differ fundamentally. For example, stroke patients partially show a dramatically reduced proprioceptive control. Here, the application of continuous movement acoustics should provide a possibility to substitute proprioceptive control rudimentally and thus support recovery in neuromotor rehabilitation. The implicit effectiveness of the acoustic information and the redundancy of focused attention should be a considerable advantage over the visual movement control. First steps towards the use of continuous kinematic movement sonification in stroke rehabilitation have already been realized [58,59]. Also, for patients with hip arthroplasty, kinematic movement sonification has the potential to accelerate the recovery, as shown by Reh et al. [60]. These are however only initial steps towards the use of real- time kinematic sonification in the neuromotor rehabilitation. Further research is clearly needed. It remains to be seen whether a prospective sonification device for the neuromotor rehabilitation of proprioceptively impaired patients will remain permanently necessary like a prosthesis or will become redundant after the treatment. It could be assumed that after a period of being exposed to an attendant auditory perception of one's own movements, the proprioception is resensitized. Auditory information might become redundant and internal feedback might be sufficient to sustain the acquired movement technique. The persisting learning effects following exposure to real-time movement sonification [14,61] are consistent with this assumption. Danna and Velay [12] studied this issue in two deafferented subjects but could not confirm the assumption. They discussed the reason for the absence of a learning effect. Motor learning might be per se impossible without proprioceptive feedback and sonification might also perspectively only serve as a prosthesis that the patients are dependent on for the rest of their life [12]. They counter that another explanation might be that the duration of the intervention has been too short to prove a learning effect and that motor learning simply takes more time in deafferented patients, possibly because most of the brain capacity is oriented towards coping with the task [12]. That would mean that the prosthesis might become dispensable one day because the patients might have regained their motor performance with the aid of movement sonification. This would offer the prospect of a sonification assisted recovery from proprioceptive deficits. A brain study from Ripollés et al. [62] supports the latter explanation. Music- supported-therapy in stroke rehabilitation results in a recovery of activation and connectivity between auditory and motor regions accompanied by an improvement of motor function [62].

In the present experiment, the hand position was coded by pitch, stereo characteristics, and spectral composition. Except for stereo characteristics, having a clear zero point with a clear spatial allocation, the parameters did not provide any natural indicator about the spatial position. For these parameters, meaning is only generated by the comparison of two target sounds. Moreover, because of the rather arbitrary choice of the limits of the parameter range, even the comparison between the target sounds at best provides information about the direction of the deviation but not about the extent of the deviation. This difficulty was compensated to some degree by the standardized familiarization protocol in which the participants gained experience with the proportions of the action area and got to know the sonification by drawing sinuous lines. Nevertheless, it remains remarkable that the participants were able to draw information from the sound. The participants were not experienced in using movement sonification and they were not informed about the systematic of the sonification. Still, they were able to use it to align target movements in space.

The reaching precision to acoustically coded targets could possibly be further enhanced. For example, the pitch range used to generate the stimuli was chosen arbitrarily. It must be considered that the choice might have been suboptimal. Moreover, there exists no reference, that spectral composition is particularly appropriate to map spatial depth. Possibly, a more informative sound characteristic can be found for this purpose. A follow-up study might focus on an expedient choice of the coding sound parameter and the optimal spectra of the used sound parameter to optimize the informational content of the alteration of the coding sound parameter.

In this study, we provide evidence that reaching space can be coded solely by artificial acoustics. Our sonification device developed in the first steps seems to be appropriate to represent target positions acoustically. The sonification of the invisible target position of the hand as instruction enabled reaching movements with respectable reaching precision. Although the participants were not informed in detail about how the sound was composed and had no previous experience with the use of movement sonification, they were able to use the kinematic acoustics to align their reaching movements. This implicit informational effect became evident, even when the possibility of an additional visual control of the reaching movement was excluded. Based on these results, a considerable benefit of a future application of an improved model of the developed sonification device in the neuromotor rehabilitation of proprioceptively impaired patients can be assumed.

Supplementary Materials: The following are available online at http://www.mdpi.com/2076-3417/10/2/429/s1, Table S1: Localization of the target points, Instruction-Audios S1–81: plane, condition, point.

Author Contributions: Conceptualization, U.F., A.O.E., G.S. and D.H.; methodology, U.F. and A.O.E.; software, H.B.; validation, U.F.; formal analysis, U.F., G.S., A.O.E., D.H., and S.G.; investigation, U.F. and D.H.; resources, A.O.E.; data curation, U.F., H.B., and D.H.; writing—original draft preparation, U.F.; writing—review and editing, U.F., G.S., D.H., S.G., H.B., and A.O.E.; visualization, U.F. and S.G.; supervision, A.O.E.; project administration, A.O.E.; funding acquisition, A.O.E. All authors have read and agreed to the published version of the manuscript.

Funding: This research was funded by the European Regional Development Fund (ERDF), project number W2-80118660.

Conflicts of Interest: The authors declare no conflict of interest.

References

1. Rossetti, Y.; Desmurget, M.; Prablanc, C. Vectorial coding of movement: Vision, proprioception, or both? *J. Neurophysiol.* **1995**, *74*, 457–463. [CrossRef] [PubMed]

2. Graziano, M.S. Is reaching eye-centered, body-centered, hand-centered, or a combination? *Rev. Neurosci.* **2001**, *12*, 175–186. [CrossRef] [PubMed]

3. Prablanc, C.; Pelisson, D.; Goodale, M.A. Visual control of reaching movements without vision of the limb. *Exp. Brain Res.* **1986**, *62*, 293–302. [CrossRef] [PubMed]

4. Oscari, F.; Secoli, R.; Avanzini, F.; Rosati, G.; Reinkensmeyer, D.J. Substituting auditory for visual feedback to adapt to altered dynamic and kinematic environments during reaching. *Exp. Brain Res.* **2012**, *221*, 33–41. [CrossRef]

5. Schmitz, G.; Bock, O. A Comparison of Sensorimotor Adaptation in the Visual and in the Auditory Modality. *PLoS ONE* **2014**, *9*, e107834. [CrossRef]

6. Boyer, E.O.; Babayan, B.M.; Bevilacqua, F.; Noisternig, M.; Warusfel, O.; Roby-Brami, A.; Hanneton, S.; Viaud-Delmon, I. From ear to hand: The role of the auditory-motor loop in pointing to an auditory source. *Front. Comput. Neurosci.* **2013**, *7*, 26. [CrossRef]

7. Schaffert, N.; Janzen, T.B.; Mattes, K.; Thaut, M.H. A review on the relationship between sound and movement in sports and rehabilitation. *Front. Psychol.* **2019**, *10*, 244. [CrossRef]

8. Effenberg, A.O. Movement sonification: Effects on perception and action. *IEEE Multimed.* **2005**, *12*, 53–59. [CrossRef]

9. Hwang, T.H.; Schmitz, G.; Klemmt, K.; Brinkop, L.; Ghai, S.; Stoica, M.; Maye, A.; Blume, H.; Effenberg, A.O. Effect-and performance-based auditory feedback on interpersonal coordination. *Front. Psychol.* **2018**, *9*, 404. [CrossRef]

10. Schaffert, N.; Mattes, K.; Effenberg, A.O. An investigation of online acoustic information for elite rowers in on-water training conditions. *J. Hum. Sport Exerc.* **2011**, *6*, 392–405. [CrossRef]

11. Schaffert, N.; Mattes, K. Effects of acoustic feedback training in elite-standard Para-Rowing. *J. Sports Sci.* **2015**, *33*, 411–418. [CrossRef]

12. Danna, J.; Velay, J.L. On the auditory-proprioception substitution hypothesis: Movement sonification in two deafferented subjects learning to write new characters. *Front. Neurosci.* **2017**, *11*, 137. [CrossRef] [PubMed]

13. Effenberg, A.O.; Schmitz, G.; Baumann, F.; Rosenhahn, B.; Kroeger, D. SoundScript-Supporting the acquisition of character writing by multisensory integration. *Open Psychol. J.* **2015**, *8*, 230–237. [CrossRef]

14. Effenberg, A.O.; Fehse, U.; Schmitz, G.; Krueger, B.; Mechling, H. Movement sonification: Effects on motor learning beyond rhythmic adjustments. *Front. Neurosci.* **2016**, *10*, 219. [CrossRef] [PubMed]

15. Sigrist, R.; Rauter, G.; Riener, R.; Wolf, P. Augmented visual, auditory, haptic, and multimodal feedback in motor learning: A review. *Psychon. Bull. Rev.* **2013**, *20*, 21–53. [CrossRef] [PubMed]

16. Beauchamp, M.S. See me, hear me, touch me: Multisensory integration in lateral occipital-temporal cortex. *Curr. Opin. Neurobiol.* **2005**, *15*, 145–153. [CrossRef]

17. Bidet-Caulet, A.; Voisin, J.; Bertrand, O.; Fonlupt, P. Listening to a walking human activates the temporal biological motion area. *Neuroimage* **2005**, *28*, 132–139. [CrossRef]

18. Scheef, L.; Boecker, H.; Daamen, M.; Fehse, U.; Landsberg, M.W.; Granath, D.O.; Mechling, H.; Effenberg, A.O. Multimodal motion processing in area V5/MT: Evidence from an artificial class of audio-visual events. *Brain Res.* **2009**, *1252*, 94–104. [CrossRef]

19. Bangert, M.; Altenmüller, E.O. Mapping perception to action in piano practice: A longitudinal DC-EEG study. *BMC Neurosci.* **2003**, *4*, 26. [CrossRef]

20. Gazzola, V.; Aziz-Zadeh, L.; Keysers, C. Empathy and the somatotopic auditory mirror system in humans. *Curr. Biol.* **2006**, *16*, 1824–1829. [CrossRef]

21. Kaplan, J.T.; Iacoboni, M. Multimodal action representation in human left ventral premotor cortex. *Cogn. Process.* **2007**, *8*, 103–113. [CrossRef] [PubMed]

22. Lahav, A.; Saltzman, E.; Schlaug, G. Action representation of sound: Audiomotor recognition network while listening to newly acquired actions. *J. Neurosci.* **2007**, *27*, 308–314. [CrossRef] [PubMed]

23. Schmitz, G.; Mohammadi, B.; Hammer, A.; Heldmann, M.; Samii, A.; Münte, T.F.; Effenberg, A.O. Observation of sonified movements engages a basal ganglia frontocortical network. *BMC Neurosci.* **2013**, *14*, 32. [CrossRef]

24. Murgia, M.; Hohmann, T.; Galmonte, A.; Raab, M.; Agostini, T. Recognising one's own motor actions through sound: The role of temporal factors. *Perception* **2012**, *41*, 976–987. [CrossRef]

25. Young, W.; Rodger, M.; Craig, C.M. Perceiving and reenacting spatiotemporal characteristics of walking sounds. *J. Exp. Psychol. Hum. Percept. Perform.* **2013**, *39*, 464–476. [CrossRef]

26. Agostini, T.; Righi, G.; Galmonte, A.; Bruno, P. The relevance of auditory information in optimizing hammer throwers performance. In *Biomechanics and Sports*; Pascolo, P.B., Ed.; Springer: Vienna, Austria, 2004; pp. 67–74.

27. Vinken, P.M.; Kröger, D.; Fehse, U.; Schmitz, G.; Brock, H.; Effenberg, A.O. Auditory coding of human movement kinematics. *Multisens. Res.* **2013**, *26*, 533–552. [CrossRef]

28. Fehse, U.; Weber, A.; Krüger, B.; Baumann, J.; Effenberg, A.O. Intermodal Action Identification. Presented at the International Conference on Multisensory Motor Behavior: Impact of Sound [Online], Hanover, Germany, 30 September–1 October 2013; p. 4. Available online: http://www.sonification-online.com/wp-content/uploads/2013/12/Postersession_neu.pdf (accessed on 12 October 2019).

29. Auvray, M.; Hanneton, S.; O'Regan, J.K. Learning to perceive with a visuo-auditory substitution system: Localisation and object recognition with 'The Voice'. *Perception* **2007**, *36*, 416–430. [CrossRef] [PubMed]

30. Levy-Tzedek, S.; Hanassy, S.; Abboud, S.; Maidenbaum, S.; Amedi, A. Fast, accurate reaching movements with a visual-to-auditory sensory substitution device. *Restor. Neurol. Neurosci.* **2012**, *30*, 313–323. [CrossRef]

31. Làdavas, E. Multisensory-based Approach to the recovery of unisensory deficit. *Ann. N. Y. Acad. Sci.* **2008**, *1124*, 98–110. [CrossRef]

32. Bevilacqua, F.; Boyer, E.O.; Françoise, J.; Houix, O.; Susini, P.; Roby-Brami, A.; Hanneton, S. Sensori-motor learning with movement sonification: Perspectives from recent interdisciplinary studies. *Front. Neurosci.* **2016**, *10*, 385. [CrossRef]

33. Scholz, D.S.; Wu, L.; Pirzer, J.; Schneider, J.; Rollnik, J.D.; Großbach, M.; Altenmüller, E.O. Sonification as a possible stroke rehabilitation strategy. *Front. Neurosci.* **2014**, *8*, 332. [CrossRef] [PubMed]

34. Rosati, G.; Oscari, F.; Spagnol, S.; Avanzini, F.; Masiero, S. Effect of task-related continuous auditory feedback during learning of tracking motion exercises. *J. Neuroeng. Rehabil.* **2012**, *9*, 79. [CrossRef] [PubMed]

35. Thaut, M.H.; McIntosh, G.C.; Rice, R.R.; Miller, R.A.; Rathbun, J.; Brault, J.M. Rhythmic auditory stimulation in gait training for Parkinson's disease patients. *Mov. Disord. Off. J. Mov. Disord. Soc.* **1996**, *11*, 193–200. [CrossRef] [PubMed]

36. Thaut, M.H.; McIntosh, G.C.; Rice, R.R. Rhythmic facilitation of gait training in hemiparetic stroke rehabilitation. *J. Neurol. Sci.* **1997**, *151*, 207–212. [CrossRef]

37. Thaut, M.H.; Kenyon, G.P.; Hurt, C.P.; McIntosh, G.C.; Hoemberg, V. Kinematic optimization of spatiotemporal patterns in paretic arm training with stroke patients. *Neuropsychologia* **2002**, *40*, 1073–1081. [CrossRef]

38. Ghai, S.; Ghai, I.; Effenberg, A.O. Effect of rhythmic auditory cueing on aging gait: A systematic review and meta-analysis. *Aging Dis.* **2018**, *9*, 901–923. [CrossRef]

39. Ghai, S.; Ghai, I.; Effenberg, A.O. Effect of rhythmic auditory cueing on gait in cerebral palsy: A systematic review and meta-analysis. *Neuropsychiatr. Dis. Treat.* **2018**, *14*, 43–59. [CrossRef]

40. Ghai, S.; Ghai, I.; Schmitz, G.; Effenberg, A.O. Effect of rhythmic auditory cueing on parkinsonian gait: A systematic review and meta-analysis. *Sci. Rep.* **2018**, *8*, 506. [CrossRef]

41. Thaut, M.H. The discovery of human auditory-motor entrainment and its role in the development of neurologic music therapy. *Prog. Brain Res.* **2015**, *217*, 253–266. [CrossRef]

42. Schauer, M.; Mauritz, K.H. Musical motor feedback (MMF) in walking hemiparetic stroke patients: Randomized trials of gait improvement. *Clin. Rehabil.* **2003**, *17*, 713–722. [CrossRef]

43. Young, W.R.; Shreve, L.; Quinn, E.J.; Craig, C.; Bronte-Stewart, H. Auditory cueing in Parkinson's patients with freezing of gait. What matters most: Action-relevance or cue-continuity? *Neuropsychologia* **2016**, *87*, 54–62. [CrossRef] [PubMed]

44. Maulucci, R.A.; Eckhouse, R.H. A real-time auditory feedback system for retraining gait. In Proceedings of the 2011 Annual International Conference of the IEEE Engineering in Medicine and Biology Society, Boston, MA, USA, 30 August–3 September 2011; pp. 5199–5202. [CrossRef]

45. Robertson, J.V.; Hoellinger, T.; Lindberg, P.; Bensmail, D.; Hanneton, S.; Roby-Brami, A. Effect of auditory feedback differs according to side of hemiparesis: A comparative pilot study. *J. Neuroeng. Rehabil.* **2009**, *6*, 45. [CrossRef] [PubMed]

46. Scholz, D.S.; Rohde, S.; Nikmaram, N.; Brückner, H.P.; Großbach, M.; Rollnik, J.D.; Altenmüller, E.O. Sonification of arm movements in stroke rehabilitation—A novel approach in neurologic music therapy. *Front. Neurol.* **2016**, *7*, 106. [CrossRef] [PubMed]

47. Rodger, M.W.; Young, W.R.; Craig, C.M. Synthesis of walking sounds for alleviating gait disturbances in Parkinson's disease. *IEEE Trans. Neural Syst. Rehabil. Eng.* **2014**, *22*, 543–548. [CrossRef] [PubMed]

48. Melara, R.D.; O'Brien, T.P. Interaction between synesthetically corresponding dimensions. *J. Exp. Psychol. Gen.* **1987**, *116*, 323–336. [CrossRef]

49. Küssner, M.B.; Tidhar, D.; Prior, H.M.; Leech-Wilkinson, D. Musicians are more consistent: Gestural cross-modal mappings of pitch, loudness and tempo in real-time. *Front. Psychol.* **2014**, *5*, 789. [CrossRef]

50. Online Evaluation of Amusia—Public. Available online: http://www.brams.org/amusia-public (accessed on 12 October 2019).

51. Peretz, I.; Gosselin, N.; Tillmann, B.; Cuddy, L.L.; Gagnon, B.; Trimmer, C.G.; Paquette, S.; Bouchard, B. On-line identification of congenital amusia. *Music Percept.* **2008**, *25*, 331–343. [CrossRef]

52. Fong, D.T.P.; Chan, Y.Y. The use of wearable inertial motion sensors in human lower limb biomechanics studies: A systematic review. *Sensors* **2010**, *10*, 11556–11565. [CrossRef]

53. Helten, T.; Brock, H.; Müller, M.; Seidel, H.P. Classification of trampoline jumps using inertial sensors. *Sports Eng.* **2011**, *14*, 155–164. [CrossRef]

54. Brock, H.; Schmitz, G.; Baumann, J.; Effenberg, A.O. If motion sounds: Movement sonification based on inertial sensor data. In *Procedia Engineering*; Drahne, P., Sherwood, J., Eds.; Elsevier: Amsterdam, The Netherlands, 2012; Volume 34, pp. 556–561. [CrossRef]

55. Effenberg, A.O.; Schmitz, G. Acceleration and deceleration at constant speed: Systematic modulation of motion perception by kinematic sonification. *Ann. N. Y. Acad. Sci.* **2018**, *1425*, 52–69. [CrossRef]

56. Ghez, C.; Gordon, J.; Ghilardi, M.F. Impairments of reaching movements in patients without proprioception. II. Effects of visual information on accuracy. *J. Neurophysiol.* **1995**, *73*, 361–372. [CrossRef] [PubMed]

57. Graziano, M.S. Where is my arm? The relative role of vision and proprioception in the neuronal representation of limb position. *Proc. Natl. Acad. Sci. USA* **1999**, *96*, 10418–10421. [CrossRef] [PubMed]

58. Schmitz, G.; Kroeger, D.; Effenberg, A.O. A mobile sonification system for stroke rehabilitation. In Proceedings of the Presented at the 20th International Conference on Auditory Display, New York, NY, USA, 22–25 June 2014. [CrossRef]

59. Schmitz, G.; Bergmann, J.; Effenberg, A.O.; Krewer, C.; Hwang, T.H.; Müller, F. Movement sonification in stroke rehabilitation. *Front. Neurol.* **2018**, *9*, 389. [CrossRef] [PubMed]

60. Reh, J.; Hwang, T.H.; Michalke, V.; Effenberg, A.O. Instruction and real-time sonification for gait rehabilitation after unilateral hip arthroplasty. In Proceedings of the Human Movement and Technology, Book of Abstracts—11th Joint Conference on Motor Control and Learning, Biomechanics and Training, Darmstadt, Germany, 28–30 September 2016; Wiemeyer, J., Seyfarth, A., Kollegger, G., Tokur, D., Schumacher, C., Hoffmann, K., Schöberl, D., Eds.; Shaker: Aachen, Germany, 2016.

61. Baudry, L.; Leroy, D.; Thouvarecq, R.; Chollet, D. Auditory concurrent feedback benefits on the circle performed in gymnastics. *J. Sports Sci.* **2006**, *24*, 149–156. [CrossRef]

62. Ripollés, P.; Rojo, N.; Grau-Sánchez, J.; Amengual, J.L.; Càmara, E.; Marco-Pallarés, J.; Juncadella, M.; Vaquero, L.; Rubio, F.; Duarte, E.; et al. Music supported therapy promotes motor plasticity in individuals with chronic stroke. *Brain Imaging Behav.* **2016**, *10*, 1289–1307. [CrossRef]

Article

Ankle Joint Dynamic Stiffness in Long-Distance Runners: Effect of Foot Strike and Shoes Features

Alessandro Garofolini [1,*], Simon Taylor [1], Patrick Mclaughlin [1], Karen J Mickle [1] and Carlo Albino Frigo [2]

[1] Institute for Health and Sport (IHeS), Victoria University, Melbourne 3011, Australia; simon.taylor@vu.edu.au (S.T.); patrick.mclaughlin@vu.edu.au (P.M.); karen.mickle@vu.edu.au (K.J.M.)
[2] Dipartimento di Elettronica, Informazione e Bioingegneria, Politecnico di Milano, 20133 Milan, Italy; carlo.frigo@polimi.it
* Correspondence: alessandro.garofolini1@live.vu.edu.au

Received: 3 September 2019; Accepted: 27 September 2019; Published: 1 October 2019

Abstract: Foot strike mode and footwear features are known to affect ankle joint kinematics and loading patterns, but how those factors are related to the ankle dynamic properties is less clear. In our study, two distinct samples of experienced long-distance runners: habitual rearfoot strikers (n = 10) and habitual forefoot strikers (n = 10), were analysed while running at constant speed on an instrumented treadmill in three footwear conditions. The joint dynamic stiffness was analysed for three subphases of the moment–angle plot: early rising, late rising and descending. Habitual rearfoot strikers displayed a statistically ($p < 0.05$) higher ankle dynamic stiffness in all combinations of shoes and subphases, except in early stance in supportive shoes. In minimal-supportive shoes, both groups had the lowest dynamic stiffness values for early and late rising (initial contact through mid-stance), whilst the highest stiffness values were at late rising in minimal shoes for both rearfoot and forefoot strikers (0.21 ± 0.04, 0.24 ± 0.06 (Nm/kg/°·100), respectively). In conclusion, habitual forefoot strikers may have access to a wider physiological range of the muscle torque and joint angle. This increased potential may allow forefoot strikers to adapt to different footwear by regulating ankle dynamic stiffness depending upon the motor task.

Keywords: running; biomechanics; footwear; joint work; loading

---

1. Introduction

There is an ongoing debate on whether the foot strike pattern of long-distance runners plays a role in defining performance and injury risk in this population [1–3]. Experienced long-distance runners are able to change their foot strike pattern during a competition [4] or if they are asked to [5]. Their ability to adopt a different foot strike pattern has been often interpreted as a sign of adaptability. The concepts of "adaptability" and "ability to adopt different execution patterns", however, are not equivalent [6]. Adaptability refers to the level of organisation embodied by the human locomotor control system [7]; it represents the richness of motor behaviours that equally can accomplish the task-goal [8]. In contrast, the ability to adopt different execution patterns refers to the ability to change joint kinematics (and kinetics) without necessarily meeting the task-goal. It is unknown if runners who adopt different execution patterns (i.e., rearfoot strikers versus forefoot strikers) have developed a different level of adaptability.

During the stance phase of running, the ankle plays a dominant role in storing and generating energy for propulsion [9,10]. The mode of foot/ground initial contact may affect the subsequent joint angle time course and the associated joint stiffness. According to Günther and Blickhan [11], the foot strike angle, stiffness and running velocity are crucial parameters for coordination of body movement dynamics. The concept of dynamic joint stiffness [12,13], defined "quasi-stiffness" by Latash and

Zatsiorsky [14], can be used to characterize the ankle behaviour during the stance phase of running [15]. Here, the ankle exhibits a first loading state in which the internal plantarflexor moment rises during dorsiflexion, and the periarticular joint structures absorb energy. It follows an unloading state in which the plantarflexion moment decreases while the joint plantarflexes, and the periarticular joint structures produce energy. The level of stiffness (that is the variation of joint moment per unit of joint angle variation) can depend on both (i) structural adaptations of the muscle–tendon units surrounding this joint and (ii) neural adaptations that control instantly the characteristics of these muscle–tendon units [16–18]. For instance, long-term adaptations in muscle and tendon architecture in the lower limb, such as shorter gastrocnemius medialis fascicles [19], thicker Achilles tendon [20] and stiffer foot arch [21], were found in habitual forefoot strikers, who usually land with a plantar-flexed ankle. Such adaptations could lead to a different load distribution in the muscle–tendon unit [22], in which the role of the elastic components is increased and the muscle fibres contract at a slower rate, which is advantageous for maximal power output and efficiency [23]. Together, both the structural and the neural adaptations contribute to defining the dimensionality of the system (degrees of freedom of the neural control system), that is the number of structures (muscles) that can be actively controlled and can be used to regulate the ankle dynamic stiffness efficiently, according to the mechanical requirements [24].

Ankle dynamic stiffness can be computed as the slope of the tangent to the moment–angle curve [12]. Using similar approaches, previous studies investigated dynamic ankle stiffness during running [9–11]. To our knowledge, Hamill and Gruber [5] were the only researchers testing change in ankle joint stiffness in two groups of runners with distinct foot strike patterns. Participants were classified as either rearfoot or forefoot strikers based on the presence of an impact peak on the vertical ground reaction force and on the ankle angle at landing. Although using these criteria runners may have been misclassified [25], according to Hamill and Gruber [5], habitual forefoot strikers exhibited a more compliant ankle and absorbed more (negative) work than habitual rearfoot strikers when running with their preferred foot strike pattern (forefoot); however, no differences were found with habitual rearfoot strikers running with a forefoot strike pattern (nonpreferred mode).

It is common for studies concerning running and ankle stiffness to simplify the loading phase of the moment–angle loop by representing the linear slope from initial foot contact to peak moment [5,15,26–28] (Figure 1, dashed line). This approach overlooks the potentially meaningful details occurring within the loading phase. For instance, at initial foot contact, the ankle joint moment increases nonlinearly with the change in angle (red portion in Figure 1). This state represents the ankle joint response to foot/ground initial loading. Thereafter, the ankle dorsiflexes slowly, while the ankle moment increases fast (blue portion in Figure 1).

This is a less compliant condition, representing the loading of the passive structures of the muscle–tendon units, and can be seen as a different state [29]. Another interesting phase that deserves attention is the unloading phase (light-blue portion in Figure 1), where the movement of the joint reverses to plantarflexion and the joint moment decreases. To the best of our knowledge, no studies have investigated the stance phase of running by considering these three task-relevant subphases of the moment–angle loop, which we expect to yield a more sensitive insight of the differences between habitual rearfoot and forefoot strikers.

The aim of this study was to investigate if foot strike loading technique has an effect on the ankle moment–angle dynamics during the stance phase of running. We expected forefoot strikers to have lower dynamic stiffness during the loading phase, based on previous findings [5]. We also expected forefoot strikers to have a higher proportion of negative work relative to positive work because of their loading technique that allows them to store and return elastic potential energy in the foot–ankle structure. To test whether habitual foot strike loading technique compromises the control of ankle stiffness, shoes with different assistive constructs were considered. We expected differences in ankle stiffness and work ratio between groups to be greater in minimally assistive shoes, due to the unfamiliarity of rearfoot strikers to this condition. Because we expected forefoot strikers have a greater intrinsic foot–ankle adaptability to external loading (i.e., greater ability to control ankle stiffness), we expected their (forefoot strikers) ankle stiffness to have a higher time-dependency. That is, ankle stiffness during unloading will depend more on ankle stiffness during loading in forefoot strikers.

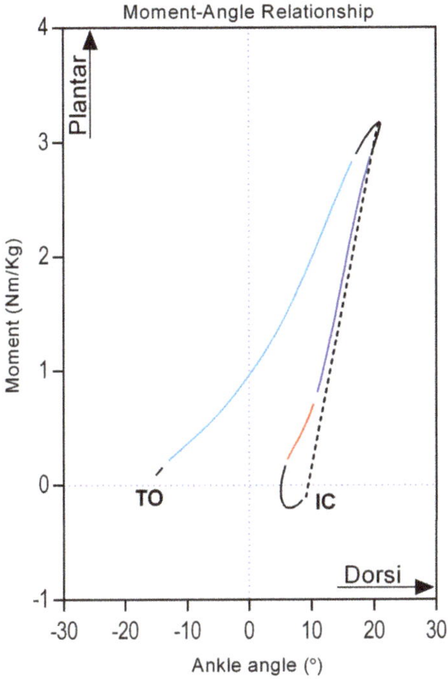

Figure 1. Example of moment–angle loop for the ankle joint. IC = initial foot contact; TO = toe off. Dashed line represents commonly computed slope for ankle stiffness during the loading phase. Red line represents foot/ground initial loading; blue line represents loading phase; light-blue line represents unloading phase.

2. Materials and Methods

2.1. Participants

Forty male long-distance runners gave their personal consent to take part in this study. Participants were excluded if they had not been running for at least 5 years, with an average of at least 40 km/week, and had not been free of neurological, cardiovascular or musculoskeletal problems within the previous six months. A number of 21 runners were found eligible. One subject was unable to complete the study protocol, which resulted in a tested sample of 20 subjects (age: 31.2 ± 6.9 yrs, height: 1.77 ± 0.07 cm, weight: 73.4 ± 7.9 kg). Participants were classified as rearfoot strikers (RFS, n = 10) or forefoot strikers (FFS, n = 10) based on their habitual mode of foot/ground initial contact. To classify their foot strike loading type, the participants were asked to run on an instrumented treadmill (AMTI Pty, Watertown, MA, USA) at their preferred speed, wearing their habitual running shoes. After a standardized 7 min of progressive warm up (starting from 3 min at 6 km/h, followed by 2 min at 8 km/h, then 2 min at 10 km/h) and accommodation period (i.e., quantitatively assessed stable foot strike angle—around 2 min), participants ran for 3 min at their preferred running speed, which was identified from the protocol suggested by Jordan and Challis [30]. Habitual foot strike mode was assessed on the basis of data collected on the last minute of running (~60 steps). The ankle internal moment of the dominant leg was analysed to this purpose within a short period, from initial ground contact to the time at which the vertical ground reaction force exceeded a threshold corresponding to body weight. Runners displaying an internal plantarflexor moment for at least 90% of this period were classified as forefoot strikers (FFS); conversely, those who displayed an internal dorsiflexor moment for at least 90% of the analysed period were classified as rearfoot strikers (RFS). This foot strike classification method was shown to perform best among other conventional methods [25].

2.2. Experimental Protocol

Tests were performed on an instrumented treadmill (Advanced Mechanical Technology Inc., Watertown, MA, USA) that collects ground reaction forces [31] at a sampling rate of 1000 Hz. Three-dimensional kinematics data of the lower extremities was recorded at a sampling rate of 250 Hz from a 14-camera VICON B-10 system (Oxford Metrics Ltd., Oxford, UK). Kinematic and ground reaction force data were synchronised using a VICON MX-Net control box and collected through Nexus 2.6 software (Vicon Motion Systems Ltd., Oxford, UK). A biomechanical model was reconstructed from 45 retroreflective markers, placed on proper landmarks of body segments (Appendix A).

After completing a standardized and progressive 7 min warm-up, participants repeated three times a 5 min running test, with different shoes for each trial; the three shoe models were characterized by their different minimalist indexes (MI). The minimalist index is a classification that takes into account structure, flexibility, pronation support and other footwear features, and ranges from 0% (maximum assistance) to 100% (least interaction with the foot) [32]. The shoes adopted in our experiments were classified at low-MI (Mizuno® Wave Rider 21, MI = 18%), medium-MI (Mizuno® Wave Sonic, MI = 56%) and high-MI (Vibram® Five fingers, MI = 96%). The order of presentation was pseudorandom, which means that combinations were balanced within each group and equal between groups. Testing speed was fixed for all participants at 11 km/h.

2.3. Data Analysis

Three-dimensional kinematics and kinetic data were analysed in Visual3D software (C-Motion, Inc, Rockville, MD, USA). A digital low-pass Butterworth filter (4th order, zero lag) was used to smooth raw kinematic and kinetic data with cut-off frequency of 15 and 35 Hz, respectively. The ankle joint angle was calculated as the relative angle between the foot and the shank longitudinal axes, and subtracted to the ankle joint angle computed during subject's standing calibration posture (reference angle). Joint moments were computed around the ankle flexion/extension axis (the axis connecting the medial and lateral malleoli) using Newton–Euler inverse dynamics approach and normalized to body mass. Stance time was defined by gait events of initial and terminal foot contact (IC and TC) that were determined by a vertical ground reaction force threshold of 20 N. Stance time was normalised to 101 data points. The ankle (internal) moment was plotted as a function of the corresponding ankle angle (moment–angle plot), and the resultant curve was subdivided into three functionally relevant phases: early rising (ERP), late rising (LRP) and descending-phase (DP), similar to Crenna and Frigo [12]. The ERP was defined as the period between an ascending threshold of 0.2 peak plantarflexion moment and the first identified abrupt change in the statistical properties of the signal (i.e., mean and slope) going forward. The LRP was defined as the period between a threshold of 0.95 ascending moment to the first change in the statistical properties of the signal (i.e., mean and slope) going backwards.

The slope of the best regression line interpolating the angle–moment curve in each phase represents the average ankle joint stiffness (K_{ankle}) in each functionally relevant phase (Figure S1). The area under the rising component and the descending component of the curve was integrated using a trapezoidal approximation. This gives the work absorbed (W_{abs}, during the loading phase) and the work produced (W_{prod}, during the unloading phase), respectively. The net work (W_{net}) produced was computed as the difference between W_{prod} and W_{abs}. Finally, the work ratio ($W_{ratio} = W_{abs}/W_{prod}$) was computed and was considered as a measure of muscle efficiency.

2.4. Statistical Analysis

An initial check for normal distribution (Shapiro–Wilk test) of the dependent variables and homogeneity of variance (Levene's test) was performed. A three-way repeated-measures ANOVA was used to test the effect of the between-factor *Group* (RFS, FFS) and within-factors *Shoe* (LOW, MED and HIGH MI index), and *Slope* (ERP, LRP and DP) on K_{ankle}. A two-way repeated measure ANOVA was used to test the effect of the between-factor *Group* (RFS, FFS) and within-factors *Shoe* for dependent variables W_{abs}, W_{prod}, W_{net} and W_{ratio}. If ANOVA was significant, a posthoc multiple comparison Tukey's test was used to determine where the differences were. Pearson Correlation coefficient (r) was calculated for all couples of dependent variables, while the coefficient of determination (r^2) was estimated from a linear regression model run among ERP, LRP, and DP. In case of non-normal distribution of data, the equivalent nonparametric tests (Kruskal–Wallis test, Dunn's multiple comparisons test, Spearman correlation) were used. All statistical analyses were performed using SPSS (version 25.0. Armonk, NY, USA: IBM Corp.). Statistical significance was set at $p < 0.05$, with multiple pairwise comparisons corrected with Bonferroni adjustment method. Magnitude of changes (effect sizes) was assessed using partial Eta squared (η_p^2) for the ANOVA, and Hedges' *g* for pairwise (posthoc) comparisons.

3. Results

No main effect of group was found for K_{ankle} ($p = 0.164$; $\eta_p^2 = 0.105$), but the main effects for shoe type ($p = 0.008$; $\eta_p^2 = 0.272$) and slope ($p < 0.001$; $\eta_p^2 = 0.889$) were obtained (Table S1). Posthoc analysis revealed that K_{ankle} was 12% higher in med-MI compared with high-MI shoes ($p = 0.007$; $g = 0.706$). Table S2 and Figure 2 show mean and SD for K_{ankle} in the three subphases of stance and among the three shoe conditions. Significant differences were found among all subphases: ERP–LRP (0.176 ± 0.01; 0.215 ± 0.01 Nm/kg/°·100) $p = 0.001$; $g = 3.9$; ERP–DP (0.176 ± 0.01; 0.091 ± 0.01 Nm/kg/°·100) $p < 0.001$; $g = 9.5$; LRP–DP (0.215 ± 0.01; 0.091 ± 0.01 Nm/kg/°·100) $p = 0.001$; $g = 12.4$. Overall K_{ankle} was highest when wearing med-MI shoes (although not statistically different from low-MI shoes; $p = 0.246$); K_{ankle} was highest during the late rising phase (LRP) and lowest during the unloading phase (DP).

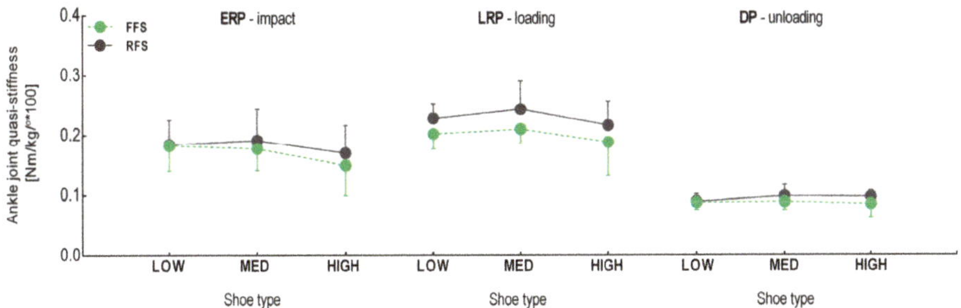

Figure 2. Mean and SD values for ankle joint dynamic stiffness of FFS and RFS for the three phases of stance, in the three shoe conditions. ERP early rising phase, LRP late rising phase, DP descending phase. Shoes conditions are termed as low-MI (LOW), medium-MI (MED) and high-MI (HIGH). MI: minimalist indexes. FFS: forefoot strikers. RFS: rearfoot strikers

Runners in high-MI shoes exhibited a lower stiffness (more compliant ankle) during the impact phase (ERP) and late rising phase (LRP); during the unloading phase (DP), low-MI shoes allowed the most compliant ankle. There was a *Shoe* by *Slope* interaction effect ($p = 0.008$; $\eta_p^2 = 0.221$; Table S1) for K_{ankle} (Figure 2, Table S2). Pairwise multiple comparisons showed that during the impact phase (ERP), K_{ankle} in high-MI shoes was lower compared with that in both low-MI and med-MI shoes (−15%, $p = 0.013$, $g = 1.2$; and −16%, $p = 0.003$, $g = 1.25$, respectively). During the late rising phase (LRP), K_{ankle} was the highest in med-MI shoes (0.227 ± 0.01 Nm/kg/°·100), but only statistically different from high-MI shoes (+12%, $p = 0.011$, $g = 1.58$). During the unloading phase (DP), differences between shoes were only significant for low-MI compared with med-MI shoes (−6%, $p = 0.009$, $g = 0.5$).

Figure 3 compares mean moment–angle loops for RFS and FFS. While curves are similar in low-MI shoes, (Figure 3, top) the base (ankle range of motion) is shifted toward the left for FFS. This is also true for medium-MI (Figure 3, middle), and high-MI shoes (Figure 3, bottom).

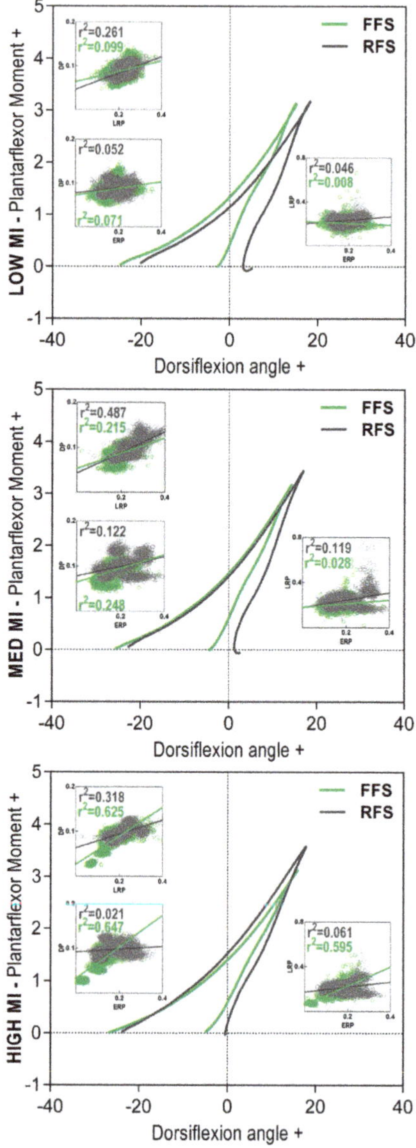

Figure 3. Ankle moment–angle plot. Group mean profiles comparison for low-MI, medium-MI and high-MI shoes. Insets report linear regression lines among early rising phase (ERP), late rising phase (LRP) and descending phase (DP).

Overall, runners exhibiting high K_{ankle} during the late rising phase (LRP) also have high K_{ankle} during the unloading phase (DP) (Table 1). For FFS, the correlation between K_{ankle} in the impact phase (ERP) and in late rising phase (LRP) increased with shoes' MI, with the highest correlation ($r_s = 0.95$; $p < 0.01$) in high-MI shoes. A similar trend was reported for correlations between K_{ankle} in impact phase (ERP) and in unloading (DP), and between K_{ankle} in late rising phase (LRP) and in unloading (DP), with highest values in the high-MI condition ($r_s = 0.84$, $p < 0.01$; $r_s = 0.89$, $p < 0.01$, respectively). Values were only significant in high-MI shoe conditions; this means that FFS in high-MI shoes with high K_{ankle} during impact phase will also have high K_{ankle} during the loading and unloading phases. For RFS, correlations between K_{ankle} in impact phase (ERP) and in late rising phase (LRP) and correlations between K_{ankle} in impact phase (ERP) and in unloading (DP) vary irrespectively to the shoe condition. The correlation between K_{ankle} in late rising phase (LRP) and in unloading (DP) increased with shoes' MI, with the highest correlation ($r_s = 0.92$; $p < 0.01$) in high-MI shoes. This means, K_{ankle} during impact has less of an effect on the subsequent subphases in RFS; instead, the late rising phase plays a central role.

In low-MI shoes, both groups presented low regression values ($r^2 \leq 0.26$, insets in Figure 3). In medium-MI shoes, K_{ankle} of RFS during the loading phase (LRP) explained 49% of the K_{ankle} variance during the unloading phase (DP), while for FFS, only 22% was explained. K_{ankle} of FFS in high-MI shoes depended on the stiffness in the previous phase: that is, stiffness during the impact phase (ERP) explained 60% of the stiffness variance during the late rising phase (LRP) and 65% of the stiffness variance during the unloading phase (DP); likewise, stiffness during the late rising phase (LRP) explained 63% of the stiffness variance during the unloading phase (DP).

We found a main effect of shoes for W_{abs} and W_{prod} ($p = 0.001$, $\eta_p^2 = 0.425$; $p < 0.001$, $\eta_p^2 = 0.517$) but no main effect of group ($p = 0.105$; $p = 0.716$) or interaction effects for *Groups* by *Shoes* were found ($p = 0.051$; $p = 0.097$) (Table S1). Figure 4 shows that W_{prod} increased significantly from low-MI to med-MI shoes (7%, $p = 0.004$, $g = 0.578$) and from med-MI to high-MI shoes (11%, $p = 0.017$, $g = 0.657$); while W_{abs} decreased as an inverse function of shoe MI index, reaching highest values in high-MI shoes (-32.58 ± 1.71 Nm/kg/°·100). The latter was significantly lower than W_{abs} in low-MI (-19%, $p = 0.002$, $g = 0.839$) and med-MI shoes (-14%, $p = 0.009$, $g = 0.674$). RFS exhibited higher W_{net} compared with FFS (24.99 ± 1.25 versus 19.47 ± 1.25; $p = 0.006$, $g = 4.420$); W_{net} increased with shoe MI index, with runners in low-MI shoes exhibiting statistically lower W_{net} (-12%; $p = 0.007$, $g = 0.456$) compared with those in med MI-shoes, and compared with those in high-MI shoes (-20%; $p = 0.028$, $g = 0.728$).

Rear foot strikers in high-MI shoes had the highest net work values (27.8 ± 8 Nm/kg/°·100) associated to increased work absorbed (+28% from LOW, $p < 0.001$ $g = 2.09$; +16% from MED, $p < 0.001$, $g = 1.60$) and produced (+30% from LOW, $p < 0.001$, $g = 2.17$; +21% from MED, $p < 0.001$, $g = 1.17$) (Figure 4); however, the work ratio (absorbed/produced) for RFS was statistically lower than for FFS (0.55 vs. 0.59, $g = 0.533$). FFS increased positive work going from LOW to MED (+5%; $p < 0.001$, $g = 0.465$) and from MED to HIGH (+6%; $p < 0.001$, $g = 0.319$); while negative work was not statistically different from LOW (28.84 ± 5.8 Nm/kg/°·100) and MED (29.21 ± 6.0 Nm/kg/°·100; $p = 0.327$), but in HIGH, negative work was higher than in both LOW (+9%; $p < 0.001$, $g = 0.342$) and MED (+8%; $p < 0.001$, $g = 0.299$); however, net work in HIGH (20.4 ± 5.5 Nm/kg/°·100) was similar ($p = 0.781$) to MED (20.2 ± 5.0 Nm/kg/°·100) and LOW (18.8 ± 6 Nm/kg/°·100).

As for the correlation between energetic (work) measures (Table 1), FFS exhibited high negative correlations values between W_{abs} and W_{prod} in all shoe conditions ($r_s \leq -0.69$), meaning that the more work they absorbed during loading, the less work they needed to produce during the unloading phase. RFS did not show such correlations; instead, they exhibited high positive correlations ($r_s \geq 0.60$) between W_{prod} and W_{net}, meaning that the net work increased as the produced work increased.

Table 1. Correlations between moment–angle loop parameters (Spearman correlation coefficient rs). * represents statistically significant correlations ($p < 0.05$); ** represents statistically significant correlations ($p < 0.01$). W_{abs}: work absorbed. W_{net}: net work. W_{prod}: work produced. W_{ratio}: work ratio.

		FFS						RFS					
		Slope LRP	Slope DP	W_{abs}	W_{prod}	W_{net}	W_{ratio}	Slope LRP	Slope DP	W_{abs}	W_{prod}	W_{net}	W_{ratio}
LOW	Slope ERP	−0.16	0.41	−0.41	0.04	−0.46	−0.39	0.19	0.20	−0.27	0.60	0.31	−0.01
	Slope LRP		0.10	0.47	−0.22	0.26	0.38		0.58	−0.03	0.44	0.36	0.35
	Slope DP			−0.52	0.09	−0.65 *	−0.71 *			−0.24	0.02	−0.41	−0.39
	W_{abs}				−0.69 *	0.37	0.64 *				−0.50	0.27	0.60
	W_{prod}					0.36	0.03					0.60	0.30
	W_{net}						0.93 **						0.88 **
MED	Slope ERP	0.41	0.67 *	−0.44	0.32	−0.09	−0.22	0.19	0.42	−0.60	0.35	0.29	−0.02
	Slope LRP		0.76 *	−0.10	0.13	−0.29	−0.21		0.67 *	−0.01	−0.07	−0.02	0.20
	Slope DP			−0.67 *	0.55	−0.27	−0.44			−0.05	−0.43	−0.47	−0.24
	W_{abs}				−0.82 **	0.08	0.50				−0.61	−0.53	0.16
	W_{prod}					0.39	−0.08					0.99 **	0.53
	W_{net}						0.86 **						0.61
HIGH	Slope ERP	0.95 **	0.84 **	−0.19	0.30	0.71 *	0.13	0.09	−0.02	0.04	0.21	0.27	0.31
	Slope LRP		0.89 **	−0.22	0.30	0.65 *	0.08		0.92 **	0.26	−0.08	0.21	0.43
	Slope DP			−0.53	0.58	0.58	0.07			0.14	−0.08	0.15	0.25
	W_{abs}				−0.93 **	−0.01	0.38				−0.52	0.19	0.66 *
	W_{prod}					0.26	−0.16					0.64 *	0.21
	W_{net}						0.61						0.79 **

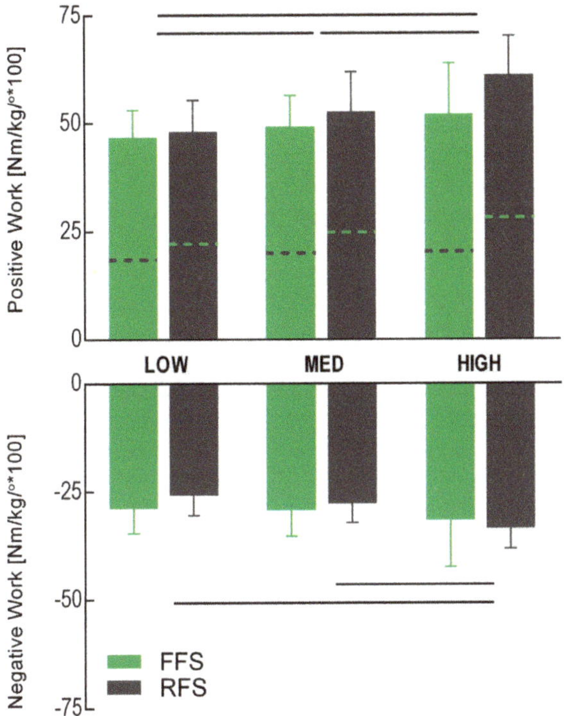

Figure 4. Mean and SD of ankle work for the three footwear conditions. Values are shown for positive and negative work for FFS and RFS. Dashed lines across the bars indicate the net work. Horizontal solid lines above and below the bar graph signify a statistically significant ($p < 0.05$) difference between footwear conditions.

4. Discussion

The purpose of this study was to explore the effect of foot strike modes and footwear features on the dynamic control of the ankle dynamics stiffness. There was no group main effect for ankle stiffness, contrary to our hypothesis that FFS had a lower ankle stiffness than RFS. Hamill, Gruber [5] investigated stiffness during the phase of stance that corresponds most closely to the LRP region of our study. By examining a main effect of group within the LRP region (ignoring ERP and DP), we have also confirmed a statistically higher (+14%; $p = 0.005$, $g = 0.725$) ankle stiffness in the RFS group. However, within the LRP, there was not a main effect of *Shoe* on ankle stiffness ($p = 0.163$). Previous studies found that changing shoe support altered the level of joint stiffness [26,33]; where ankle dynamic stiffness increased as the shoe hardness decreased [34]. While increasing stiffness may be functional in preventing excessive joint movement [35], it has been identified as a possible risk of Achilles tendon injuries in runners [36].

The rearfoot strike loading technique generated more positive (produced) work by the ankle joint. This confirms our hypothesis and is consistent with previous studies that found ankle plantar flexor muscles to store more elastic energy (negative work) during the loading phase of fast running (i.e., forefoot strike) compared with positive work during unloading [37,38]. The RFS group in our study exhibited 34% higher net work compared with FFS (Table S2 and Table 1), which correlated strongly with the work produced (Figure 3); indicating that there was more muscle energy produced compared with elastic energy stored [39]. Efficient running is achieved by efficiently storing and releasing elastic energy at each step; our results are in line with previous literature data that found

FFS to store and return more elastic energy than RFS [40–42]. Despite this energetic advantage, FFS is consistently reported to be energetically inefficient [43,44], probably because storing energy in passive structures requires muscle contraction [45]. Therefore, it may be concluded that saving and releasing energy in the plantarflexor muscles may not significantly reduce the whole-body metabolic cost of running with a forefoot strike pattern [46].

The FFS group demonstrated a time-dependent ankle stiffness across the stance phase, especially for the high-MI shoe condition, fulfilling our hypothesis. Furthermore, within the same shoe condition, the FFS group had strong correlations between ankle stiffness (K_{ankle}) during both impact and loading phases and net work (W_{net}). By controlling ankle stiffness, the work around the ankle was modulated, probably to achieve a functional redistribution of loading along the lower limb joints [10,47]. Furthermore, Figure 3 indicates that the K_{ankle} of FFS running in minimally supportive shoes is constant through the impact, late rising (loading) and unloading subphases, suggesting that foot strike at landing is a determinant for ankle dynamic stiffness not only at impact, but also during the loading and unloading phases. A similar correlation has been found between the initial joint stiffness and maximal stiffness during the stance phase of hopping [48]. One possible explanation for a constant ankle stiffness is that in that configuration (ankle plantarflexion with minimal support) the ankle–foot complex can express its spring-like function [49–51]; while increasing shoe support may introduce a level of instability that requires a trade-off between the task-goals of energy recycling and stable locomotion [52].

Shoe characteristics influenced the control of ankle dynamic stiffness. Both groups were able to reduce ankle dynamic stiffness during impact and loading phase when wearing high-MI shoes (Figure 2, Table S2). However, both groups also increased the work produced and absorbed, so that the total net work done around the ankle during stance increased as a function of the shoe MI index (Figure 4, Table S2). Control and modulation of these loads need a certain level of adaptability of both the musculoskeletal and neuronal systems [53]. This may explain the high risk of certain injuries when changing from low- to high-MI shoes [54] or from RFS to FFS patterns [55].

Limitations

We acknowledge that the energy absorbed or produced at the foot/ankle is not only associated to flexion/extension, but also to foot/shoes deformation [56]. In addition, in this study, analysis was limited to the ankle joint. Indeed, adding analysis on the work done around knee and hip would have validated our assumption on leg-level force stabilization. Other limitations are the assumed symmetry between dominant and nondominant leg. The modulation of joint dynamic stiffness and the redistribution of joint work may vary if significant asymmetry exist [57].

5. Conclusions

In this study, we investigated the effect of habitual rearfoot strike loading pattern, and the assistance of shoes, on ankle stiffness control. Our results suggested that RFS has reduced adaptability when compared with FFS, but the constraint of this ability is dependent on the shoe worn. These findings reiterate the idea that functional changes at joint level are important to define the redistribution of load along the lower-limb kinetic chain in order to solve leg-level force control. Shoes with a low MI may limit the ability to utilize the spring-like function of the ankle–foot complex, while shoes with high MI may promote the exploitation of the system redundancy. However, further studies are warranted to confirm the effect of shoes on ankle neuromuscular adaptations.

Supplementary Materials: The following are available online at http://www.mdpi.com/2076-3417/9/19/4100/s1, Figure S1: Example of ankle moment–angle relationship for a FFS subject and a RFS subject for the normalized stance phase from initial contact to toe-off. Table S1: Primary statistical results for differences between *Groups*, *Shoes*, and *Slopes* for mean ankle stiffness (K_{ankle}), work produced (W_{prod}), work absorbed (W_{abs}), work net (W_{net}), and work ratio (W_{ratio}). Table S2: Mean and (SD) for *Groups, Shoes,* and *Slopes* for mean ankle stiffness, work produced (W_{prod}), work absorbed (W_{abs}), work net (W_{net}), and work ratio (W_{ratio}).

Author Contributions: A.G.: Conceptualization, Methodology, Formal Analysis, Investigation, Data curation, Software, Visualization, Writing—Original draft preparation. S.T.: Conceptualization, Methodology, Supervision, Reviewing and Editing. P.M.: Methodology, Supervision, Reviewing and Editing. K.J.M.: Methodology, Supervision, Reviewing and Editing. C.A.F.: Writing-Reviewing and Editing.

Funding: This research received no external funding.

Acknowledgments: The authors gratefully acknowledge Mizuno Footwear Company for their financial support.

Conflicts of Interest: The authors declare no competing interests relevant to the content of this study.

Appendix A

Biomechanical Model

A set of retroreflective markers arranged in cluster setup were used to track 3D position of body segments, while landmark-derived virtual markers and movement-derived virtual markers were used to calibrate the position and orientation of the lower body skeletal system. Semirigid clusters of 4–5 markers were attached to lower-body segments so that the location of the cluster centroid was minimally affected by muscular contraction and related mass deformation. To minimize effects of skin movement artefact [58,59], we secured the semirigid clusters over extra-long neoprene bands made of antimigration material that wrapped and fastened on the thigh and shank segments. Individual trunk and pelvis retroreflective markers were placed over the 7th cervical vertebrae, sterno-clavicular notch, 10th thoracic vertebrae and posterior- and anterior-superior iliac spines. Virtual markers were used to identify medial and lateral epicondyles of the femur and medial and lateral malleoli. A custom version of the IOR multisegment foot model [60] was adopted for the foot marker setup. Retroreflective markers were placed on calcanei, first metatarsal bases and heads, second metatarsal bases and heads, navicular bones and base and heads of the 5th metatarsals.

To fix the 9.5 mm reflective markers on the foot, we removed the internal screw from the markers and replaced with a 6 mm diameter × 1.5 mm long Rare Earth Magnet fixed with superglue. After identifying the foot anatomical landmarks, we applied a similar magnet on the skin, fixed with topical skin adhesive glue. Participants performed testing in socks and shoes. All shoes were modified, with the circular holes cut at anatomical landmarks. Foot markers were attached to magnets that were preglued to the skin of the participants, ensuring repeatable marker location associated with reattachment process between footwear conditions (Figure A1).

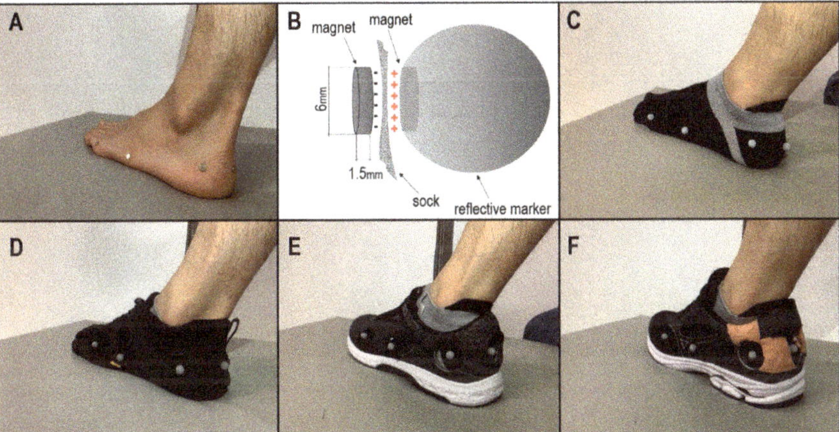

Figure A1. (**A**) Magnets glued to bony landmarks; (**B**) Schematic representation of magnets interaction; (**C**) markers placed over the sock, maintaining the same position; (**D–F**) markers position in the three shoe conditions: Vibram® Five fingers (**D**), Mizuno® Wave Sonic (**E**), Mizuno® Wave Rider 21 (**F**).

Hip joint centre and knee joint axis of rotation were defined using functional movement trials according to Camomilla and Cereatti [61], and Schwartz and Rozumalski [62]. A six-degrees of freedom segment model was built for biomechanical analysis in Visual3D software (C-motion Inc., Rockville, MD, USA). Standard methods were used to calibrate segment pose from marker setup and reconstruct the subject biomechanical model in Visual3D. For joint rotations, we used a right-handed orthogonal coordinate systems, where the z-axis represented the axial direction of the segment. The x-axis lied in the frontal plane perpendicular to the z-axis. The y-axis lied on the sagittal plane in the antero-posterior direction. In Visual3D, joint angles were calculated using an x–y–z Cardan–Euler sequence representing flexion/extension, abduction/adduction and axial rotation of the thigh, shank and foot [63]. For the pelvis, the Cardan sequence was reversed (z–y–x), as recommended by Baker [64]. Joint angles were normalized to the subject static reference position, recorded as a "standing calibration trial". For the scope of this study, the segment movements of interest were those within the sagittal plane only (i.e., flexion/extension rotations).

The force signal recorded was assigned to relevant foot segment based on detection software in Visual3D. The estimated foot assigned to the force is based on the proximity between the location of the centre of mass of the foot and the transverse plane location of the centre of pressure on the force plate. Force signals were then used to compute joint moment (through inverse dynamic calculations) represented in the joint coordinate system [65].

References

1. Bramble, D.M.; Lieberman, D.E. Endurance running and the evolution of Homo. *Nature* **2004**, *432*, 345–352. [CrossRef] [PubMed]
2. Davis, I.S.; Rice, H.M.; Wearing, S.C. Why forefoot striking in minimal shoes might positively change the course of running injuries. *J. Sport Health Sci.* **2017**, *6*, 154–161. [CrossRef] [PubMed]
3. Hamill, J.; Gruber, A.H. Is changing footstrike pattern beneficial to runners? *J. Sport Health Sci.* **2017**, *6*, 146–153. [CrossRef] [PubMed]
4. Larson, P.; Higgins, E.; Kaminski, J.; Decker, T.; Preble, J.; Lyons, D.; McIntyre, K.; Normile, A. Foot strike patterns of recreational and sub-elite runners in a long-distance road race. *J. Sport Health Sci.* **2011**, *29*, 1665–1673. [CrossRef] [PubMed]
5. Hamill, J.; Gruber, A.H.; Derrick, T.R. Lower extremity joint stiffness characteristics during running with different footfall patterns. *Eur. J. Sport Sci.* **2014**, *14*, 130–136. [CrossRef] [PubMed]
6. Garofolini, A.; Taylor, S.; Mclaughlin, P.; Vaughan, B.; Wittich, E. Acute adaptability to barefoot running among professional AFL players. *Footwear Sci.* **2017**, *9*, S44–S45. [CrossRef]
7. Lacquaniti, F.; Ivanenko, Y.P.; Zago, M. Patterned control of human locomotion. *J. Physiol.* **2012**, *590*, 2189–2199. [CrossRef]
8. Todorov, E.; Jordan, M.I. Optimal feedback control as a theory of motor coordination. *Nat. Neurosci.* **2002**, *5*, 1226–1235. [CrossRef] [PubMed]
9. Jin, L.; Hahn, M.E. Modulation of lower extremity joint stiffness, work and power at different walking and running speeds. *Hum. Mov. Sci.* **2018**, *58*, 1–9. [CrossRef] [PubMed]
10. Schache, A.G.; Brown, N.A.T.; Pandy, M.G. Modulation of work and power by the human lower-limb joints with increasing steady-state locomotion speed. *J. Exp. Biol.* **2015**, *218*, 2472–2481. [CrossRef]
11. Günther, M.; Blickhan, R. Joint stiffness of the ankle and the knee in running. *J. Biomech.* **2002**, *35*, 1459–1474. [CrossRef]
12. Crenna, P.; Frigo, C. Dynamics of the ankle joint analyzed through moment–angle loops during human walking: Gender and age effects. *Hum. Mov. Sci.* **2011**, *30*, 1185–1198. [CrossRef] [PubMed]
13. Gabriel, R.C.; Abrantes, J.; Granata, K.; Bulas-Cruz, J.; Melo-Pinto, P.; Filipe, V. Dynamic joint stiffness of the ankle during walking: Gender-related differences. *Phys. Ther. Sport* **2008**, *9*, 16–24. [CrossRef] [PubMed]
14. Latash, M.L.; Zatsiorsky, V.M. Joint stiffness: Myth or reality? *Hum. Mov. Sci.* **1993**, *12*, 653–692. [CrossRef]
15. Stefanyshyn, D.J.; Nigg, B.M. Dynamic angular stiffness of the ankle joint during running and sprinting. *J. Appl. Biomech.* **1998**, *14*, 292–299. [CrossRef] [PubMed]

16. Feldman, A. Superposition of motor programs—I. Rhythmic forearm movements in man. *Neuroscience* **1980**, *5*, 81–90. [CrossRef]

17. Guissard, N.; Duchateau, J. Neural Aspects of Muscle Stretching. *Exerc. Sport Sci. Rev.* **2006**, *34*, 154–158. [CrossRef]

18. Duchateau, J.; Enoka, R.M. Neural control of lengthening contractions. *J. Exp. Biol.* **2016**, *219*, 197–204. [CrossRef]

19. Cronin, N.J.; Finni, T. Treadmill versus overground and barefoot versus shod comparisons of triceps surae fascicle behaviour in human walking and running. *Gait Posture* **2013**, *38*, 528–533. [CrossRef]

20. Lichtwark, G.A.; Cresswell, A.G.; Newsham-West, R.J. Effects of running on human Achilles tendon length–tension properties in the free and gastrocnemius components. *J. Exp. Biol.* **2013**, *216*, 4388–4394. [CrossRef]

21. Lieberman, D.E. Strike type variation among Tarahumara Indians in minimal sandals versus conventional running shoes. *J. Sport Health Sci.* **2014**, *3*, 86–94. [CrossRef]

22. Kubo, K.; Miyazaki, D.; Ikebukuro, T.; Yata, H.; Okada, M.; Tsunoda, N. Active muscle and tendon stiffness of plantar flexors in sprinters. *J. Sports Sci.* **2017**, *35*, 742–748. [CrossRef] [PubMed]

23. Lichtwark, G.; Bougoulias, K.; Wilson, A. Muscle fascicle and series elastic element length changes along the length of the human gastrocnemius during walking and running. *J. Biomech.* **2007**, *40*, 157–164. [CrossRef] [PubMed]

24. Latash, M.L. The bliss (not the problem) of motor abundance (not redundancy). *Exp. Brain Res.* **2012**, *217*, 1–5. [CrossRef] [PubMed]

25. Garofolini, A.; Taylor, S.; Mclaughlin, P.; Vaughan, B.; Wittich, E. Foot strike classification: A comparison of methodologies. *Footwear Sci.* **2017**, *9*, S129–S130. [CrossRef]

26. Sinclair, J.; Atkins, S.; Taylor, P.J. The Effects of Barefoot and Shod Running on Limb and Joint Stiffness Characteristics in Recreational Runners. *J. Mot. Behav.* **2016**, *48*, 79–85. [CrossRef]

27. Kuitunen, S.; Komi, P.V.; Kyröläinen, H. Knee and ankle joint stiffness in sprint running. *Med. Sci. Sports Exerc.* **2002**, *34*, 166–173. [CrossRef] [PubMed]

28. Powell, D.W.; Williams, D.S., 3rd; Windsor, B.; Butler, R.J.; Zhang, S. Ankle work and dynamic joint stiffness in high-compared to low-arched athletes during a barefoot running task. *Hum. Mov. Sci.* **2014**, *34*, 147–156. [CrossRef]

29. Roberts, T.J. Contribution of elastic tissues to the mechanics and energetics of muscle function during movement. *J. Exp. Biol.* **2016**, *219*, 266–275. [CrossRef]

30. Jordan, K.; Challis, J.H.; Newell, K.M. Speed influences on the scaling behavior of gait cycle fluctuations during treadmill running. *Hum. Mov. Sci.* **2007**, *26*, 87–102. [CrossRef]

31. Garofolini, A.; Taylor, S.; Lepine, J. Evaluating dynamic error of a treadmill and the effect on measured kinetic gait parameters: Implications and possible solutions. *J. Biomech.* **2019**, *82*, 156–163. [CrossRef] [PubMed]

32. Esculier, J.-F.; Dubois, B.; Dionne, C.E.; Leblond, J.; Roy, J.-S. A consensus definition and rating scale for minimalist shoes. *J. Foot Ankle Res.* **2015**, *8*, 1–9. [CrossRef] [PubMed]

33. Apps, C.; Sterzing, T.; O'Brien, T.; Lake, M. Lower limb joint stiffness and muscle co-contraction adaptations to instability footwear during locomotion. *J. Electromyogr. Kinesiol.* **2016**, *31*, 55–62. [CrossRef] [PubMed]

34. Baltich, J.; Maurer, C.; Nigg, B.M. Increased vertical impact forces and altered running mechanics with softer midsole shoes. *PLoS ONE* **2015**, *10*, e0125196. [CrossRef] [PubMed]

35. Riemann, B.L.; Myers, J.B.; Lephart, S.M. Sensorimotor system measurement techniques. *J. Athl. Train.* **2002**, *37*, 85–98. [PubMed]

36. Lorimer, A.V.; Hume, P.A. Stiffness as a Risk Factor for Achilles Tendon Injury in Running Athletes. *Sports Med.* **2016**, *46*, 1921–1938. [CrossRef]

37. Lai, A.; Schache, A.G.; Brown, N.A.; Pandy, M.G. Human ankle plantar flexor muscle–tendon mechanics and energetics during maximum acceleration sprinting. *J. R. Soc. Interface* **2016**, *13*, 20160391. [CrossRef]

38. Hof, A.; van Zandwijk, J.; Bobbert, M. Mechanics of human triceps surae muscle in walking, running and jumping. *Acta Physiol. Scand.* **2002**, *174*, 17–30. [CrossRef]

39. Biewener, A.A.; Roberts, T.J. Muscle and tendon contributions to force, work, and elastic energy savings: A comparative perspective. *Exerc. Sport Sci. Rev.* **2000**, *28*, 99–107.

40. Hasegawa, H.; Yamauchi, T.; Kraemer, W.J. Foot strike patterns of runners at the 15-km point during an elite-level half marathon. *J. Strength Cond. Res.* **2007**, *21*, 888–893.

41. Lieberman; Venkadesan, M.; Werbel, W.A.; Daoud, A.I.; D'Andrea, S.; Davis, I.S.; Mang'Eni, R.O.; Pitsiladis, Y. Foot strike patterns and collision forces in habitually barefoot versus shod runners. *Nature* **2010**, *463*, 531–535. [CrossRef] [PubMed]

42. Perl, D.P.; Daoud, A.I.; Lieberman, D.E. Effects of footwear and strike type on running economy. *Med. Sci. Sports Exerc.* **2012**, *44*, 1335–1343. [CrossRef] [PubMed]

43. Gruber, A.H.; Umberger, B.R.; Braun, B.; Hamill, J. Economy and rate of carbohydrate oxidation during running with rearfoot and forefoot strike patterns. *J. Appl. Physiol. (1985)* **2013**, *115*, 194–201. [CrossRef] [PubMed]

44. Ogueta-Alday, A.; Rodríguez-Marroyo, J.A.; García-López, J. Rearfoot striking runners are more economical than midfoot strikers. *Med. Sci. Sports Exerc.* **2014**, *46*, 580–585. [CrossRef]

45. Holt, N.C.; Roberts, T.J.; Askew, G.N. The energetic benefits of tendon springs in running: Is the reduction of muscle work important? *J. Exp. Biol.* **2014**, *217*, 4365–4371. [CrossRef] [PubMed]

46. Gruber, A.; Umberger, B.R.; Miller, R.H.; Hamill, J.H. Muscle mechanics and energy expenditure of the triceps surae during rearfoot and forefoot running. *bioRxiv* **2018**. [CrossRef]

47. Yen, J.T.; Auyang, A.G.; Chang, Y.-H. Joint-level kinetic redundancy is exploited to control limb-level forces during human hopping. *Exp. Brain Res.* **2009**, *196*, 439–451. [CrossRef] [PubMed]

48. Rapoport, S.; Mizrahi, J.; Kimmel, E.; Verbitsky, O.; Isakov, E. Constant and variable stiffness and damping of the leg joints in human hopping. *J. Biomech. Eng.* **2003**, *125*, 507–514. [CrossRef]

49. Farris Dominic, J.; Brent, J.R. Modulation of leg joint function to produce emulated acceleration during walking and running in humans. *R. Soc. Open Sci.* **2017**, *4*, 160901. [CrossRef] [PubMed]

50. Kelly, L.A.; Farris, D.J.; Lichtwark, G.A.; Cresswell, A.G. The Influence of Foot-Strike Technique on the Neuromechanical Function of the Foot. *Med. Sci. Sports Exerc.* **2018**, *50*, 98–108. [CrossRef]

51. Riddick, R.; Farris, D.J.; Kelly, L.A. The foot is more than a spring: Human foot muscles perform work to adapt to the energetic requirements of locomotion. *J. R. Soc. Interface* **2019**, *16*, 20180680. [CrossRef] [PubMed]

52. Ferris, D.P.; Louie, M.; Farley, C.T. Running in the real world: Adjusting leg stiffness for different surfaces. *Proc. R. Soc. Lond. Ser. B Biol. Sci.* **1998**, *265*, 989–994. [CrossRef] [PubMed]

53. Cronin, N.J.; Carty, C.P.; Barrett, R.S. Triceps surae short latency stretch reflexes contribute to ankle stiffness regulation during human running. *PLoS ONE* **2011**, *6*, e23917. [CrossRef] [PubMed]

54. Giuliani, J.; Masini, B.; Alitz, C.; Owens, B.D. Barefoot-simulating footwear associated with metatarsal stress injury in 2 runners. *Orthopedics* **2011**, *34*, e320–e323. [CrossRef] [PubMed]

55. Daoud, A.I.; Geissler, G.J.; Wang, F.; Saretsky, J.; Daoud, Y.A.; Lieberman, D.E. Foot Strike and Injury Rates in Endurance Runners: A Retrospective Study. *Med. Sci. Sports Exerc.* **2012**, *44*, 1325–1334. [CrossRef] [PubMed]

56. Farinelli, V.; Hosseinzadeh, L.; Palmisano, C.; Frigo, C. An easily applicable method to analyse the ankle-foot power absorption and production during walking. *Gait Posture* **2019**, *71*, 56–61. [CrossRef]

57. Exell, T.A.; Irwin, G.; Gittoes, M.J.; Kerwin, D.G. Implications of intra-limb variability on asymmetry analyses. *J. Sport Health Sci.* **2012**, *30*, 403–409. [CrossRef]

58. Taylor, W.R.; Ehrig, R.M.; Duda, G.N.; Schell, H.; Seebeck, P.; Heller, M.O. On the influence of soft tissue coverage in the determination of bone kinematics using skin markers. *J. Orthop. Res.* **2005**, *23*, 726–734. [CrossRef]

59. Leardini, A.; Chiari, L.; Della Croce, U.; Cappozzo, A. Human movement analysis using stereophotogrammetry: Part 3. Soft tissue artifact assessment and compensation. *Gait Posture* **2005**, *21*, 212–225. [CrossRef]

60. Leardini, A.; Benedetti, M.G.; Berti, L.; Bettinelli, D.; Nativo, R.; Giannini, S. Rear-foot, mid-foot and fore-foot motion during the stance phase of gait. *Gait Posture* **2007**, *25*, 453–462. [CrossRef]

61. Camomilla, V.; Cereatti, A.; Vannozzi, G.; Cappozzo, A. An optimized protocol for hip joint centre determination using the functional method. *J. Biomech.* **2006**, *39*, 1096–1106. [CrossRef] [PubMed]

62. Schwartz, M.H.; Rozumalski, A. A new method for estimating joint parameters from motion data. *J. Biomech.* **2005**, *38*, 107–116. [CrossRef] [PubMed]

63. Robertson, D.G.E.; Robertson, G.; Caldwell, G.; Hamill, J.; Kamen, P.G.; Whittlesey, S. *Research Methods in Biomech*, 2nd ed.; Human Kinetics: Champaign, IL, USA, 2013.

64. Baker, R. Pelvic angles: A mathematically rigorous definition which is consistent with a conventional clinical understanding of the terms. *Gait posture* **2001**, *13*, 1–6. [CrossRef]

65. Schache, A.G.; Baker, R. On the expression of joint moments during gait. *Gait Posture* **2007**, *25*, 440–452. [CrossRef] [PubMed]

 *applied sciences*

Article

Changes in Ground Reaction Forces, Joint Mechanics, and Stiffness during Treadmill Running to Fatigue

Zhen Luo [1,†], Xini Zhang [1,†], Junqing Wang [1], Yang Yang [1], Yongxin Xu [1] and Weijie Fu [1,2,*]

[1] School of Kinesiology, Shanghai University of Sport, Shanghai 200438, China; luozhen716@gmail.com (Z.L.); xinizhang.sus@gmail.com (X.Z.); wangjunqing.sus@gmail.com (J.W.); yangyang.sus@gmail.com (Y.Y.); xuyongxin266@gmail.com (Y.X.)

[2] Key Laboratory of Exercise and Health Sciences of Ministry of Education, Shanghai University of Sport, Shanghai 200438, China

* Correspondence: fuweijie@sus.edu.cn; Tel.: +86-21-65507368; Fax: +86-21-51253242

† Zhen Luo and Xini Zhang contributed equally to this work.

Received: 31 October 2019; Accepted: 10 December 2019; Published: 13 December 2019

Abstract: Purpose: This study aimed to determine the changes in lower extremity biomechanics during running-induced fatigue intervention. Methods: Fourteen male recreational runners were required to run at 3.33 m/s until they could no longer continue running. Ground reaction forces (GRFs) and marker trajectories were recorded intermittently every 2 min to quantify the impact forces and the lower extremity kinematics and kinetics during the fatiguing run. Blood lactate concentration (BLa) was also collected before and after running. Results: In comparison with the beginning of the run duration, (1) BLa significantly increased immediately after running, 4 min after running, and 9 min after running; (2) no changes were observed in vertical/anterior–posterior GRF and loading rates; (3) the hip joint range of motion (θ_{ROM}) significantly increased at 33%, 67%, and 100% of the run duration, whereas θ_{ROM} of the knee joint significantly increased at 67%; (4) no changes were observed in ankle joint kinematics and peak joint moment at the ankle, knee, and hip; and (5) vertical and ankle stiffness decreased at 67% and 100% of the run duration. Conclusion: GRF characteristics did not vary significantly throughout the fatiguing run. However, nonlinear adaptations in lower extremity kinematics and kinetics were observed. In particular, a "soft landing" strategy, achieved by an increased θ_{ROM} at the hip and knee joints and a decreased vertical and ankle stiffness, was initiated from the mid-stage of a fatiguing run to potentially maintain similar impact forces.

Keywords: fatiguing run; instrumented treadmill; joint range of motion; stiffness

1. Introduction

Long-distance running has many benefits, including reducing the risk of cardiovascular disease [1]. Although there is not sufficient literature demonstrating that injuries are inevitable for all runners, running-related musculoskeletal injuries are still very common. Previous studies showed that the incidence of injury per runner for 1000 h is 18.2–92.4% [2–4] or 6.8–59% [5]. Amongst them, more than 74% of long-distance runners adopt the heel strike pattern [6]. The repetitive impact forces occurred during touchdown can reach a magnitude ranging from 2- to 3-times body weight [7], and are considered as high risk factors of lower limb injury [6].

Long-term and high-intensity running inevitably induces neuromuscular fatigue, which further influences musculoskeletal system of the lower limbs. Generally, the movement control of the lower extremity decreases after fatigue, which is an important cause of running injury [8,9]. However, conclusions on the relationship among fatigue, impact force, and lower extremity biomechanics during prolonged running remain uncertain [10]. Morin [11] found that vertical ground reaction forces (vGRF) decreases with fatigue after multiple groups of repeated sprints are performed on a force treadmill. By contrast, Christina et al. [12] observed that the impact peak and loading rate of the runners increase significantly after fatigue intervention. However, Gerlach et al. [13] found no significant differences in vGRF after fatigue intervention induced by repeated sprints. Similarly, Slawinski et al. [14] reported no changes in vGRF after nine well-trained runners performed an exhaustive exercise over 2000 m on an indoor track. Collectively, multifactorial causes underlie these different responses, and further studies beyond the analysis of the GRF level are warranted to provide insight on the joint mechanics strategies and the underlying stiffness characteristics occurring during running to fatigue.

Most previous studies focused on the direct relationship between fatigue and injury, especially the biomechanical changes in the lower extremities before and after fatigue [15–19]. For instance, Radzak et al. [14] reported the biomechanical asymmetry (e.g., loading rate and knee/vertical stiffness) of the lower limbs before and after fatigue. However, the fatigued state was considered to be achieved only by increased rated perceived exertion (RPE) scores. Abt et al. [17] compared the differences in running kinematics and shock absorption before and after fatigue. However, the influence of running fatigue on kinematics remains inconclusive, which might be partly due to the nonlinear adaptations in lower extremity biomechanics. Clansey et al. [18] reported an increased vertical force loading rate along with increased hip extension and ankle plantarflexion at incremental running speeds before and after two 20-min fatiguing treadmill runs. Nevertheless, they did not determine the threshold at which point the changes may occur. Thus, more investigations are necessary to examine fatigue-induced changes in lower extremity biomechanics not only in pre- and post-fatigue conditions but during the whole progress of a fatiguing run.

In addition, none of the aforementioned studies provided a direct accessible evaluation of running fatigue. Fatigue is generally associated with lactic acid accumulation [20]. Blood lactate concentration (BLa) is an indicator of muscle metabolites and recovery [21]. In the current study, BLa was used to help determine fatigue as a consequence of an insufficient oxygen supply in the capillary blood [22].

The purpose of this study, therefore, was to determine the changes in lower extremity biomechanics, namely, vertical and anterior–posterior GRFs, loading rates, joint mechanics and stiffness, during treadmill running to fatigue. It was hypothesized that joint mechanics and stiffness with fatigue would be changed. Specifically, runners would increase their vGRFs and joint range of motion. They would also have low vertical stiffness (k_{vert}) and joint stiffness during the progress of a fatiguing run.

2. Materials and Methods

2.1. Participants

Fourteen male recreational runners (age: 27.9 ± 7.5 years; height: 175.6 ± 5.5 cm; body mass: 70.4 ± 7.1 kg; weekly running volume: 29.3 ± 12.0 km) with a minimum of 15 km/week for at least 3 months prior to the study were recruited to participate in the study. All the participants had no history of musculoskeletal injuries to the lower extremity in the previous 6 months and did not engage in strenuous exercise for 24 h before the study. Each participant signed an informed consent form before the experiments. The study was approved by the Institutional Review Board of the Shanghai University of Sports (No. 2017007).

2.2. Instrumentation

Kinematic data were collected using an eight-camera infrared 3D motion capture system (Vicon T40, Oxford Metrics, UK) at a sampling rate of 100 Hz. Forty infrared retroreflective markers, each with a diameter of 14.0 mm, were attached bilaterally to both lower extremities to define hip, knee, and ankle joints by using a plug-in gait marker set (Figure 1). GRFs were measured with a split-belt fully instrumented treadmill (Bertec Corporation, Columbus, OH, USA) at a sampling rate of 1000 Hz. The 3D kinematic and GRF data were synchronized using the Vicon system. A heart rate (HR) monitor (V800, Polar Electro-OY, Kempele, Finland) was attached to each participant's chest to continuously monitor their HR during the entire fatiguing run procedure. Blood lactate concentration (BLa) was collected at different time points that were determined from the capillary blood sampled from the right fingertip and assessed with a Biosen C-line glucose analyzer (EKF Diagnostics, Magdeburg, Germany). Specifically, after cleaning with a sterile alcohol swab, a finger prick capillary puncture was performed and approximately 10 μL of sampled whole blood was aspirated into an EKF Diagnostics glucose and lactate analyzer [23]. Four measurement time points were obtained [21,24]: upon arrival at the laboratory (pre-), immediately after running (post-0 min), 4 min after running (post-4 min), and 9 min after running (post-9 min).

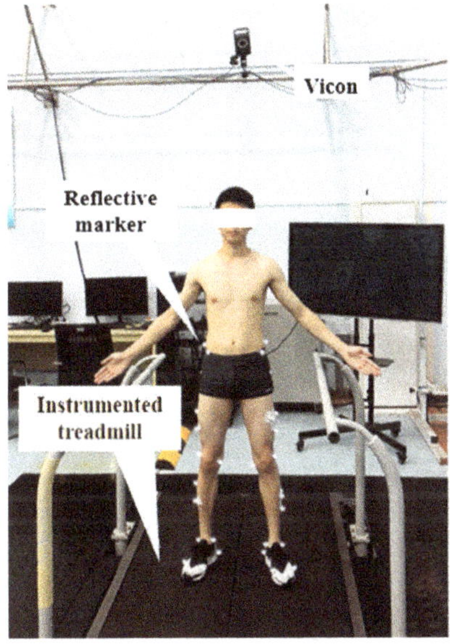

Figure 1. Marker set and experimental setup.

2.3. Running-Induced Fatigue Protocol

The participants were naked from the waist up and run at their self-selected speed for 5 min on the instrumented treadmill as a warm-up and familiarization. Standardized neutral running shoes (Nike Air Pegasus 34) were provided (Nike, Beaverton, OR, USA). Afterwards, they were required to run at 3.33 m/s [25] until they could not continue running [26]. They were considered to have experienced fatigue, and intervention was terminated when both following criteria were met: (1) the HR of the participants reached 90% of their age-calculated maximum HR, and (2) the participants could not continue running.

During the fatiguing run, the running time and the highest HR were recorded. The rated perceived exertion (RPE) using a Borg 6–20 scale was acquired immediately after running.

2.4. Data Processing

For the marker trajectories and GRF data, the measurements of at least 20 steps (15 s) were recorded intermittently every 2 min during the fatiguing run [22]. The sagittal plane kinematic data of the dominant lower extremity, defined as the preferred kicking leg, were filtered through a Butterworth fourth-order, low-pass filter at a cut-off frequency of 14 Hz with V3D software (v5, C-Motion Inc.,

Germantown, MD, USA) [27]. The kinematic variables of the hip, knee and ankle joints included the following: (1) joint angles (θ_0) at initial contact (Figure 2), (2) maximum joint extension/dorsiflexion angles ($\theta_{max\text{-}ext}$) and peak joint flexion/plantar flexion angles ($\theta_{max\text{-}flx}$) during stance, (3) joint ranges of motion (θ_{ROM}, $\theta_{ROM} = \theta_{max\text{-}ext} - \theta_{max\text{-}flx}$) during stance, (4) changes in joint angle ($\Delta\theta$, $\theta_{max\text{-}flex} - \theta_0$), and (5) maximum extension/plantarflexion angular velocity ($\omega_{max\text{-}ext}$) during the stance.

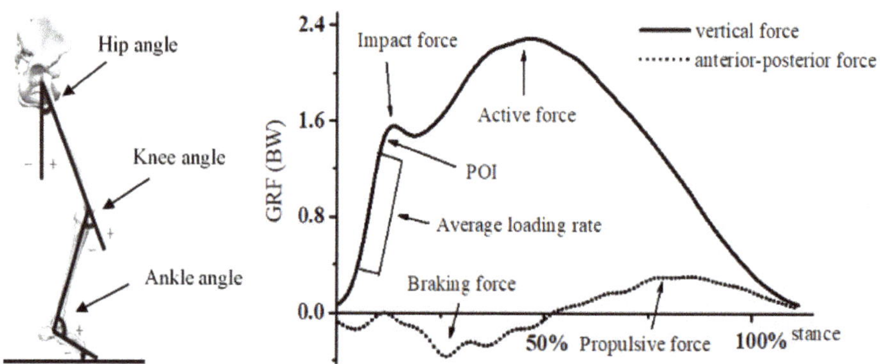

Figure 2. Scheme of lower extremity kinematics and ground reaction force variables. Note: GRF, ground reaction force; BW: body weight; POI: a point of interest.

GRF variables included (1) first and second peak vertical GRFs (F_{zmax1} and F_{zmax2}) and the occurrence time of F_{zmax1} and F_{zmax2} (t_F_{zmax1} and t_F_{zmax2}), (2) maximum and average loading rates (LR_{max} and LR_{avg}; Figure 2), (3) contact time (CT) and (4) maximum propulsive and braking GRF (F_{ymax} and F_{ymin}). Loading rate was calculated on the basis of the method described by Futrell et al. [28]. In brief, a point of interest (POI) was defined as the first point above 75% of a participant's body weight with the instantaneous loading rate less than 15 body weight/s. LR_{max} (i.e., the maximum instantaneous slope) and LR_{avg} (the average slope) were calculated from 20% to 100% and from 20% to 80% of the force at the POI, respectively (Figure 2).

Kinetic variables included the peak moments and joint stiffness [29] [k_j, Equation (1)] of hip, knee and ankle the vertical stiffness of the lower extremity [30] [k_{vert}, Equation (2)], and they were expressed as follows:

$$k_j = \frac{\Delta M}{\Delta\theta} \tag{1}$$

where ΔM is the joint moment difference between initial contact and mid-stance, and $\Delta\theta$ is the joint angle difference between initial contact and mid-stance.

$$k_{vert} = \frac{GRF_i}{\Delta y} \tag{2}$$

where GRF_i is the vertical ground reaction force at the lowest position of the center of gravity (CoG), and Δy is the maximum vertical displacement of CoG.

2.5. Statistics

All data are given as mean ± standard deviation. A power analysis was performed prior to the study to indicate the statistical power. It revealed that a sample size of 14 was sufficient to minimize the probability of Type II error for our variables of interest. A repeated measures ANOVA was performed to determine the effects of time points (pre-, post-0 min, post-4 min, and post-9 min) on BLa. Moreover, the variables from the relative time points of beginning, 33%, 67%, and end of each participant's test were included in this analysis because the participants fatigued at varying periods [31]. Thus, repeated measures analysis of variance (ANOVA) was performed to determine the effects of periods on GRFs, loading rates, joint mechanics, and stiffness (20.0, SPSS, Inc., Chicago, IL, USA). Partial eta squared (η_p^2) was used as an estimate of effect size. Post-hoc pairwise comparison tests were used to assess the changes at the different time points. The level of significance was set at 0.05.

3. Results

3.1. Running-Induced Fatigue Intervention

The intervention time to produce a fatigue state was 28.5 ± 10.4 min. The maximal HR and RPE scale observed during fatigue were 182.9 ± 7.7 bpm and 17.2 ± 0.9, respectively. Moreover, a main effect of time points was observed for BLa ($p < 0.001$, $\eta_p^2 = 0.76$). Specifically, a significant increase in BLa was observed immediately after running (+329.9%, $p < 0.001$), 4 min after running (+251.6%, $p < 0.001$), and 9 min after running (+135.9%, $p = 0.006$) compared with that in the time point corresponding to the pre-running (Figure 3).

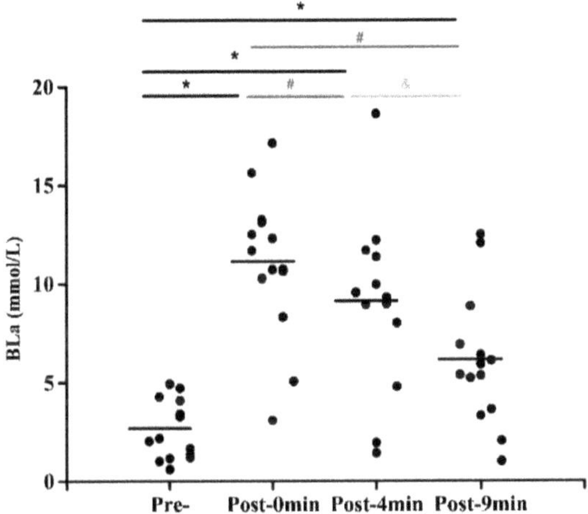

Figure 3. Blood lactate concentration (BLa) at time points corresponding to (**1**) upon arrival at the laboratory (pre-), (**2**) immediately after running (post-0 min), (**3**) 4 min after running (post-4 min), and (**4**) 9 min after running (post-9 min). *, #, and & significantly different from the pre-, post-0 min, and post-4 min with $p < 0.05$, respectively.

3.2. Ground Reaction Force

No significant differences were observed in F_{zmax1}, t_F_{zmax1}, LR_{max}, LR_{avg}, F_{zmax2}, t_F_{zmax2}, F_{ymax}, and F_{ymin} at four time points corresponding to the beginning, 33%, 67%, and 100% of the run duration (Table 1). CT increased significantly at 67% of the run duration compared to the beginning ($p < 0.05$).

Table 1. Ground reaction forces (GRF) characteristics at time points corresponding to the beginning, 33%, 67%, and 100% of the run duration.

Parameter	Beginning	33%	67%	100%	*p*-Value	η_p^2
F_{zmax1} (BW)	1.93 ± 0.21	1.87 ± 0.22	1.90 ± 0.21	1.88 ± 0.22	0.45	0.07
t_F_{zmax1} (s)	0.03 ± 0.01	0.03 ± 0.01	0.03 ± 0.00	0.03 ± 0.01	0.11	0.14
LR_{max} (BW/s)	120.88 ± 25.71	112.58 ± 24.20	118.80 ± 26.14	121.62 ± 27.15	0.04	0.19
LR_{avg} (BW/s)	102.34 ± 22.72	99.58 ± 20.58	102.82 ± 23.09	106.64 ± 25.64	0.16	0.12
F_{zmax2} (BW)	2.48 ± 0.20	2.45 ± 0.14	2.52 ± 0.15	2.51 ± 0.16	0.25	0.10
t_F_{zmax2} (s)	0.09 ± 0.01	0.09 ± 0.00	0.09 ± 0.01	0.09 ± 0.01	0.85	0.02
CT (s)	0.22 ± 0.01	0.22 ± 0.01	0.23 ± 0.01 *	0.23 ± 0.01	0.01	0.25
F_{ymax} (BW)	0.32 ± 0.03	0.32 ± 0.03	0.32 ± 0.03	0.32 ± 0.03	0.89	0.02
F_{ymin} (BW)	-0.47 ± 0.06	-0.48 ± 0.06	-0.49 ± 0.08	-0.48 ± 0.07	0.76	0.03

Note: F_{zmax1}, first peak vGRF; t_F_{zmax1}, the occurrence time of F_{zmax1}; LR_{max}, maximum loading rate; LR_{avg}, average loading rate; F_{zmax2}, second peak vGRF; t_F_{zmax2}, the occurrence time of F_{zmax2}; CT, contact time; F_{ymax}, maximum propulsive GRF; and F_{ymin}, maximum braking GRF. * indicates significant differences from the values obtained at the beginning ($p < 0.05$).

3.3. Joint Mechanics

In comparison with joint range of motion (θ_{ROM}) at the beginning, θ_{ROM} of the knee joint significantly increased at 33% and 67% of the run duration, and θ_{ROM}, $\Delta\theta$, θ_0, and $\theta_{max\text{-}ext}$ of the hip joint significantly increased at 33%, 67%, and 100% of the run duration (Figure 4 and Table 2). No significant differences were observed in θ_0, θ_{max}, $\omega_{max\text{-}ext}$, and M_{max} of the three joints, and θ_{ROM} and $\Delta\theta$ of the ankle joint amongst four time points (Table 2).

Table 2. Kinematics and joint moment of hip, knee, and ankle joints at time points corresponding to the beginning, 33%, 67%, and 100% of the run duration.

Joint	Parameter	Beginning	33%	67%	100%	*p*-Value	η_p^2
Ankle	θ_0 (°)	2.11 ± 2.80	0.87 ± 3.57	0.92 ± 3.32	0.84 ± 3.41	0.05	0.18
	$\theta_{max\text{-}ext}$ (°)	19.40 ± 3.09	19.49 ± 3.52	19.88 ± 3.08	19.90 ± 2.98	0.25	0.10
	$\theta_{max\text{-}flex}$ (°)	-21.61 ± 4.01	-21.07 ± 2.75	-22.06 ± 3.90	-21.73 ± 2.77	0.51	0.06
	θ_{ROM} (°)	41.01 ± 4.06	40.56 ± 3.18	41.94 ± 3.49	41.63 ± 2.37	0.24	0.12
	$\Delta\theta$ (°)	23.72 ± 4.14	21.94 ± 3.49	22.98 ± 3.46	22.57 ± 3.17	0.23	0.10
	$\omega_{max\text{-}ext}$ (°/s)	584.90 ± 56.63	577.63 ± 48.96	584.13 ± 46.71	567.19 ± 64.55	0.77	0.03
	M_{max} (N·m/kg)	-3.69 ± 0.42	-3.65 ± 0.53	-3.71 ± 0.48	-3.76 ± 0.61	0.83	0.02
Knee	θ_0 (°)	-17.08 ± 5.81	-18.15 ± 4.90	-18.89 ± 5.09	-18.67 ± 5.00	0.04	0.24
	$\theta_{max\text{-}ext}$ (°)	-14.16 ± 4.30	-13.15 ± 5.27	-13.56 ± 5.06	-14.54 ± 5.30	0.09	0.15
	$\theta_{max\text{-}flex}$ (°)	-37.63 ± 5.34	-38.60 ± 5.56	-37.92 ± 8.63	-39.47 ± 5.54	0.37	0.08
	θ_{ROM} (°)	23.47 ± 4.17	25.45 ± 3.17 *	25.97 ± 2.76 *	24.93 ± 2.59	<0.01	0.32
	$\Delta\theta$ (°)	20.56 ± 3.80	20.45 ± 3.45	19.03 ± 7.36	20.81 ± 3.30	0.46	0.06
	$\omega_{max\text{-}ext}$ (°/s)	492.84 ± 78.22	460.21 ± 91.37	479.65 ± 95.40	462.48 ± 83.61	0.12	0.14
	M_{max} (N·m/kg)	-1.77 ± 0.48	-1.79 ± 0.38	-1.84 ± 0.60	-1.91 ± 0.56	0.69	0.04
Hip	θ_0 (°)	26.93 ± 8.88	29.48 ± 3.83 *	29.70 ± 9.68 *	30.74 ± 9.98 *	<0.01	0.47
	$\theta_{max\text{-}ext}$ (°)	27.78 ± 8.35	30.05 ± 8.58 *	30.92 ± 8.64 *	30.47 ± 9.69 *	<0.01	0.45
	$\theta_{max\text{-}flex}$ (°)	-9.24 ± 8.22	-9.8 ± 9.41	-9.87 ± 8.34	-10.60 ± 10.18	0.28	0.09
	θ_{ROM} (°)	37.02 ± 3.00	39.88 ± 2.84 *	40.79 ± 3.10 *	40.07 ± 4.27 *	<0.01	0.44
	$\Delta\theta$ (°)	36.17 ± 3.83	39.31 ± 2.84 *	39.57 ± 3.13 *	41.34 ± 4.38 *	<0.01	0.41
	$\omega_{max\text{-}ext}$ (°/s)	307.81 ± 30.56	311.94 ± 23.54	316.99 ± 28.24	308.58 ± 29.92	0.62	0.04
	M_{max} (N·m/kg)	2.97 ± 0.50	3.23 ± 0.86	3.22 ± 0.86	3.45 ± 0.74	0.09	0.15

Note: θ_0, angle at initial contact; $\theta_{max\text{-}ext}$, maximum extension/dorsiflexion angle during the stance phase; $\theta_{max\text{-}flx}$, maximum flexion/plantar flexion angle during the stance phase; θ_{ROM}, joint range of motion; $\Delta\theta$, $\theta_{max\text{-}flx}-\theta_0$; $\omega_{max\text{-}ext}$, maximum extension/plantarflexion angular velocity during the stance phase; M_{max}, peak joint moment maximum. * indicates significant differences from the values at the beginning ($p < 0.05$).

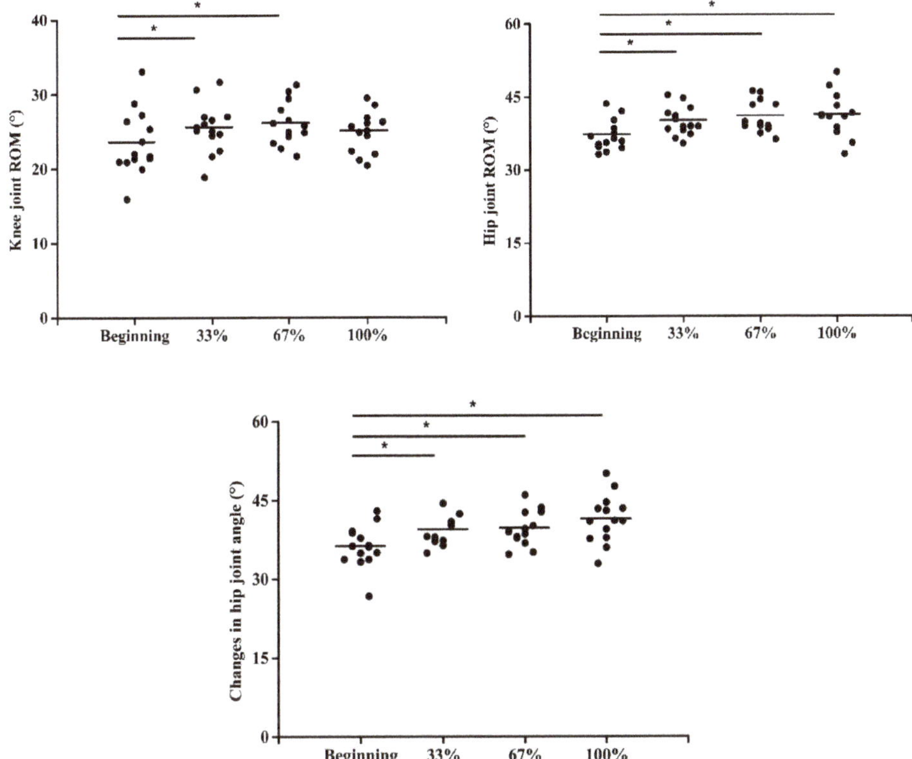

Figure 4. Knee joint range of motion (ROM), hip joint ROM, and changes in angle ($\Delta\theta$) of the hip joint at time points corresponding to the beginning, 33%, 67%, and 100% of the run duration. * significantly different from the beginning with $p < 0.05$.

3.4. Stiffness

In comparison with k_{vert} at the beginning, k_{vert} showed a trend towards a decrease at 67% and 100% of the run duration, respectively (Table 3). Meanwhile, significant main effects of time points were observed for Δy and k_{ankle} (Table 3). Specifically, k_{ankle} significantly increased at 33%, 67%, and 100% of the run duration (Figure 5).

Table 3. Vertical stiffness and joint stiffness at time points corresponding to the beginning, 33%, 67%, and 100% of the run duration.

Parameter	Beginning	33%	67%	100%	*p*-Value	η_p^2
k_{vert} (N/kg/m)	67.41 ± 10.54	65.65 ± 6.97	63.05 ± 8.23	61.32 ± 6.67	0.06	0.17
Δy (m)	0.038 ± 0.004	0.038 ± 0.004	0.040 ± 0.005	0.041 ± 0.004	0.03	0.20
k_{hip} (N·m/kg/°)	0.55 ± 0.35	0.62 ± 0.22	0.56 ± 0.28	0.77 ± 0.45	0.07	0.16
k_{knee} (N·m/kg/°)	0.04 ± 0.03	0.06 ± 0.03 *	0.05 ± 0.03	0.06 ± 0.03	<0.01	0.33
k_{ankle} (N·m/kg/°)	0.21 ± 0.03	0.19 ± 0.03	0.20 ± 0.03 *#	0.20 ± 0.04 *#&	0.03	0.20

Note: k_{vert}, vertical stiffness; Δy, changes in the vertical displacement of the center of gravity; k_{joint}, joint stiffness. *, #, and & significantly different from the pre-, post-0 min, and post-4 min with $p < 0.05$, respectively.

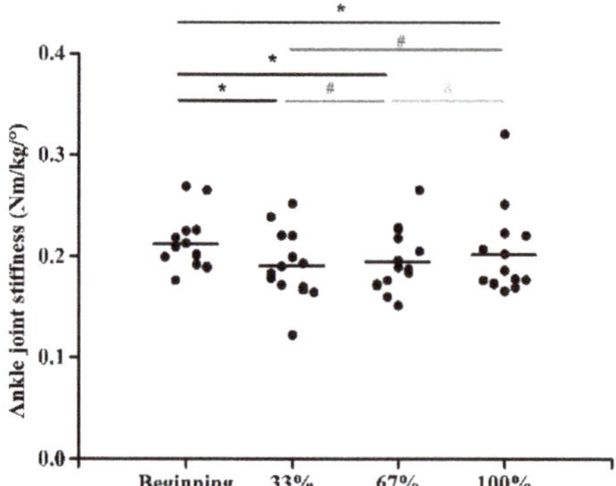

Figure 5. Stiffness of the ankle joint at time points corresponding to the beginning, 33%, 67%, and 100% of the run duration. *, #, and & significantly different from the beginning, 33%, and 67% of the run duration with $p < 0.05$, respectively.

4. Discussion

This study aimed to determine the changes in lower extremity biomechanics, i.e., vertical and anterior–posterior GRFs, loading rates, joint mechanics and stiffness, during treadmill running to fatigue. Our findings supported the hypotheses that an increased range of motion at the hip and knee joints and a decreased vertical and ankle stiffness of the lower extremity were observed during the progress of a fatiguing run. However, GRF characteristics, i.e., vertical/propulsive/braking GRF and peak loading rates, did not vary significantly throughout the fatiguing run.

4.1. Fatigue Intervention

In this experiment, after the fatigue intervention, the blood lactic acid was changed from 2.5 ± 1.5 mmol/L (at rest) to 11.1 ± 3.6 mmol/L (immediately after fatigue), which was consistent with a previous study (from 1.9 ± 0.8 mmol/L to 12.4 ± 3.8 mmol/L) [32]. Specifically, in comparison with the rest state, the blood lactic acid at the ninth minute still significantly increased, and the intervention intensity could be considered the standard of running fatigue. The average intervention time of this experiment was 28.5 min, the maximum heart rate reached 182.9 bpm, and the RPE scale was 17.2 (corresponding to "very hard"). These values were supported by our previous study [33,34].

4.2. Ground Reaction Force

In our study, ground reaction forces, such as first and second peaks, had no significant difference before and after fatigue. Our observation was consistent with the findings of m [35] and Abt et al. [18], who used the same test speed or a faster speed in their experiments. It implied that the impact forces as well as the active forces would not change significantly throughout the whole fatiguing run. Generally, repetitive and passive impact forces in running are considered to be one of the main causes of the overuse injury of the lower extremities; these forces elicit a comprehensive fatigue effect and inhibit musculoskeletal remodeling and repair [36]. According to this mechanism, one of the key factors causing running overuse injury is fatigue, and the repeatability of running is one of the potential factors that may cause fatigue [18]. In addition, the loading rate is considered a sensitive impact force parameter in the run-and-jump motion [4,37,38], which is more reflective of the relationship between running impact and running injury compared with the impact peak. However, in our study, the average loading rate, along with vertical GRFs, did not vary significantly during the whole fatigue process. This observation also supported that of Zadpoor et al. [39]. In summary, fatigue intervention had no significant effect on the impact force and loading rate during a fatiguing run because the impact and loading rate were closely related to the strike pattern. This finding indicated that no linear relationship was found between the impact and loading rate and the muscle fatigue of lower extremities without changing the running strike pattern. Three joints and the surrounding muscles would adapt, transmit, and further attenuate the impact force, avoiding damage caused by repeated impacts when individuals were running for a long time.

4.3. Joint Mechanics

In this study, θ_{ROM} of the knee joint increased significantly at 33% and 66% of the run duration. However, no changes were observed for the maximum knee flexion. Abt et al. [18] required the male participants to run on a treadmill at 3.3 m/s speed until exhaustion. Similarly, no significant differences existed for maximum knee and ankle flexion after fatigue. On the other hand, θ_{ROM} and $\Delta\theta$ of the hip joint at 33%, 67% and 100% of the run duration significantly increased compared with those at the beginning. This finding supported the results of Sara et al. [40] possibly because θ_{ROM} of hip and knee joints, as the large joints of the lower extremities, increased, leading to a decrease in Δy during the stance phase rather than relying solely on the single joint of the knee joint. It is speculated that the kinematic changes after running fatigue may be a compensation mechanism to reduce the possibility of injury rather than the result of fatigue [9]. θ_{ROM} of the hip and knee joints decreased starting from the middle stage of the fatiguing run. This finding seems to be another evidence of the abovementioned explanation. Generally, approximately 70–80% of the impact force is absorbed by the knee joint during running, and the resolution of this impact force is crucial to prevent overuse damage [41]. To absorb these impact forces, the human body properly coordinates the joint activities of the lower limbs through the regulation of joint activities in the musculoskeletal system [31,42]. The current findings indicated that the hip and knee kinematic chains of human lower extremities played an important role in response to the progress of fatigue but did not passively attenuate the impact.

4.4. Stiffness

In the progress of a fatiguing run, k_{vert} showed a trend towards a decrease at 67% and 100% of the running period compared with that of the pre-fatigue period. Meanwhile, Δy significantly increased at the middle and late stages. However, no differences were observed in vGRF. These findings might explain the nonlinear relationship between vGRF and k_{vert}. Furthermore, the stiffness of the ankle joints significantly reduced at 33%, 67%, and 100% of the running period compared with that of the pre-fatigue period.

Generally, vertical stiffness is considered an important factor in the spring-mass model of the musculoskeletal system, which is closely related to injuries [43]. Previous studies [31] suggested that k_{vert} decreases during fatigue if an individual is fatigued at a constant speed, and changes in k_{vert} are inversely related to Δy rather than vGRF during the stance phase. These results were supported by our findings. Goodwin et al. [44] showed that the reduction in k_{vert} during running was related to the range of motion (ROM) of the larger joint. To maintain the stability of the lower extremity and reduce the cumulative impact injuries, the human body automatically reduces Δy and k_{vert}. This strategy achieves a "soft landing" within a certain joint ROM. Previous studies [31,45] found that the "soft landing" was related to the decreased stiffness of the ankle joints after fatigue, which was in accordance with our findings. Moreover, k_{vert} is also related to running economy and energy efficiency. Another study [46] has shown that the energy utilization rate increased with k_{vert}. In the later stage of medium/high-intensity long-term exercise, k_{vert} might decrease as oxygen uptake increases, and the energy utilization and running economy gradually decrease. Future studies should focus on the effects of fatigue on the running economy of the joints of the lower extremities after runners' transition from a rearfoot strike to a forefoot strike. Future studies should also investigate the differences in running economy between different strike patterns.

4.5. Limitations

In this study we required participants to run on the treadmill, but one cannot be certain that their running pattern was not affected by factors such as the adaptation of the lower extremity to the running belt transfer speed and the instability of treadmill itself. Furthermore, we did not collect surface electromyographic data to simplify the design by focusing on GRFs and joint mechanics and by limiting the experimental devices that were attached to the runners. Finally, the role of gender should also be taken into account in the future study.

5. Conclusions

GRF characteristics, i.e., vertical/propulsive/braking GRF and peak loading rates, did not vary significantly throughout the fatiguing run. However, nonlinear adaptations in lower extremity kinematics and kinetics were observed. In particular, a "soft landing" strategy, achieved by an increased range of motion at the hip and knee joints and a decreased vertical and ankle stiffness of the lower extremity, was initiated from the mid-stage (e.g., 33% and 67%) of a fatiguing run to potentially maintain similar impact forces. These findings provide preliminary evidence suggesting the hip and knee kinematic chains played an important role in response to the progress of running to fatigue which may require additional attention during training.

Author Contributions: Z.L., X.Z., contributed equally. Conceptualization, W.F.; methodology, Z.L.; formal analysis, Z.L., X.Z., J.W., and Y.Y.; investigation, Z.L., X.Z., J.W., Y.Y., and W.F.; resources, W.F.; data curation, Z.L., X.Z.; writing—original draft preparation, Z.L., X.Z., J.W., and Y.X.; writing—review and editing, X.Z. and W.F.; project administration, W.F.; funding acquisition, W.F.

Funding: This work was supported by the National Natural Science Foundation of China (11772201, 11572202); Talent Development Fund of Shanghai Municipal (2018107); Shanghai Committee of Science and Technology (17080503300); the National Key Technology Research and Development Program of the Ministry of Science and Technology of China (2019YFF0302100); the "Dawn" Program of Shanghai Education Commission (19SG47), China.

References

1. Williams, P.T. Dose of exercise and health benefit. *Arch. Intern. Med.* **1997**, *157*, 1774. [CrossRef] [PubMed]
2. Saragiotto, B.T.; Yamato, T.P.; Hespanhol Junior, L.C.; Rainbow, M.J.; Davis, I.S.; Lopes, A.D. What are the main risk factors for running-related injuries? *Sports Med.* **2014**, *44*, 1153–1163. [CrossRef] [PubMed]
3. Van Middelkoop, M.; Kolkman, J.; Van Ochten, J.; Bierma-Zeinstra, S.M.; Koes, B.W. Risk factors for lower extremity injuries among male marathon runners. *Scand. J. Med. Sci. Sports* **2008**, *18*, 691–697. [CrossRef] [PubMed]
4. van Gent, R.N.; Siem, D.; van Middelkoop, M.; van Os, A.G.; Bierma-Zeinstra, S.M.; Koes, B.W. Incidence and determinants of lower extremity running injuries in long distance runners: A systematic review. *Br. J. Sports Med.* **2007**, *41*, 469–480. [CrossRef] [PubMed]
5. Buist, I.; Bredeweg, S.W.; Bessem, B.; van Mechelen, W.; Lemmink, K.A.; Diercks, R.L. Incidence and risk factors of running-related injuries during preparation for a 4-mile recreational running event. *Br. J. Sports Med.* **2010**, *44*, 598–604. [CrossRef] [PubMed]
6. Bonacci, J.; Saunders, P.U.; Hicks, A.; Rantalainen, T.; Vicenzino, B.G.; Spratford, W. Running in a minimalist and lightweight shoe is not the same as running barefoot: A biomechanical study. *Br. J. Sports Med.* **2013**, *47*, 387–392. [CrossRef]
7. Lieberman, D.E.; Venkadesan, M.; Werbel, W.A.; Daoud, A.I.; D'Andrea, S.; Davis, I.S.; Mang'eni, R.O.; Pitsiladis, Y. Foot strike patterns and collision forces in habitually barefoot versus shod runners. *Nature* **2010**, *463*, 531–535. [CrossRef]
8. Lieber, R.L. Biomechanical response of skeletal muscle to eccentric contractions. *J. Sport Health Sci.* **2018**, *7*, 294–309. [CrossRef]
9. Derrick, T.R.; Dereu, D.; McLean, S.P. Impacts and kinematic adjustments during an exhaustive run. *Med. Sci. Sports Exerc.* **2002**, *34*, 998–1002. [CrossRef]

10. Zadpoor, A.A.; Nikooyan, A.A. The effects of lower extremity muscle fatigue on the vertical ground reaction force: A meta-analysis. *Proc. Inst. Mech. Eng. H* **2012**, *226*, 579–588. [CrossRef]
11. Morin, J.B.; Samozino, P.; Edouard, P.; Tomazin, K. Effect of fatigue on force production and force application technique during repeated sprints. *J. Biomech.* **2011**, *44*, 2719–2723. [CrossRef] [PubMed]
12. Christina, K.A.; White, S.C.; Gilchrist, L.A. Gilchrist. Effect of localized muscle fatigue on vertical ground reaction forces and ankle joint motion during running. *Hum. Mov. Sci.* **2001**, *20*, 257–276. [CrossRef]
13. Gerlach, K.E.; White, S.C.; Burton, H.W.; Dorn, J.M.; Leddy, J.J.; Horvath, P.J. Kinetic changes with fatigue and relationship to injury in female runners. *Med. Sci. Sports Exerc.* **2005**, *37*, 657–663. [CrossRef] [PubMed]
14. Slawinski, J.; Heubert, R.; Quievre, J.; Billat, V.; Hanon, C. Changes in spring-mass model parameters and energy cost during track running to exhaustion. *J. Strength Cond. Res.* **2008**, *22*, 930–936. [CrossRef] [PubMed]
15. Radzak, K.N.; Putnam, A.M.; Tamura, K.; Hetzler, R.K.; Stickley, C.D. Asymmetry between lower limbs during rested and fatigued state running gait in healthy individuals. *Gait Posture* **2017**, *51*, 268–274. [CrossRef] [PubMed]
16. Kuhman, D.; Melcher, D.; Paquette, M.R. Ankle and knee kinetics between strike patterns at common training speeds in competitive male runners. *Eur. J. Sport Sci.* **2016**, *16*, 433–440. [CrossRef]
17. Brown, A.M.; Zifchock, R.A.; Hillstrom, H.J.; Song, J.; Tucker, C.A. The effects of fatigue on lower extremity kinematics, kinetics and joint coupling in symptomatic female runners with iliotibial band syndrome. *Clin. Biomech.* **2016**, *39*, 84–90. [CrossRef]
18. John, P.; Abt, A.S.M.L.; Sell, T.C. Running kinematics and shock absorption do not change after brief exhaustive running. *J. Strength Cond. Res.* **2011**, *25*, 1479–1485.
19. Clansey, A.C.; Hanlon, M.; Wallace, E.S.; Lake, M.J. Effects of fatigue on running mechanics associated with tibial stress fracture risk. *Med. Sci. Sports Exerc.* **2011**, *44*, 1917–1923. [CrossRef]
20. Jan Hoff, Y.S. Arnstein finstad, eivind wang, and jan helgerud increased blood lactate level deteriorates running economy in world class endurance athletes. *J. Ofstrength Cond. Res.* **2016**, *30*, 1373–1378. [CrossRef]
21. Hsu, W.-C.; Tseng, L.-W.; Chen, F.-C.; Wang, L.-C.; Yang, W.-W.; Lin, Y.-J.; Liu, C. Effects of compression garments on surface EMG and physiological responses during and after distance running. *J. Sport Health Sci.* **2017**. [CrossRef]
22. Weist, R.; Eils, E.; Rosenbaum, D. The Influence of Muscle Fatigue on Electromyogram and Plantar Pressure Patterns as an Explanation for the Incidence of Metatarsal Stress Fractures. *Am. J. Sports Med.* **2004**, *32*, 1893–1898. [CrossRef] [PubMed]
23. Tanner, R.K.; Fuller, K.L.; Ross, M.L. Evaluation of three portable blood lactate analysers: Lactate Pro, Lactate Scout and Lactate Plus. *Eur. J. Appl. Physiol.* **2010**, *109*, 551–559. [CrossRef] [PubMed]
24. Devlin, J.; Paton, B.; Poole, L.; Sun, W.; Ferguson, C.; Wilson, J.; Kemi, O.J. Blood lactate clearance after maximal exercise depends on active recovery intensity. *J. Sports Med. Phys. Fit.* **2014**, *54*, 271–278.
25. Maas, E.; De Bie, J.; Vanfleteren, R.; Hoogkamer, W.; Vanwanseele, B. Novice runners show greater changes in kinematics with fatigue compared with competitive runners. *Sports Biomech.* **2018**, *17*, 350–360. [CrossRef]
26. Koblbauer, I.F.; van Schooten, K.S.; Verhagen, E.A.; van Dieen, J.H. Kinematic changes during running-induced fatigue and relations with core endurance in novice runners. *J. Sci. Med. Sport* **2014**, *17*, 419–424. [CrossRef]
27. Fu, W.; Fang, Y.; Gu, Y.; Huang, L.; Li, L.; Liu, Y. Shoe cushioning reduces impact and muscle activation during landings from unexpected, but not self-initiated, drops. *J. Sci. Med. Sport* **2017**. [CrossRef]
28. Futrell, E.E.; Jamison, S.T.; Tenforde, A.S.; Davis, I.S. Relationships between Habitual Cadence, Footstrike, and Vertical Loadrates in Runners. *Med. Sci. Sports Exerc.* **2018**, *50*, 1837–1841. [CrossRef]
29. Hamill, J.; Moses, M.; Seay, J. Lower extremity joint stiffness in runners with low back pain. *Res. Sports Med.* **2009**, *17*, 260–273. [CrossRef]
30. Liu, Y.; Peng, C.H.; Wei, S.H.; Chi, J.C.; Tsai, F.R.; Chen, J.Y. Active leg stiffness and energy stored in the muscles during maximal counter movement jump in the aged. *J. Electromyogr. Kinesiol.* **2006**, *16*, 342–351. [CrossRef]
31. Dutto, D.J.; Smith, G.A. Changes in spring-mass characteristics during treadmill running to exhaustion. *Med. Sci. Sports Exerc.* **2002**, *34*, 1324–1331. [CrossRef]

32. Stajer, V.; Vranes, M.; Ostojic, S.M. Correlation between biomarkers of creatine metabolism and serum indicators of peripheral muscle fatigue during exhaustive exercise in active men. *Res. Sports Med.* **2018**. [CrossRef] [PubMed]

33. Zhang, X.; Xia, R.; Dai, B.; Sun, X.; Fu, W. Effects of Exercise-Induced Fatigue on Lower Extremity Joint Mechanics, Stiffness, and Energy Absorption during Landings. *J. Sports Sci. Med.* **2018**, *17*, 640–649. [PubMed]

34. Avela, J.; Kyröläinen, H.; Komi, P.V.; Rama, D. Reduced reflex sensitivity persists several days after long-lasting stretch-shortening cycle exercise. *J. Appl. Physiol.* **1999**, *86*, 1292–1300. [CrossRef] [PubMed]

35. Bazuelo-Ruiz, B.; Dura-Gil, J.V.; Palomares, N.; Medina, E.; Llana-Belloch, S. Effect of fatigue and gender on kinematics and ground reaction forces variables in recreational runners. *Peer J.* **2018**, *6*, e4489. [CrossRef] [PubMed]

36. Hreljac, A.; Marshall, R.N.; Hume, P.A. Evaluation of lower extremity overuse injury potential in runners. *Med. Sci. Sports Exerc.* **2000**, *32*, 1635–1641. [CrossRef]

37. Pohl, M.B.; Hamill, J.; Davis, I.S. Biomechanical and anatomic factors associated with a history of plantar fasciitis in female runners. *Clin. J. Sport Med.* **2009**, *19*, 372–376. [CrossRef]

38. Milner, C.E.; Ferber, R.; Pollard, C.D.; Hamill, J.; Davis, I.S. Biomechanical factors associated with tibial stress fracture in female runners. *Med. Sci. Sports Exerc.* **2006**, *38*, 323–328. [CrossRef]

39. Zadpoor, A.A.; Nikooyan, A.A. The relationship between lower-extremity stress fractures and the ground reaction force: A systematic review. *Clin. Biomech.* **2011**, *26*, 23–28. [CrossRef]

40. Winter, S.; Gordon, S.; Watt, K. Effects of fatigue on kinematics and kinetics during overground running: A systematic review. *J. Sports Med. Phys. Fit.* **2017**, *57*, 887–899.

41. Kim, W.; Voloshin, A.S.; Johnson, S.H. Modeling of heel strike transients during running. *Hum. Mov. Sci.* **1994**, *13*, 221–244. [CrossRef]

42. Mercer, J.A.; Bates, B.T.; Dufek, J.S.; Hreljac, A. Characteristics of shock attenuation during fatigued running. *J. Sports Sci.* **2003**, *21*, 911–919. [CrossRef] [PubMed]

43. Jordan, M.J.; Aagaard, P.; Herzog, W. A comparison of lower limb stiffness and mechanical muscle function in ACL-reconstructed, elite, and adolescent alpine ski racers/ski cross athletes. *J. Sport Health Sci.* **2018**, *7*, 416–424. [CrossRef] [PubMed]

44. Jonathan, S.; Goodwin, J.T.B.; Todd, A.; Schwartz, D.S. Blaise Williams III. Clinical predictors of dynamic lower extremity stiffness during running. *J. Orthop. Sports Phys. Ther.* **2018**, *49*, 98–104.

45. Kuitunen, S.; Komi, P.V.; Kyröläinen, H. Knee and ankle joint stiffness in sprint running. *Med. Sci. Sports Exerc.* **2002**, *34*, 166–173. [CrossRef]

46. Perl, D.P.; Daoud, A.I.; Lieberman, D.E. Effects of footwear and strike type on running economy. *Med. Sci. Sports Exerc.* **2012**, *44*, 1335–1343. [CrossRef]

Article

Behavioral Dynamics of Pedestrians Crossing between Two Moving Vehicles

Soon Ho Kim [1], Jong Won Kim [2], Hyun-Chae Chung [3], Gyoo-Jae Choi [4] and MooYoung Choi [1,*

[1] Department of Physics and Astronomy and Center for Theoretical Physics, Seoul National University, Seoul 08826, Korea; soonhokim@snu.ac.kr
[2] Department of Healthcare Information Technology, Inje University, Gimhae 50834, Korea; jongwonkim@inje.ac.kr
[3] Department of Sports and Exercise Science, Kunsan National University, Gunsan 54150, Korea; hcx@kunsan.ac.kr
[4] School of Automotive and Mechanical Engineering, Kunsan National University, Gunsan 54150, Korea; gjchoi@kunsan.ac.kr
* Correspondence: mychoi@snu.ac.kr; Tel.: +82-2-880-6615

Received: 25 December 2019; Accepted: 20 January 2020; Published: 26 January 2020

Abstract: This study examines the human behavioral dynamics of pedestrians crossing a street with vehicular traffic. To this end, an experiment was constructed in which human participants cross a road between two moving vehicles in a virtual reality setting. A mathematical model is developed in which the position is given by a simple function. The model is used to extract information on each crossing by performing root-mean-square deviation (RMSD) minimization of the function from the data. By isolating the parameter adjusted to gap features, we find that the subjects primarily changed the timing of the acceleration to adjust to changing gap conditions, rather than walking speed or duration of acceleration. Moreover, this parameter was also adjusted to the vehicle speed and vehicle type, even when the gap size and timing were not changed. The model is found to provide a description of gap affordance via a simple inequality of the fitting parameters. In addition, the model turns out to predict a constant bearing angle with the crossing point, which is also observed in the data. We thus conclude that our model provides a mathematical tool useful for modeling crossing behaviors and probing existing models. It may also provide insight into the source of traffic accidents.

Keywords: pedestrian behavior; human locomotion; human dynamics; traffic safety

1. Introduction

Pedestrians make up a large portion of traffic accident fatalities, particularly in areas of high population density [1,2]. Exploring the behavioral dynamics of road crossing may provide insight into the fundamental source of accidents [2–4]. The task of crossing a road involves a goal-directed movement, as the pedestrian desires to reach the other side of the street subject to the avoidance condition due to passing vehicles. This relates the problem to studies on human behavior during tasks of avoidance [3,5] as well as interception [6–8].

The constant bearing-angle model has gained attention as a possible strategy that humans employ in order to intercept moving objects [6–8]. This walking strategy is tied to how human locomotion is visually controlled [9–12]. In addition, studies involving movements through gaps have employed the concept of affordance, the possibilities for action constrained by the environment and physical conditions of the actor [13–15]. Statistical analyses of pedestrian inter-vehicle gap acceptance rates, which depend on the pedestrian's perception of affordance, have also been reported [16]; these studies, however, do not typically provide a dynamic model of action.

The purpose of this study is to develop a model of road crossing that can be used to analyze data and test hypotheses. Specifically, we examine how pedestrians cross a street between two moving vehicles. An experiment is constructed in which human participants cross a street within a virtual environment with a range of experimental conditions. A mathematical model is developed which accurately describes typical crossing patterns. The model is applied to experimental data via minimization of the root-mean-square deviation (RMSD) with respect to the model parameters. When the best fit parameters are found, the model equations give us the position of the pedestrian as a function of time.

Discussing the meaning of each parameter of the model, we examine how the average value of each parameter varies depending on crossing conditions. Specifically, we consider the effects of adjusting the gap size and the initial distance as well as the pedestrian's age, vehicle speed, and vehicle type. We further make use of our model to derive an inequality among the parameters that must be satisfied for successful crossing to occur, and, accordingly, to describe the affordance [13,15] of the crossing situation. The possibility of intercepting the gap between the cars has been mathematically modeled with the environmental factors (e.g., speed of cars, length of lanes) and the walker's capabilities such as walking speed and response time. It is also observed in the data that the bearing angle tends to be constant with respect to the crossing point, which is obtained analytically from the model equations.

In Section 2, we describe the experimental procedure and lay out our model. The concept of affordance and bearing angle are also formulated using the model. In Section 3, we draw conclusions from the fitting parameters of the model. The results are visualized and analyzed in the affordance and bearing-angle viewpoints. In Section 4, we discuss the implications of our results and we conclude in Section 5.

2. Methods

2.1. Data Collection

Sixteen children (of age 12.2 ± 0.8 yrs, i.e., mean age 12.2 years and standard deviation 0.8 years), sixteen young adults (of age 22.8 ± 2.6 yrs), and fourteen elderly people (of age 54.1 ± 4.9 yrs) with normal or corrected-to-normal vision were recruited for this experiment. Informed written consent was obtained from all individual participants. The experiment was conducted in accordance with the Declaration of Helsinki, and the protocol was approved by the Kunsan National University Research Board. Each subject was placed on a customized treadmill (of dimensions 0.67 m wide, 1.26 m long, and 1.10 m high) with four magnetic counters that track movements. A Velcro belt connected to the treadmill is worn to decrease vertical and lateral movements, and a handrail is placed for stability. The treadmill turns with minimal friction as the participants walk, and magnetic counters track rotations (left panel, Figure 1). Each participant wore an Oculus Rift (Menlo Park, CA, USA) DK1 virtual reality headset connected to a standard desktop PC. The headset portrayed a realistic view of a street with a crosswalk in 1280×800 resolution 3D stereoscopic visual images that respond realistically according to the participant's movements. Practice trials were performed prior to the experiments to familiarize each participant with the treadmill and virtual environment before the experiment. (see [4] for details).

A diagram of the virtual crossing situation is given in Figure 2. The participant, standing at rest before a car lane with a "ready" signal shown, was instructed to press a button when a "go" signal appears, and then to look at her/his left side, from which two vehicles, one in front of the other, were moving at equal constant speed v_c. The participant was instructed to cross between the two vehicles if she/he believed she/he could cross successfully. If a collision with a vehicle occurred, the simulations halted. The experimenter recorded whether there was a successful crossing, a collision, or no crossing.

Figure 1. Images of experimental setup. Left panel: photograph of student participating in an experiment. Middle panel: cartoon crosswalk used for calibration. Right panel: Screenshot of the virtual crosswalk used in the experiment.

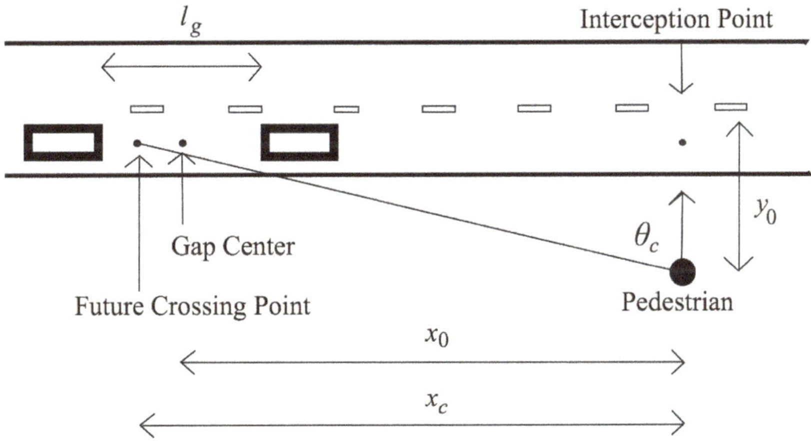

Figure 2. Top-down diagram of the crossing environment at time $t = 0$. The two boxes on the road depict two vehicles facing the right, which move forward at a constant speed. The circle depicts the pedestrian, who begins at rest and is facing in the direction the arrow. Labeled are experimental parameters l_g, y_0, and the bearing angle to the crossing point θ_c.

The experiment was designed to examine the participants' responses to a single gap, and so only two vehicles were present. We examine the responses of the participants to changing gap characteristics by varying experimental parameters. These include pedestrian starting position y_0, gap time t_g (or gap length l_g), vehicle speed v_c, and the vehicle type (sedan or bus). Details of age groups and a full list of experimental parameters are given in Table 1.

The gap center is defined to be the midpoint of the gap between the vehicles. While the width of the gap and speed of the vehicles are varied, the initial position x_0 of the gap center is always set in such a way that the gap center reaches the interception point at time $t = 4$ s. Accordingly, we have $x_0 = -v_c \times 4$ s.

Table 1. Table summarizing age groups and experimental parameters.

Participant Group	Number of Participant
Children	16
Young adults	16
Elderly	14
Experimental Parameter	**Values Used**
y_0 (initial distance)	$-3.5, -4.5, -5.5, -6.5$ m
t_g (gap time)	2.5, 3, 4 s
v_c (vehicle speed)	30, 60 km/h
vehicle type	sedan, bus
total configurations	48

2.2. Model

The following model, called the simple crossing model, was used to analyze crossing data. The velocity is assumed to follow a logistic function of time in the form

$$v(t) = v_{\max} \frac{\exp[(t - t_a)/\tau]}{1 + \exp[(t - t_a)/\tau]}, \tag{1}$$

which, upon integration, results in the position as a function of time:

$$y(t) = y_0 + v_{\max}\tau \log\{1 + \exp[(t - t_a)/\tau]\}. \tag{2}$$

Equation (2) is plotted in Figure 3a (red line). Constants t_a, τ, $v_{\max}$ are fitting parameters whose meanings can be understood as follows: The measurement begins at time $t = 0$. Assuming $t_a - 2\tau > 0$, we have the initial position and velocity of the pedestrian, $y(t{=}0) \approx y_0\,(< 0)$ and $v(t{=}0) \approx 0$, respectively. Then, the pedestrian accelerates smoothly until the maximum velocity $v_{\max}$ is reached. The parameter τ then serves as a measure for the duration of this acceleration, the midpoint between which is given by t_a. Note that, at time $t = t_a - 2\tau$, the velocity in Equation (1) becomes $v(t = t_a{-}2\tau) = v_{\max}\, e^{-2}/(1 + e^{-2}) \approx 0.1\, v_{\max}$. While Equation (1) never gives $v = 0$ exactly, in practice, we may define $t_d \equiv t_a - 2\tau$ to be the time at which the pedestrian begins to accelerate forward. If preferable, one may take alternatively $t_d \equiv t_a - 3\tau$, which corresponds to $v(t = t_d) \approx 0.01\, v_{\max}$.

A second model, called the two-step crossing model, is used to analyze crossings that have more than one acceleration event and thus do not fit the simple crossing model (Figure 3b). The two-step crossing model is discussed in the Appendix A.

Each piece of data classified as a simple crossing is fit to Equation (2) by minimizing RMSD with respect to the fitting parameters. We probe the effects of gap characteristics by examining how the distributions of parameters change with the variation of certain features of the gap, and discuss the results in Section 3.2. Those data displaying the two-step pattern are fitted separately to the extended model, and the results are discussed in the Appendix A.

2.3. Affordance

Affordance stands for the range of possible actions that the environment offers to the acting agent. In the crossing task, the affordance is determined by how long the gap overlaps with the participant's walking trajectory. Assuming the simple crossing model (i.e., Equations (1) and 2)), the affordance of the gap is described by the inequality

$$t_f - \tau \log[e^{(-y_0 - w/2)/v_{\max}\tau} - 1] < t_a$$
$$< t_b - \tau \log[e^{(-y_0 + w/2)/v_{\max}\tau} - 1]. \tag{3}$$

Here, $t_f \equiv |x_0 + l_g/2|v_c^{-1}$ corresponds to the time at which the back bumper of the leading vehicle passes the intersection point and $t_b \equiv |x_0 - l_g/2|v_c^{-1}$ corresponds to the time at which the front bumper of the trailing vehicle passes the point, while w denotes the width of the vehicles and equals 1.5 m in our experiment. t_f is hence manifested in Figure 3a by the time coordinate of the right side of the box to the left (2.5 s), while t_b is by that of the left side of the box to the right (5.5 s). Equation (3) thus describes the condition under which the pedestrian's trajectory passes between the two boxes in Figure 3a.

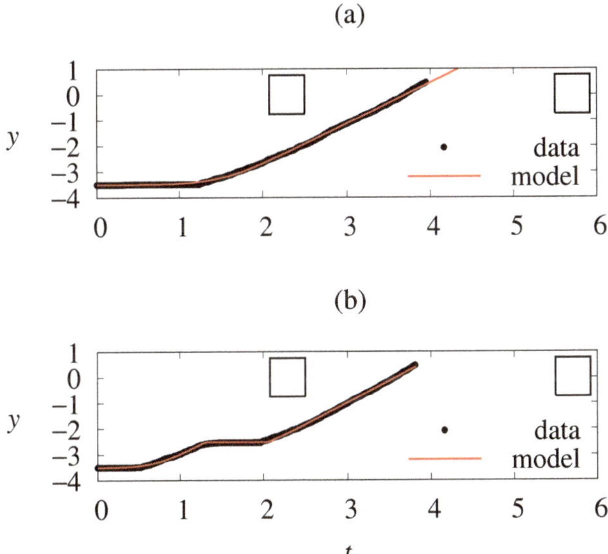

Figure 3. Pedestrian position-time plots illustrating typical crossing patterns. Black traces indicate example data; red traces indicate the corresponding model fits. Left and right boxes indicate the temporal and spatial area that the leading and the trailing vehicles occupy, respectively. Intersecting one of the box lines would indicate a collision. In both examples, the conditions are described by $y_0 = -3.5$ m, $t_g = 3.0$ s, and $v_c = 30$ km/h, with the vehicle type set to be a sedan. (**a**) an example of the simple cross with a single acceleration event followed by constant speed walking; (**b**) an example of the two-step cross with two acceleration events.

In general typical values of τ are small than the time scale of crossing, e.g., compared with $(-y_0 \pm w/2)/v_{\max}$. (Note that $y_0 < 0$ in our coordinate system.) Accordingly, we may take the limit $\tau \to 0$, and reduce Equation (3) to

$$t_f - \frac{1}{v_{\max}}\left(-y_0 - \frac{w}{2}\right) < t_a < t_b - \frac{1}{v_{\max}}\left(-y_0 + \frac{w}{2}\right). \tag{4}$$

This provides a simpler inequality involving two fitting parameters.

2.4. Bearing Angle

Dynamics of interceptive movement are often described in terms of the bearing angle [7,8], which refers to the angle between the velocity vector of the human subject and the line of sight between the subject and the object she/he hopes to intercept. In brief, this model asserts that people intercept a moving object by choosing such a trajectory that the bearing angle is kept constant.

Our case of crossing a road may be cast into an interception task: The pedestrian must "intercept" the empty gap between the vehicles [17]. We may hence apply the bearing angle approach to our crossing experiment and model. One difficulty with this approach is that the gap is not a point but a moving area. As an obvious choice, we may simply use the gap center $x_g(t)$, with respect to which the bearing angle is $\theta_g(t) = \arctan\left[x_g(t)/y(t)\right]$. However, the pedestrian may not cross the gap center; it is thus more appropriate to examine the bearing angle with respect to the point within the moving gap that the pedestrian actually crosses. With t^* denoting the crossing time, we have $y(t^*) = 0$ and let the position of the gap center at the crossing time be Δx, i.e., $x_g(t^*) = \Delta x$. We then define the crossing point,

$$x_c(t) = x_g(t) - \Delta x = x_0 - \Delta x + v_c t = v_c(t - t^*),\tag{5}$$

and consider the angle with respect to x_c:

$$\theta_c(t) = \arctan\left[\frac{x_c(t)}{y(t)}\right].\tag{6}$$

Taking the time derivative of Equation (6) results in

$$\dot{\theta}_c = \frac{x_c y}{x_c^2 + y^2}\left(\frac{\dot{x}_c}{x_c} - \frac{\dot{y}}{y}\right).\tag{7}$$

Assuming that y follows the simple crossing model (i.e., Equation (2)), $|\dot{y}|$ is small when $t < t_a$. Considering the signs of variables (especially, $x_c < 0$ and $\dot{x}_c > 0$), we thus have that $\dot{\theta}_c < 0$, indicating a decreasing bearing angle. When $t > t_a + 2\tau$, the speed approaches the maximum: $\dot{y} \approx v_{\max}$, so that we have

$$\frac{\dot{x}_c}{x_c} - \frac{\dot{y}}{y} \approx \frac{v_c}{v_c(t - t^*)} - \frac{v_{\max}}{v_{\max}(t - t^*)} = 0,\tag{8}$$

which, upon substituting into Equation (7), yields $\dot{\theta}_c = 0$ or a constant bearing angle. The model thus predicts that the bearing angle should decrease at the first stage of crossing and remain constant thereafter. The constant value θ_c^* that the bearing angle approaches can be estimated by

$$\lim_{\Delta t \to 0} \theta_c(t^* - \Delta t) = \lim_{\Delta t \to 0} \arctan\left(\frac{v_c \Delta t}{v_{\max} \Delta t}\right)$$

$$= \arctan\left(\frac{v_c}{v_{\max}}\right).\tag{9}$$

3. Results

3.1. Data Analysis

Tables 2 and 3 show the percentage of successful crossings in this group and the proportions of two-step crossings to the total successful crossings. The success rate drops significantly when the gap length is made small at 20.8 m but still stays above 80%. The highest proportion of two-step crossings occurs when the gap is the shortest and the walking distance is the furthest.

Table 2. Proportion of successful crossings to all crossing attempts for several values of y_0 and l_g, when $v_c = 30$ km/h and vehicle type is sedan.

	Successful Crossings		
l_g (m) y_0 (m)	20.8	25	33.3
−3.5	88 %	100 %	100 %
−4.5	100 %	100 %	100 %
−5.5	100 %	94 %	100 %
−6.5	82 %	94 %	100 %

Table 3. Proportion of two-step crossings to all successful crossings for several values of y_0 and l_g, when $v_c = 30$ km/h and vehicle type is sedan.

	Two-Step Crossings		
l_g (m) y_0 (m)	20.8	25	33.3
−3.5	42 %	25 %	6 %
−4.5	31 %	0 %	0 %
−5.5	0 %	26 %	0 %
−6.5	15 %	6 %	0 %

Equation (2) was fit to simple crossings with an average RMSD of 0.068 m. The low RMSD values indicate that the model accurately describes the majority of crossings. Two-step crossings were also found to be accurate, and are discussed in the Appendix A. Examples of the model equations fit to simple and two-step crossing time series are given in Figure 3a,b, respectively.

3.2. Behavioral Response to Gap Features

Restricting the analysis to simple crossings, we consider the variations of the parameters to changing crossing conditions. Experimental parameters y_0 and l_g affect directly the affordance of the gap by changing the temporal window of the gap or the distance the pedestrian needs to traverse to reach the gap. Effects of the experimental parameters on the three fitting parameters $v_{\max}, t_a,$ and τ have been examined; only t_a has turned out to respond significantly. Figure 4 shows the distribution of t_a obtained for several values of y_0 and l_g. It is observed that t_a generally increases as y_0 approaches zero. This can be understood intuitively as follows: Recall that y_0 denotes the distance the pedestrian must traverse to reach the gap. The larger the distance, the earlier they must begin walking. However, when the initial position is farther, namely, when y_0 is made larger, this trend disappears and t_a tends to stay at slightly over one second ($t_a \gtrsim 1\,\mathrm{s}$). This is likely to result from the minimum response time. Namely, the pedestrian may not cross earlier than the earliest timing at which they can reasonably begin to walk. On the other hand, an increase in the gap size appears to lower t_a. This indicates that, when the gap is accessible earlier, the pedestrian tends to cross earlier. The distributions of the other two parameters $v_{\max}$ and τ have also been examined. While the average value of $v_{\max}$ tends generally to increase with y_0, the trend is not statistically significant. No significant trends have been observed for τ.

Contrary to y_0 and l_g, the vehicle speed v_c and the vehicle type are manipulated without changing the gap affordance. These experimental parameters affect the visual perception of the gap without changing its temporal window of availability. Figure 5 displays the effects of the vehicle speed and type on t_a when the gap time t_g is set to be 3 s and y_0 to be −3.5 m. Doubling the vehicle speed results in a significant increase in t_a. Moreover, in several cases, buses resulted in a greater value of t_a than sedans did.

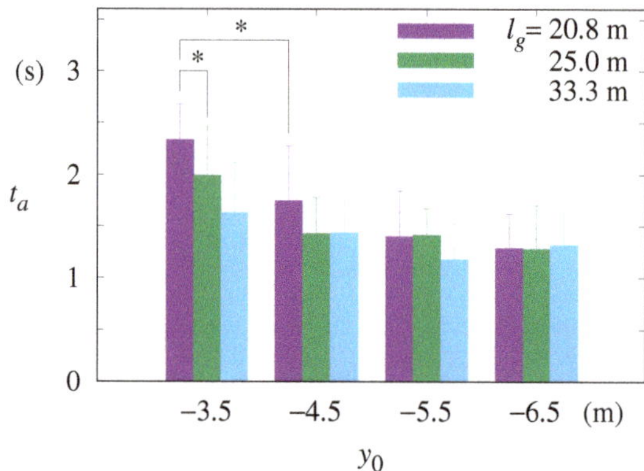

Figure 4. Distributions of parameter t_a for varying y_0 and l_g. Here, $v_c = 30$ km/h and vehicle type is sedan. Columns indicate the average values of t_a in the data for given experimental conditions while error bars represent standard deviations. Pairs of samples, marked with asterisks, are presumed to belong to different distributions ($p < 0.05$) according to the Mann-Whitney U test. (Note here that not all such pairs are marked.)

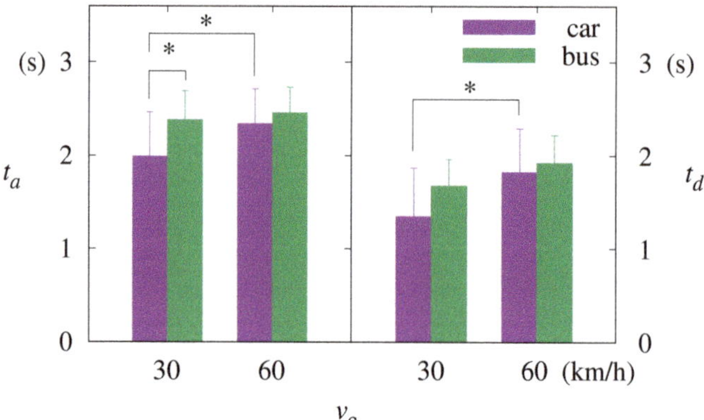

Figure 5. Distributions of parameters t_a (left) and t_d (right) for varying vehicle speeds v_c and for two vehicle types. Other parameters are set to $t_g = 3$ s and $y_0 = -3.5$ m. Columns indicate the average values in the data for given experimental conditions while error bars represent standard deviations. Pairs of samples, marked with asterisks, are presumed to belong to different distributions ($p < 0.05$) according to the Mann–Whitney U test.

On the other hand, when the same comparison is made for data with $y_0 < -3.5$ m, there arises no significant shift in t_a or t_d upon changing the vehicle type. For $y_0 < -4.5$ m, no significant shift is observed upon changing the vehicle speed as well. This suggests that, when the initial distance is sufficiently far, pedestrian's judgement of the gap is hardly affected by the vehicle type or speed.

Finally, we examine differences among age groups. According to the Mann–Whitney U test, the difference in the distribution of $v_{\max}$ is found to be significant ($p < 0.05$) when either the young adult group or the elderly group is compared with the child group. Both the young adult and elderly groups consistently have higher average values of $v_{\max}$ across all crossing conditions, by about 0.3 m/s.

While children have generally slightly lower values of t_a, perhaps a sign of earlier start up times to compensate for their lower speeds, the differences are not found to be statistically significant. The young adult and elderly groups do not show significant differences.

3.3. Parameters Fall within Affordance

Due to the accuracy of the simple crossing model, we expect Equations (3) and (4) to hold for the fitting parameters derived from the data. Figure 6a presents the affordance boundaries in the 3D parameter space. The curved surfaces depict the boundaries specified by Equation (3), and each data point plots the parameters corresponding to a single crossing. For comparison, data for two-step crossings (crosses) as well as simple crossings (circles) are displayed. It is observed that the circles, corresponding to simple crossings indeed lie within the volume between the boundaries. In contrast, most of the crosses are located outside, above the higher surface. This indicates that the subject is on route to collide with the leading vehicle and therefore deceleration is necessary.

(a)

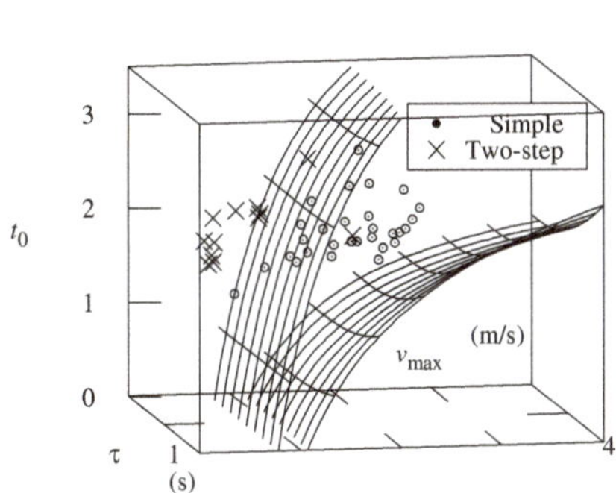

(b)

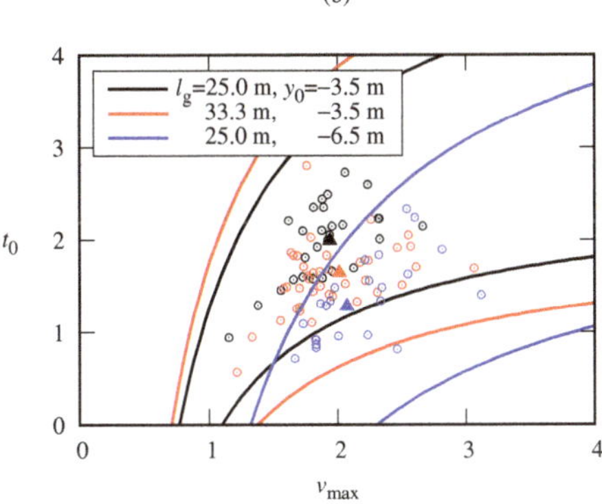

Figure 6. Data plotted with boundaries representing the affordance of the gap. (**a**) data plotted with surfaces in the three-dimensional parameter space (τ, v_{max}, t_a) defined by Equation (3). Data points for $l_g = 25$ m and $y_0 = -3.5$ m are plotted for all age groups; dots represent simple crossings and crosses represent two-step crossings; (**b**) data plotted with boundaries on the two-dimensional plane (v_{max}, t_a) defined by Equation (4). To illustrate the effects of a shift in affordance, data and boundaries for $l_g = 25$ m and $y_0 = -3.5$ m (black) are plotted with those for l_g changed to 33.3 m and for y_0 changed to -6.5 m are shown in red and in blue, respectively. Triangles indicate average parameter values.

Plotting Equation (4) on the 2D parameter plane (v_{max}, t_a) corresponds to the projection of the 3D plot in Figure 6a onto the plane defined by $\tau = 0$. This results in Figure 6b, where two more cases have been included in addition to the case of the gap length $l_g = 25\,$m and the initial position $y_0 = -3.5\,$m presented in Figure 6a. Namely, to probe how the distribution of parameter values shifts with l_g and y_0, we consider the data for a larger gap $l_g = 33.3\,$m and for a farther initial position $y_0 = -6.5\,$m. Accordingly, Figure 6b presents data for three sets of the gap length and initial position together with the corresponding boundaries (lines instead of surfaces in Figure 6a) determined by Equation (4). Specifically, the cases of $(l_g = 33.3\,$m$, y_0 = -3.5\,$m$)$ and $(l_g = 25\,$m$, y_0 = -6.5\,$m$)$ are plotted in red and in blue, respectively, as well as the case of $(l_g = 25\,$m$, y_0 = -3.5\,$m$)$ plotted in black. It is observed that, as the gap is widened, the average behavior (designated by red triangles) shifts toward smaller t_a and larger v_{max}. This reflects the tendency of the pedestrian to cross early before the gap center when possible. On the other hand, in the case that the initial position becomes farther from the intersection point, the pedestrian must compensate by either beginning to walk earlier or walking faster. Data in blue indeed exhibit on average a decrease in t_a and a slight increase in v_{max} (which is, however, not statistically significant). We remark that Figure 6b corresponds directly to Figures 3C and 3D of [13] while Figure 6a is a generalization.

3.4. Bearing Angle Analysis

Finally, we examine the bearing angle of the data and compare it with the model predictions. Figure 7 shows the bearing angle as a function of time for two sets of data (colored lines). The bearing angle tends to decrease in the first few seconds of crossing and to remain constant thereafter, as predicted by Equation (9). In addition, the analytical results given by Equation (6) are plotted with the average parameter values (black line). Both the theory (analytical result from the model) and the experiment (result computed from data) show that a constant bearing angle is held once the pedestrian starts moving at a nearly constant speed. It is shown in Figure 7 that the time interval during which the constant angle is observed is significantly shorter for $y_0 = -3.5\,$m (red) than for $y_0 = -6.5\,$m (cyan). This reflects the smaller value of t_a in the latter case, when the initial position is farther from the interception point.

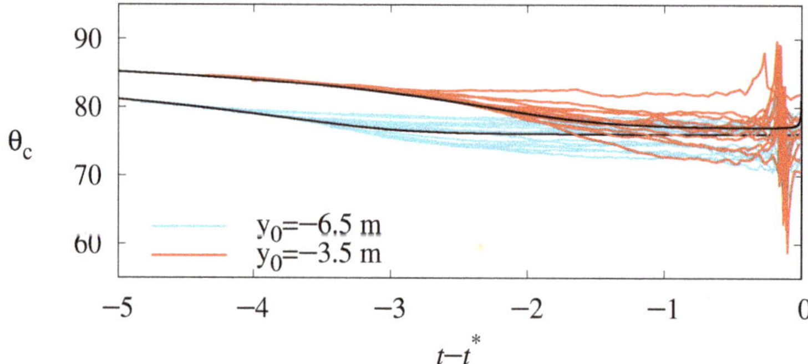

Figure 7. Time evolution of the bearing angle θ_c, defined by Equation (6). The gap length is $l_g = 25\,$m and the initial position is $y_0 = -3.5\,$m and $-6.5\,$m for red and cyan lines, respectively. Colored lines are computed from data, while black lines show the analytical results using average parameter values. The time axis is given in terms of time before crossing $t - t^*$, so that the zero point is equal to the intersection time of the run (i.e., the time at which $y(t) = 0$).

We note that the fluctuations of the data in Figure 7 are due to measurement error, which is magnified immediately before the pedestrian meets the crossing point. This can be seen in Equation (5) by considering that $y(t) \rightarrow 0$ as $t \rightarrow t^*$, causing the error in the argument of the arctan function to become magnified.

4. Discussion

In this study, we have proposed a model for pedestrian crossing and utilized it to extract information from experimental data. The model fit the data with high accuracy, allowing for applications of different methods. In particular, the model allows us to visualize the affordance of each gap and see whether the data lies within it. The model also predicts a constant bearing angle, which has been observed in the data.

The fitting parameters of the model, t_a, $v_{\max}$ and τ, provide a physically intuitive interpretation of the data. Analysis has revealed that pedestrians respond to shifting gap affordances primarily by timing their accelerations, rather than changing their walking speeds, as shown by the distributions of t_a. Moreover, shifts in t_a have been observed in response to the speed of the gap and the size of the surrounding vehicles, even if the gap affordance remains the same, indicating that these environmental factors can change the visual perception of the gap. However, this trend disappeared when the initial distance was greater, suggesting that a greater distance from the road tends to offer a more accurate visual perception of the gap. It has also been observed that children's lower walking velocities indirectly shrank the affordances of the gaps and in certain situations failed to compensate.

Due to the simplicity of the equations, this methodology offers a versatile method to analyze pedestrian behavior. While the velocity equation does not necessarily need to follow a logistic function in particular, the high accuracy with which the equation fits the data and its ease of manipulation makes it an ideal tool for such an analysis.

It should be noted that the accuracy of affordance judgments in virtual environments has been questioned in previous studies [18]. In addition, it is not straightforward to measure walking speeds on treadmills, and a different method was proposed [19]. These factors should be considered when interpreting the results, but we expect them to affect neither qualitative results nor the efficacy of the model. Moreover, it would be desirable to include more general walking scenarios where the pedestrian is not constrained to walk in a straight line. This is left for future study.

5. Conclusions

We have utilized a newly developed model for crossing behavior to extract the control parameter used for adjusting to changing gaps, visualizing crossings within the gap affordance, and testing the bearing angle hypothesis. The model thus serves as a useful tool with which pedestrian behavior can be understood. It can also be used in the context of previous modeling frameworks and may further be developed to extract elements of crossing behavior which leads to collision.

Author Contributions: Conceptualization, M.C., J.W.K. and H.-C.C.; methodology, M.C. and J.W.K.; software, S.H.K.; visualization, S.H.K.; investigation, H.-C.C. and G.-J.C.; writing–original draft preparation, S.H.K.; writing–review and editing, J.W.K. and M.C.; All authors have read and agreed to the published version of the manuscript.

Funding: This research was funded by National Research Foundation of Korea through the Basic Science Research Program (Grant No. 2019R1F1A1046285) and by Korea Institute for Advancement of Technology and Ministry of Trade, Industry, and Energy (Grant No. 10044775).

Conflicts of Interest: The authors declare no conflict of interest.

Appendix A. Analysis of Two-Step Crossings

In order to model two-step crossings in which there are two acceleration events (Figure 3b), we extend the model in the following way: We first take the acceleration equation

$$\ddot{y} = \frac{\dot{y}}{\tau}\left(1 - \frac{\dot{y}}{v_{\max}}\right), \tag{A1}$$

which is equivalent to Equations 1 and 2 with appropriate initial conditions. In the two-step crossing, after acceleration (i.e., at time $t > t_a + 2\tau$), the pedestrian will decelerate at a point y_s and stop until she/he accelerates again at time t_s. This behavior may be described by adding two terms in Equation (A1), which leads to

$$\ddot{y} = \frac{\dot{y}}{\tau}\left(1 - \frac{\dot{y}}{v_{\max}}\right) - r_s \dot{y} \exp\left[-\frac{(y - y_s)^2}{\sigma_s^2}\right]\theta(t_s - t)$$
$$+ v_s\delta(t - t_s), \tag{A2}$$

where the second and the third terms of the right-hand side represent repulsion and impulse force, respectively. The repulsion is centered at position y_s with range σ_s; y_s may be interpreted as the point beyond which the pedestrian perceives to be unsafe, due to the incoming traffic. Accordingly, y_s is the position of the flat region of the curve in Figure 3b, i.e., $y_s = -2.3\,\mathrm{m}$ in this example. The Heaviside step function $\theta(t_s - t)$ effectively "turns off" the repulsion force at time t_s, thus removing the potential for collision after the vehicle has passed. The parameter r_s adjusts the overall strength of the repulsion. The impulse term is necessary for the model to undergo sharp acceleration from rest, so that the pedestrian starts to walk again at time t_s. In the example of Figure 3b, the time at which the subject begins the second acceleration is given by $t_s = 2.0\,\mathrm{s}$. The magnitude v_s of the impulse is a fraction of $v_{\max}$, and determines how quickly the model regains the maximum velocity.

We fit Equation (A2) to the data for the case $y_0 = -3.5\,\mathrm{m}$ and $l_g = 25\,\mathrm{m}$, yielding $y_s = -2.29 \pm 0.22\,\mathrm{m}$. This implies that participants who performed two-step crossings walked forward about 1.2 m before stopping, which amounts to about 1.5 m from the path of the vehicles. We also have the impulse magnitude $v_s/v_{\max} = 0.67 \pm 0.22$ and time $t_s = 2.41 \pm 0.26\,\mathrm{s}$, which corresponds to the time for the pedestrian to start walking again. This range includes the time (about 2.5 s) at which the leading vehicle passes the interception point. Other parameters, repulsion strength r_s and range σ_s, which suffer from large fluctuations due to the very limited sample number, are obtained as $r_s = 520 \pm 480\,\mathrm{s}^{-1}$ and $\sigma_s = 0.26 \pm 0.24\,\mathrm{m}$. However, the sample size of two-step crossings was insufficient to derive statistically meaningful results. A possible behavioral interpretation of this type of crossing is as exchanging the additional energy required for deceleration and acceleration in favor of more safety or security.

References

1. OECD. *Road Infrastructure, Inclusive Development and Traffic Safety in Korea*; OECD Publishing: Paris, France, 2016.
2. Hamed, M.M. Analysis of pedestrians' behavior at pedestrian crossings. *Saf. Sci.* **2001**, *38*, 63–82. [CrossRef]
3. Te Velde, A.F.; Van Der Kamp, J.; Savelsbergh, G.J.P. Five- to twelve-year-olds' control of movement velocity in a dynamic collision avoidance task Arenda. *Br. J. Dev. Psychol.* **2008**, *26*, 33–50. [CrossRef]
4. Azam, M.; Choi, G.-J.; Chung, H.-C. Perception of Affordance in Children and Adults While Crossing Road between Moving Vehicles Muhammad. *Psychology* **2017**, *8*, 1042–1052. [CrossRef]
5. Fajen, B.R.; Warren, W.H. Behavioral Dynamics of Steering, Obstacle Avoidance, and Route Selection Brett. *J. Exp. Psychol. Hum.* **2003**, *29*, 343–362. [CrossRef] [PubMed]
6. Chapman, S. Catching a Baseball. *Am. J. Phys.* **1968**, *36*, 868–870. [CrossRef]
7. Bastin, J.; Craig, C.; Montagne, G. Prospective strategies underlie the control of interceptive actions. *Hum. Mov. Sci.* **2006**, *25*, 718–732. [CrossRef] [PubMed]
8. Fajen, B.R.; Warren, W.H. Behavioral dynamics of intercepting a moving target. *Exp. Brain Res.* **2007**, *180*, 303–319. [CrossRef] [PubMed]
9. Gibson, J.J. The Ecological Approach to Visual Perception. In *Visual Guidance of Intercepting a Moving Target on Foot*; Houghton Mifflin Harcourt (HMH): Boston, MA, USA, 1979; p. 127.
10. Patla, A.E. Understanding the roles of vision in the control of human locomotion. *Gait Posture* **1997**, *5*, 54–69. [CrossRef]

11. Warren, W.H.; Kay, B.A.; Zosh, W.D.; Duchon, A.P.; Sahuc, S. Optic flow is used to control human walking. *Nat. Neurosci.* **2001**, *4*, 213–216. [CrossRef] [PubMed]

12. Fajen, B.R.; Warren, W.H. Visual guidance of intercepting a moving target on foot. *Perception* **2004**, *33*, 689–715. [CrossRef] [PubMed]

13. Fajen, B.R. Guiding locomotion in complex, dynamic environments. *Front. Behav. Neurosci.* **2013**, *7*, 85. [CrossRef] [PubMed]

14. Plumert, J.M.; Kearney, J.K.; Cremer, J.F. Children's perception of gap affordances: Bicycling across traffic-filled intersections in an immersive virtual environment. *Child Dev.* **2004**, *75*, 1243–1253. [CrossRef] [PubMed]

15. Fajen, B.R.; Matthis, J.S. Direct perception of action-scaled affordances: The shrinking gap problem. *J. Exp. Psychol. Hum. Percept. Perform* **2011**, *37*, 1442–1457. [CrossRef] [PubMed]

16. Yannis, G.; Papadimitriou, E.; Theofilatos, A. Pedestrian gap acceptance for mid-block street crossing. *Transp. Plan. Technol.* **2013**, *36*, 450–462. [CrossRef]

17. Chihak, B.J.; Plumert, J.M.; Ziemer, C.J.; Kearney, J.K. Synchronizing self and object movement: How child and adult cyclists intercept moving gaps in a virtual environment. *J. Exp. Psychol. Hum. Percept. Perform* **2010**, *36*, 1535–1552. [CrossRef] [PubMed]

18. Creem-Regehr Sarah, H.; Gill, D.M.; Pointon, G.D.; Bodenheimer, B.; Stefanucci, J. Mind the gap: Gap affordance judgments of children, teens, and adults in an immersive virtual environment. *Front. Robot. AI* **2019**, *6*, 96. [CrossRef]

19. Yoon, J.; Park, H.S.; Damiano, D.L. A novel walking speed estimation scheme and its application to treadmill control for gait rehabilitation. *J. Neuroeng. Rehabil.* **2012**, *9*, 62. [CrossRef] [PubMed]

Article

Functional Roles of Saccades for a Hand Movement

Yuki Sakazume, Sho Furubayashi and Eizo Miyashita *

School of Life Science and Technology, Tokyo Institute of Technology, Tokyo 152-8550, Japan;
sakazume.y.aa@m.titech.ac.jp (Y.S.); furubayashi.s.aa@m.titech.ac.jp (S.F.)
* Correspondence: miyashita.e.aa@m.titech.ac.jp; Tel.: +81-45-924-5573

Received: 6 March 2020; Accepted: 27 April 2020; Published: 28 April 2020

Abstract: An eye saccade provides appropriate visual information for motor control. The present study was aimed to reveal the role of saccades in hand movements. Two types of movements, i.e., hitting and circle-drawing movements, were adopted, and saccades during the movements were classified as either a leading saccade (LS) or catching saccade (CS) depending on the relative gaze position of the saccade to the hand position. The ratio of types of the saccades during the movements was heavily dependent on the skillfulness of the subjects. In the late phase of the movements in a less skillful subject, CS tended to occur in less precise movements, and precision of the movement tended to be improved in the subsequent movement in the hitting. While LS directing gaze to a target point was observed in both types of the movements regardless of skillfulness of the subjects, LS in between a start point and a target point, which led gaze to a local minimum variance point on a hand movement trajectory, was exclusively found in the drawing in a less skillful subject. These results suggest that LS and some types of CS may provide positional information of via-points in addition to a target point and visual information to improve precision of a feedforward controller in the brain, respectively.

Keywords: eye movements; reaching arm movements; eye-hand coordination; visual information; movement precision; movement segment; forward model; control model of the brain

---

1. Introduction

Complex movement is considered to have consecutive segments of movement [1,2]. A connecting point of each segment can be called a via-point [3]. Thus, previous studies have focused on hitting or reaching movements with the hand as an elemental movement or a movement segment to reveal behavioral characteristics [4,5] and to investigate neuronal activity patterns [6,7]. Such studies often include a drawing movement [8]. Investigation of generation mechanism of the via-point may lead to understanding a complex movement.

In daily life, movements depend heavily on visual information, which is used for feedforward and feedback controls. For example, goal-directed movements of the hand are usually accompanied by an eye saccade that precisely directs an individual's gaze to an observed target location in case of stational target [9–13] or a predicted target location in case of moving target [14,15] before the hand starts to move. This type of saccades has been thought to provide visual information about the observed or predicted target location to guide a hand movement. Feedback control of the reaching movements also relies substantially on visual information [16–19]. Additionally, another type of saccade was identified during the performance of a simple line-drawing movement in which a sequence of small eye saccades closely followed the trajectory of a pencil [20]. The author of this study suggested that this type of saccade contributed to feedback control. Therefore, at least two types of saccades may be associated with hand movement: a saccade that directs an individual's gaze to a target position, which has been observed during a reaching movement, and a saccade that directs an individual's gaze to a hand position, which has been observed during the drawing of a simple line.

The goal of the present study was to reveal the relationship between the two types of saccades and hand movements, especially its feedforward control, and to get an insight into the role of visual information acquired by the saccades on movement control of the hand. We adopted hitting and circle-drawing movements as discrete and continuous movements, respectively. These two types of movements were selected instead of reaching and circle-tracking movements to increase the relative contribution of feedforward control to feedback control. Saccades during the movements were quantitatively classified as either a leading saccade (LS) or catching saccade (CS) depending on the relative gaze position of the saccade to a cursor position that represented the hand position: LS and CS directed the gaze to the cursor position in the direction of the cursor movement and to around the current cursor position, respectively. Precision of the hand movements was analyzed in relation to the saccades.

2. Materials and Methods

Two Japanese monkeys (*Macaca fuscata*; male: Monkey H, 6.7 kg, female: Monkey U, 6.6 kg) were used in the present study. Using a robotic arm manipulandum, the monkeys were trained to execute hitting and drawing tasks. During the tasks, hand and gaze positions of the monkey were recorded. All experimental procedures were performed in accordance with the Guidelines for Proper Conduct of Animal Experiments of Science Council of Japan and approved by the Committee for Animal Experiment at Tokyo Institute of Technology.

2.1. Subjects and Apparatus

To fixate the head position during the tasks, a head holder was installed on each monkey. Inducted by ketamine (10 mg/kg, intramuscular injection (i.m.)) and xylazine (2 mg/kg, i.m.), the monkey was deeply anesthetized with pentobarbital (25 mg/kg for an initial dose and 12.5 mg/kg for a supplementary dose when required by intravenous injection), a surgical operation was performed under aseptic conditions while monitoring heart rate and blood saturated oxygen concentration. The head holder was installed on the monkey skull and fixed with dental cement after cortical screws were implanted into the skull.

After an appropriate recovery period from the surgical operation, the monkeys were well trained to perform behavioral tasks (Figure 1). The monkey sat comfortably in a primate chair with its head fixated. RANARM, a robotic arm manipulandum for normal and altered reaching movements [21], was used to control the tasks and record the hand position of the monkeys, and a 19-inch computer monitor was set 61 cm away from the monkeys' eyes. A cursor on the monitor provided visual feedback regarding the current spatial location of a gripper of RANARM that the monkey held in its hand. Gaze position was estimated by measuring the position and shape of the pupil at a sampling rate of 1000 Hz with EyeLink (SR Research, Ottawa, ON, Canada).

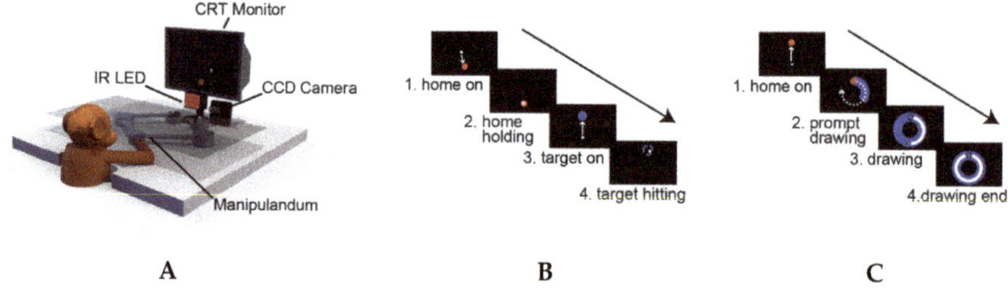

Figure 1. Experimental setup and time consequence of the tasks are illustrated. (**A**) A monkey directs a cursor on a monitor by manipulating a robotic arm with its right hand on a horizontal plane. The hand position and its eye position are recorded during the tasks. (**B**) Hitting task: monkeys are required to move the cursor from the bottom red home point to the top blue target point within a designated amount of time.

(**C**) Circle drawing task: a prompt shows the required direction (blue: clockwise, green: counter-clockwise (not shown)) of drawing while the monkey holds the cursor within the start point. Eight evenly spaced start points are used. The monkey can begin drawing at any time and without any time restrictions on movement time.

2.2. Behavioral Tasks

Two tasks were executed by each monkey with the right hand: a hitting task and a circle-drawing task.

In the hitting task, a red home point appeared 0.100 m below the center of the monitor at the start of each trial. The monkey was required to move the cursor to the home point and then hold it there for a random period of time that ranged from 0.50 to 1.00 s. Next, a blue target point appeared 0.075 m above the center of the monitor, and the monkey was required to move the cursor to the target point within a designated amount of time. If the monkey hit the target point in the allotted time, the target point disappeared, and the monkey received a drop of water as a reward. If the monkey did not hit the target in time, the trial was considered a failure. To ensure quick and precise hand movements, the size of the home and target points, and the required task time (reaction time plus movement time) were regulated so that the ratio of successful trials to failure trials was equal to approximately 1.

In the circle-drawing task, a red start point (radius: 0.030 m) appeared at one of eight predetermined positions located at 45° intervals around a circle (radius: 0.100 m) in the center of the monitor. The monkey was required to move the cursor (radius: 0.010 m) into the start point and to hold the cursor there for 1 sec. Then, a prompt consisting of a circle was quickly drawn beginning from the start point and using different colors to identify the direction of movement (blue: clockwise, green: counterclockwise). During this period, the monkey was required to hold the cursor within the start point and was then able to begin its drawing movement at any time after the prompt drawing was completed. There were no time restrictions on the movement, and the trajectory of the cursor during hand movement was recorded. Three indices were used to determine a successful trial: 1) the drawing direction of the monkey was the same as that of the prompt drawing; 2) the center of cursor passed within 0.070 m of the center of the prompt trajectory; and 3) the cursor returned to the start point. The monkey got a drop of water as a reward after a successful trial.

2.3. Data Recording and Analysis

Gaze and hand positional data were outputted to a home-made data recording system using LabView (National Instruments, Austin, TX, USA) through a 16-bit analog to digital converter and saved as a binary file. Data were analyzed using MATLAB (The MathWorks, Natick, MA, USA).

2.3.1. Saccade

The eye saccades were detected from the gaze positional data that were filtered with a fourth-order Butterworth low-pass filter at 50 Hz. The onset and offset of the saccades were determined using the following criteria: the gaze speed exceeded a threshold criterion of 61°/s or it was lower than a threshold criterion of 3°/s after the saccade onset. We classified the saccades into two types depending on the relative gaze position at offset of the saccade to the hand position: a leading saccade (LS) and a catching saccade (CS). LS and CS directed the gaze to the cursor position in the direction of the cursor movement and to around the current cursor position, respectively. We analyzed CS in the hitting task and both LS and CS in the circle-drawing task from a viewpoint of relationships between saccades and precision of the hand movement.

2.3.2. Hand Movement

The cursor positional data that corresponded to the hand position were obtained using RANARM and were filtered with a fourth-order Butterworth low-pass filter at 10 Hz. Precision of the hand movement was evaluated.

In the hitting task, it was evaluated at three different points during the movement where the hand acceleration was maximum, its velocity was maximum, and its acceleration was minimum, each of which represents the initial, middle, and terminal phases of the movement. Movement precision was defined based on the positional variance of the cursor's x-coordinate. At the acceleration maximum point, another index, which was the directional variance of tangential movement direction (Dm) of the cursor, was adopted. We expected that Dm was more robust to variation of the start position of the cursor and was a more appropriate index for a very early phase of the movement.

In the circle-drawing task, the position of the cursor was expressed in polar coordinates, and precision of the hand movement was defined as the positional variance of the cursor in the radial direction at a given phase. As another index of the precision, we also adopted curvature of the hand movement, which was calculated as follows: $\left| \dot{x}\ddot{y} - \ddot{x}\dot{y} \right| \left(\dot{x}^2 + \dot{y}^2 \right)^{-3/2}$, where x and y are the cursor's coordinates of its position in Cartesian coordinates with the horizontal direction and the center of the monitor as its x-axis and origin, respectively.

3. Results

3.1. Hitting Task

Monkeys had to control its hand quickly and precisely to execute the task. For *Monkey H*, the radii of the home and target points were set at 0.012 and 0.014 m, respectively, and the task time was designated at 0.75 s. This subject completed a total of 780 trials, of which 401 (52%) were successful. For *Monkey U*, the radii of the home and target points were both set at 0.006 m, and the task time was designated at 0.54 s. This subject completed a total of 560 trials, of which 403 (58%) were successful.

Although success ratio of the task was more or less the same in the two monkeys, movement characteristics of the hand in the hitting task varied depending on the monkey. *Monkey U* showed better performance than *Monkey H* with smaller target diameter, shorter movement time (0.47 ± 0.04 s vs 0.33 ± 0.03 s) (mean ± standard deviation [SD]), higher peak hand velocity (0.54 ± 0.05 m/s vs 0.90 ± 0.07 m/s) (mean ± SD), all of which indicate more precise feedforward control of the hand.

In the following analysis, trials that did not reach 0.125 m in y-coordinate within the designated task time were excluded; 762 trials in Monkey H and 556 trials in Monkey U were analyzed as a result. In almost all of the hitting task trials, both monkeys made a saccade directly to the target point upon its appearance and prior to the onset of hand movement. Subsequently, two different types of gaze control were observed. In one type, the monkey kept its gaze on the target point until the cursor hit it (Figure 2A), and in the other type, the monkey returned its gaze from the target to the cursor and made a couple of saccades around the cursor while following its movement (Figure 2B). LS and CS that execute these two types of gaze control were analyzed. We defined LS and CS as follows: LS was a saccade that directed the gaze in the range from 0.135 to 0.200 m in y-coordinate; CS was one that was not LS and difference of gaze position directed by that and the cursor position was within ± 0.050 m in y-coordinate.

The LS was observed in both monkeys in more than 90% of the trials. Inversely correlated with the performance of the task, interestingly, the number of the CS in Monkey H was prominently higher than that in Monkey U (Table 1). Therefore, the CS in Monkey H was analyzed in the following analysis. In the trial with the CS in Monkey H, the subject executed the CS along the hand trajectory (Figure 3A). The gaze points following the CS in Monkey H seemed bimodally distributed in the direction of cursor movement with a border point at 0.11 m from the center of the start point (Figure 3B), which roughly corresponded to the cursor velocity maximum point. We tentatively classified the CS into two types; the CS in hand acceleration phase (CSa) and the CS in hand deceleration phase (CSd).

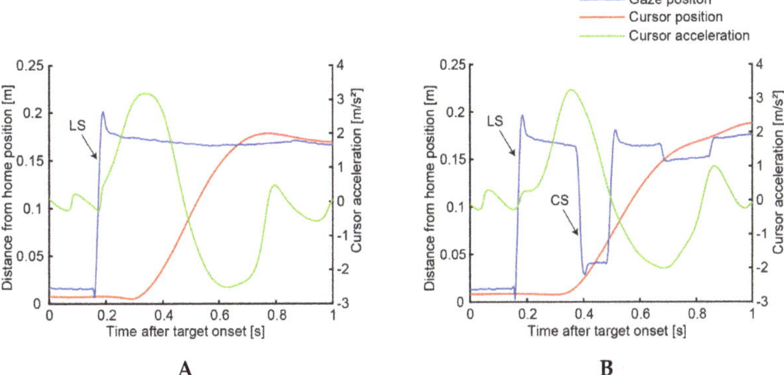

Figure 2. Two types of saccades during the hitting task are depicted. (**A**) leading saccade (LS) occurs almost simultaneously at the onset of acceleration of the cursor. (**B**) catching saccade (CS) follows LS in this trial.

Table 1. Occurrence probability of each type of saccades during the hitting task.

Subject	Number of Trials	Saccade [Times/Trial] (Number of Trials)				
		Total	LS	Total CS	CSa [1]	CSd [2]
Monkey H	762	2.51 (762)	1.77 (762)	0.35 (242)	0.27 (199) [3]	0.09 (67) [3]
Monkey U	556	1.49 (556)	1.16 (540)	0.01 (8)	0.01(8)	0.00 (0)

[1] CSa: CS in cursor acceleration phase. [2] CSd: CS in cursor deceleration phase. [3] Both CSa and CSd were observed in 24 trials.

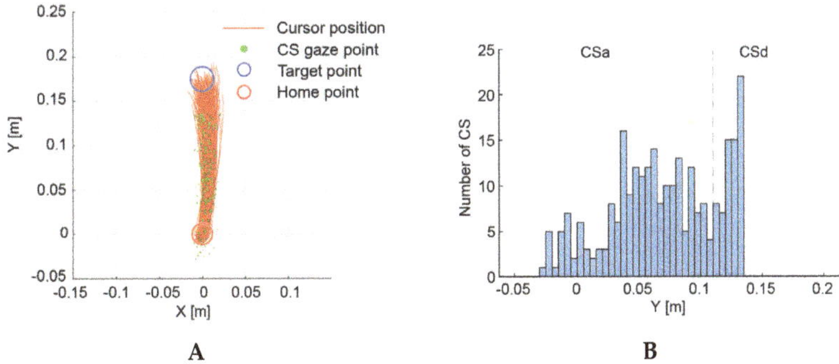

Figure 3. Cursor trajectories of trials with CS and distribution of gaze points directed by CS. (**A**) Cursor trajectories of the hitting trials with CS and gaze points directed by CS in *Monkey H*. (**B**) Distribution of gaze points directed by CS in y-coordinate. Broken line indicates the border point (0.11 m) of CS in hand acceleration phase (CSa) and CS in hand deceleration phase (CSd).

To analyze the relationship between the CS and precision of the hand movement, the hitting task trials were grouped into three; those without the CS, those with CSa, and those with CSd, and precision of the hand movement among the groups was compared. The trials with CSd showed a tendency towards larger variance of the movement than the other two groups from the middle phase of the movement and showed significantly larger variance than those without the CS both at the middle (F-test, $p = 0.0144$) and deceleration (F-test, $p = 0.0158$) phases of the movement (Figure 4A). Although there was no significant difference in the variance of Dm among the three groups, we found that only the subsequent trials to ones with CSd showed a tendency towards smaller variance than the current trials (Figure 4B). This was indicative of influence of CSd on improvement of precision of the subsequent movement in its initial phase, i.e. feedforward control component. Like in the task performance indices, e.g. size of the target and peak hand velocity, precision of the movement by *Monkey U* was superior to that by *Monkey H* in all the movement phases.

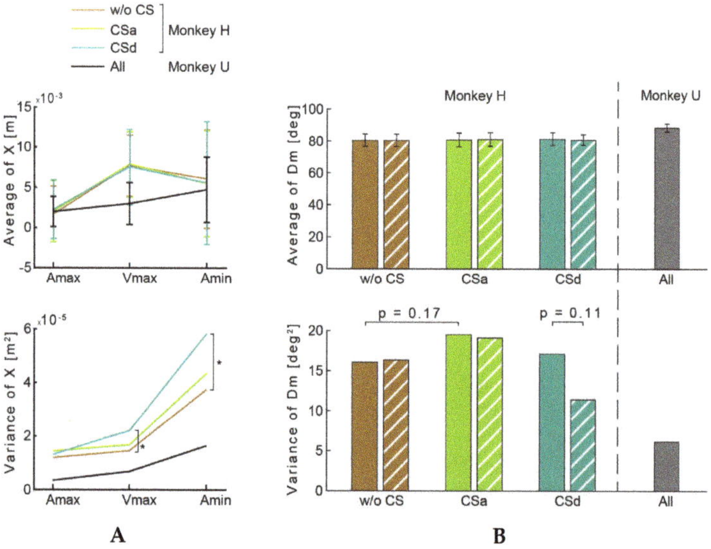

Figure 4. Effects of the CS on accuracy and precision of the hand movement in the hitting task are shown. Trials are grouped into three depending on the types of saccades during the trial in Monkey H; those without (w/o) CS (520 trials), with CSa (199 trials), and with CSd (67 trials). As a comparison, the result of all trials in Monkey U (556 trials) is also shown. (**A**) Top and bottom panels: average and variance of the cursor position in x-coordinate are shown, respectively (* $p < 0.05$). They are evaluated at cursor acceleration maximum (Amax), velocity maximum (Vmax), and acceleration minimum (Amin) point during the hand movement. (**B**) Top and bottom panels: average and variance of movement direction (Dm) are shown, respectively. The current and subsequent trials are shown as filled boxes and white oblique-lined boxes, respectively. An error bar represents a standard deviation in all panels.

3.2. Circle-Drawing Task

Following a training period, both monkeys were able to smoothly draw a circle from all eight start points in both directions of rotation at success rates up to around 90%. In total, 2,000 trials (two rotation directions × eight start points × 125 trials) were recorded from each monkey. All of the following analyses were applied to the successful trials in which cursor trajectory was within ± 2 SD range from the average and the saccades that directed the gaze within ± 3 SD range from the average trajectory of the cursor.

Monkey H performed the task with a movement time of 0.90 ± 0.14 s and 0.97 ± 0.11 s (mean ± SD) in the counterclockwise and clockwise directions, respectively. *Monkey U* did it with a movement time of 0.77 ± 0.08 s and 0.71 ± 0.07 s (mean ± SD) in the counterclockwise and clockwise directions, respectively.

Similar to the hitting task, both monkeys made several saccades throughout the drawing task (Figure 5A). We focused on the saccades that led the gaze on the way to the target. Therefore, we excluded the saccade within the final 45° range of the phase and plotted the distribution of phase difference of the saccade against the cursor (Figure 5B). In Monkey H, the distribution seemed to be bimodal with peaks around at 0° and −100° in both rotation directions. In contrast, in *Monkey U*, it seemed to be unimodal with a peak around at 0° in both rotation directions. We defined LS and CS in the drawing task as a saccade that had the phase difference smaller than −35° and within ±35°, respectively. More than 40% of the saccades were the LS in Monkey H (46% in the counterclockwise rotation direction, 44% in the clockwise rotation direction). In contrast, the CS dominated in Monkey U (45% in the counterclockwise rotation direction, 76% in the clockwise rotation direction) (Figure 5C) (Table 2).

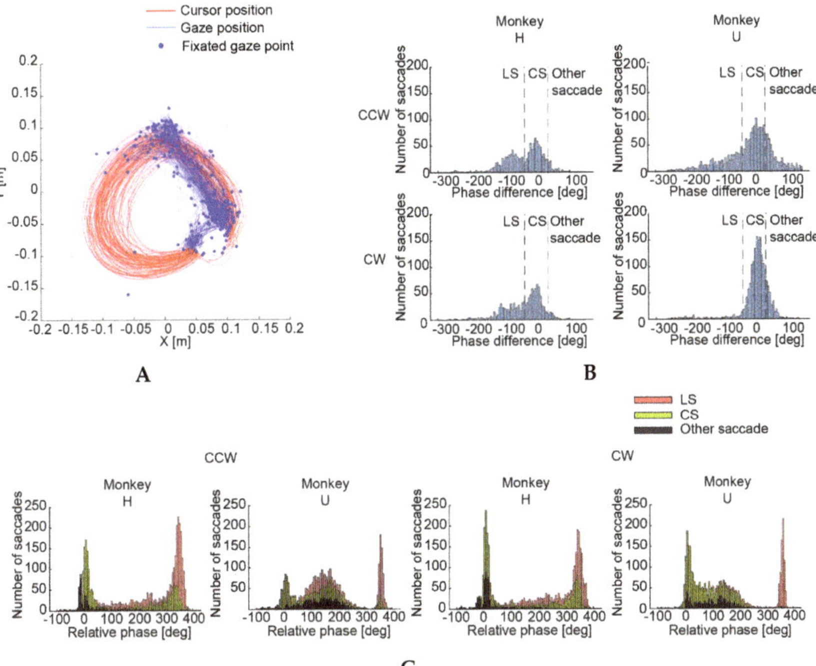

Figure 5. Two types of saccades during the circle-drawing task are depicted. (**A**) Example trials of the circle-drawing task from the 315° start point toward the clockwise direction in *Monkey H*. (**B**) Distributions of gaze points directed by saccades relative to the hand position are shown. Broken lines indicate the borders (–35° and 35°) of LS, CS, and other saccades. (**C**) Distributions of gaze points directed by LS, CS, and the other saccade are plotted in relation to a start point (a positive sign is assigned to the phase when the gaze point is in the cursor movement direction). Bin width is 5° in all the panels.

Table 2. Occurrence probability of each type of saccades during the circle-drawing task.

Subject	Direction [1]	Trial	Saccade [Times/Trial] (Number of Trials) All/Midway [2]		
			Total [3]	LS	CS
Monkey H	CCW	743	5.22 (741)/ 1.81 (723)	2.33 (735)/ 0.83 (457)	2.27 (694)/ 0.86 (447)
	CW	751	4.77 (751)/ 1.72 (648)	1.78 (714)/ 0.75 (439)	2.22 (668)/ 0.88 (441)
Monkey U	CCW	749	4.70 (749)/ 3.04 (749)	1.70 (691)/ 0.89 (531)	1.80 (638)/ 1.35 (624)
	CW	769	4.60 (769)/ 2.29 (769)	1.10 (679)/ 0.16 (110)	2.60 (744)/ 1.73 (729)

[1] CCW: counterclockwise, CW: clockwise. [2] Midway: saccades except the final 45° range of the phase. [3] Total = LS + CS + Other saccades.

We then analyzed the relationship of these saccades and precision of the hand movement. In Monkey H, variance of the movement tended to be small at a specific phase around 30°–60° irrespective of the eight different start points and two different rotation directions, toward which point the LS tended to be directed (Figure 6A). When the LS was aligned at the local minimum phase of the variance in the range of 0°–100°, it clearly showed that the LS directed the gaze 0° (counterclockwise rotation direction) and −30° (clockwise rotation direction) in advance to the local minimum point of the variance (Figure 6B). In Monkey U, the number of the LS was relatively small (29% in the counterclockwise rotation direction, 7% in the clockwise rotation direction), and no obvious relationship of the LS and precision of the hand movement was found.

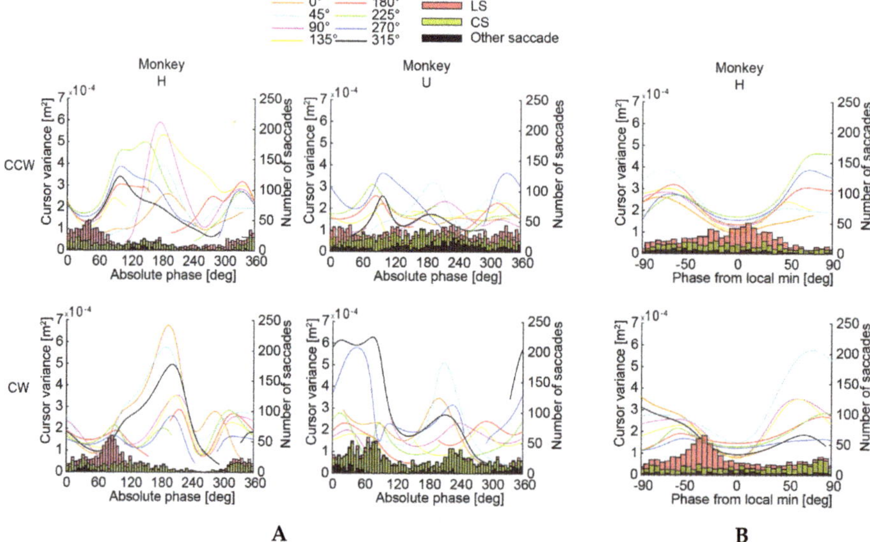

Figure 6. Relationships of the cursor positional variance and gaze points directed by saccades in between (within 45°–315° phase range relative to the start point) the start and target points in the circle-drawing task are shown. (**A**) The relationships are plotted in absolute phase (phase in polar coordinates with its origin at the center of the monitor). Cursor variance and number of each saccade are indicated by the solid lines and bars, respectively. Note that the cursor variance gets local minimum at the phase of around 30°–60° for almost all starting point regardless of the rotational directions in *Monkey H*. (**B**) Gaze points directed by saccades are plotted against the local minimum phase of the cursor variance in *Monkey H* (a negative sign is assigned to the phase when the gaze leads the cursor movement). Bin width is 5° in all the panels.

Finally, the CS was analyzed in terms of precision of the drawing movement. We applied the same method as in the hitting task; precision of the movement was compared between trials with CSd within the phase range of 60° to the final target (CSt) and ones without it. As an index of the movement precision, curvature of the cursor was evaluated at the local minimum of the acceleration of the cursor. In *Monkey H*, the trials with CSt showed significantly higher variance of the curvature than those without CSt in the counterclockwise (F-test, $p = 1.6 \times 10^{-7}$) and clockwise (F-test, $p = 3.1 \times 10^{-3}$) rotation directions, respectively (Figure 7). The analysis was not applied to Monkey U because there were a few CSt found in both rotation directions.

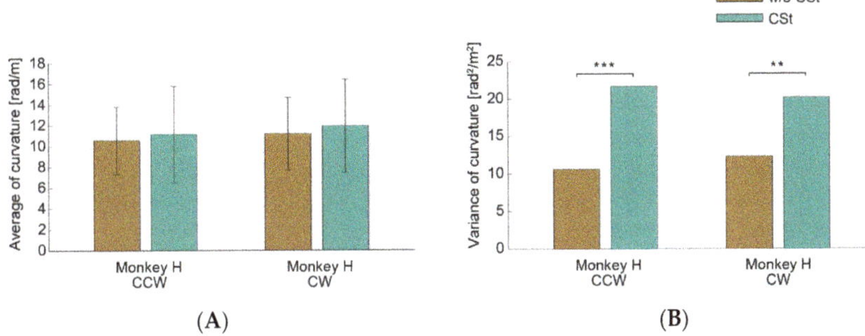

Figure 7. Effects of CS close to the target (CSt) on the cursor movement accuracy and precision during the circle-drawing task Monkey H are shown. They are evaluated at the cursor acceleration minimum point close to the target (**A**) Average of curvature of the cursor trajectory. An error bar represents a standard deviation. (**B**) Variance of the curvature. ** $p < 0.01$. *** $p < 0.001$.

4. Discussion

The relationship between saccades and hand movement has been studied using both pointing and drawing movements. For example, goal-directed pointing tasks were used to assess the type of saccades that preceded the onset of hand movement [9–15]. This type of saccades has been thought to provide visual information about the observed or predicted target location to guide a hand movement. Additionally, a simple line-drawing task was used to identify another type of saccade in which the gaze closely followed a pencil trajectory with a sequence of small saccades [20]. This type of saccade may be related to the feedback control of hand movement.

In the present study, saccades during the movements were quantitatively classified as either a leading saccade (LS) or catching saccade (CS) depending on the relative gaze position of the saccade to a cursor position that represented the hand position, and relationships between the saccades and feedforward control component of hand movements were analyzed.

4.1. Functional Roles of the Saccades

We found CS during the hitting task in addition to the circle-drawing task. Although CS during the circle-drawing task was predictable considering that a similar type of saccade was observed in a previous study using a line-drawing task [20], CS during the hitting task was a novel finding.

CS during the hitting task was exclusively observed in the monkey that showed lower precision of the movement than the other. While CS was found both in the hand acceleration phase and in the hand deceleration phase (CSd), precision of the middle and late phases of the movement in the trials with CSd was significantly lower than that in the trials without CSd. Since the subject had to heavily rely on feedforward control to accomplish the task because of its designated short movement time, the observed precision is considered to largely reflect that of the feedforward control component. Taking the following two results into account, therefore, we assume that CSd during the hitting task may arise from the imprecision of the feedforward control component to improve its precision. Firstly, the trials with CSd were significantly less precise than those without CSd at the middle and deceleration phases of the movement. Secondly, the initial movement direction, which represents the feedforward control component, in the subsequent trial tended to be more precise than that in the current trial with CSd, while no significant difference of the precision was found between trials without CSd. This assumption may support the previous study that showed a faster adaptation rate in a novel task in a trial with continuous visual feedback of the cursor position than in a trial with post-trial knowledge of task performance [22]. Although we could not detect any evidence suggesting contribution of CS to online error correction but find those suggesting contribution of CSd on offline error correction ("error" is used in terms of not systematic error but accidental error), it should be further carefully investigated which saccade, i.e. LS or CS, provides visual information for the online feedback error correction [17,23,24].

In the drawing task, two types of CS were also observed like in the hitting task; one occurred in early phase of the movement (CSe) and the other in late phase of the movement, especially within a phase close to the target (CSt). While the monkey that showed higher precision in the hitting task had not CSt but exclusively CSe, the one that showed lower precision had both CSt and CSe. Like CSd in the hitting task, CSt in the circle-drawing task was observed exclusively in the monkey that showed lower precision in the hitting task. Furthermore, precision at the deceleration local minimum phase of the movement close to the target (curvature was used as an index) in the trials with CSt was significantly lower than that in the trials without CSt, which coincides with the finding in the hitting task. Taking this analogy between CSd and CSt into account, we assume that CSt during the circle-drawing task may also arise from the imprecision of the feedforward controller to improve its precision. As for CSe, we presume that CSe may have a role on precise control of the hand movement because CSe was predominantly observed in the monkey that executed more precise control of the hitting task although we could not find any valuable index to evaluate the function of CSe.

The present study also found LS during the circle-drawing task that directed the gaze to a point in between the start and target points although there was no explicit presentation of a via-point. This type of gaze control has been reported in walking in natural terrain [25], in which a saccade directed the gaze to a future point of foot placement about 1.5 ms in advance. In the present study, the LS (LSm) was always followed by a fixated gaze until the cursor passed the area and the gaze points of LSm were concentrated at a point with 40° or 0° phase lead depending on the rotation direction where the positional variance of the cursor reached a local minimum. These findings suggest that the gaze point by LSm is directed to close to an internally set via-point. These points were set in working coordinates, i.e. external coordinates because the local minimum points of the variance of the cursor were more or less consistent irrespective to the start point of the drawing movement. LSm was observed in the monkey that executed less precise control of the hitting task, suggesting that the skillful monkey regarded the circle-drawing movement as single segment of movement and the less skillful monkey divided the movement into multiple segments setting via-points.

4.2. The Saccades and the Control Model of an Arm Movement

An optimal feedback control model was proposed as one for pointing or reaching arm movements [26] and has successfully explained various characteristics of those behaviors [27–30]. According to the model, an optimal controller produces a control input to muscles based on the estimated state of an arm, i.e., a predicted online state that may be corrected by observed information. Furthermore, a forward model generates the predicted state using the control input. Thus, the optimal controller is able to serve as a controller for a segmental movement, and to control an object even without feedback information if the forward model is accurate and precise enough.

Based on the minimum intervention principle of the model, it predicts that variance of trajectories of movement becomes minimum at an aimed point, i.e. a via-point in multiple segmental movement. Therefore, the local minimum points of the variance of the cursor in the circle-drawing task can be reasonably regarded as a via-point. In the case of movements that are more complicated than the circle-drawing, it is possible that the brain internally sets via-points [3] toward which LS directs an individual's gaze and, in turn, sequentially feeds information to some type of neurological optimal controller.

The model does not inherently implement a mechanism that adaptively changes the forward model depending on a change of a control object. Since there has been accumulated evidence suggesting that the brain can adaptively change its feedforward control signal depending on a change of an environment [31–34], the forward model in the brain must adaptively changes. For the adaptive change of the forward model, some types of CS, e.g., CSd in the hitting task and CSt in the circle-drawing task may provide visual information.

Finally, a higher-level mechanism that governs the way of adaptation or learning has been proposed as meta-learning, which is mainly investigated in the framework of reinforcement learning [35,36]. An apparent different strategy to the drawing task found in two subjects in the present study might be a resultant outcome of a kind of meta-learning because the subjects had acquired the skill to perform the task through a reward-based training process. If this is the case, we suggest that precision or reliability of the forward model should be considered as an important factor to determine a strategy in the meta-learning. In other words, a more reliable forward model may lead to a strategy to perform a complex movement with a smaller number of moment segments.

Author Contributions: Conceptualization, E.M.; Formal analysis, Y.S. and S.F.; Investigation, Y.S.; Methodology, E.M.; Supervision, E.M.; Visualization, Y.S. and S.F.; Writing—original draft, E.M.; Writing—review & editing, Y.S., S.F. and E.M. All authors have read and agree to the published version of the manuscript.

Funding: This work was supported in part by Grant-in-Aid for Scientific Research on Innovative Areas, "The study on the neural dynamics for understanding communication in terms of complex hetero systems (No. 4103)" (211200112).

Acknowledgments: The authors would like to thank Tomohiro Nabe (MEng), Shinichiro Murakami (MEng), and Zhiwei He (MEng), who were master course students at Tokyo Institute of Technology, for their help of the animal experiment.

Conflicts of Interest: The authors declare no conflict of interest.

References

1. Morasso, P.; Mussa Ivaldi, F.A. Trajectory formation and handwriting: A computational model. *Biol. Cybern.* **1982**, *45*, 131–142. [CrossRef] [PubMed]
2. Viviani, P.; Cenzato, M. Segmentation and coupling in complex movements. *J. Exp. Psychol. Hum. Percept. Perform.* **1985**, *11*, 828–845. [CrossRef] [PubMed]
3. Wada, Y.; Kawato, M. A via-point time optimization algorithm for complex sequential trajectory formation. *Neural Netw.* **2004**, *17*, 353–364. [CrossRef]
4. Morasso, P. Spatial Control of Arm Movements. *Exp. Brain Res.* **1981**, *42*, 223–227. [CrossRef]
5. Gomi, H.; Kawato, M. Human arm stiffness and equilibrium-point trajectory during multi-joint movement. *Biol. Cybern.* **1997**, *76*, 163–171. [CrossRef] [PubMed]
6. Georgopoulos, A.P.; Kalaska, J.F.; Caminiti, R.; Massey, J.T. On the Relations between the Direction of Two-Dimensional Arm Movements and Cell Discharge in Primate Motor Cortex. *J. Neurosci.* **1982**, *2*, 1527–1537. [CrossRef] [PubMed]
7. Scott, S.H.; Kalaska, J.F. Changes in motor cortex activity during reaching movements with similar hand paths but different arm postures. *J. Neurophysiol.* **1995**, *73*, 2563–2567. [CrossRef] [PubMed]
8. Klein Breteler, M.D.; Hondzinski, J.M.; Flanders, M. Drawing sequences of segments in 3D: Kinetic influences on arm configuration. *J. Neurophysiol.* **2003**, *89*, 3253–3263. [CrossRef]
9. Abrams, R.A.; Meyer, D.E.; Kornblum, S. Eye-hand coordination: Oculomotor control in rapid aimed limb movements. *J. Exp. Psychol. Hum. Percept. Perform.* **1990**, *16*, 248–267. [CrossRef]
10. Frens, M.A.; Erkelens, C.J. Coordination of hand movements and saccades: Evidence for a common and a separate pathway. *Exp. Brain Res.* **1991**, *85*, 682–690. [CrossRef]
11. Bekkering, H.; Adam, J.J.; van den Aarssen, A.; Kingma, H.; Whiting, H.T. Interference between saccadic eye and goal-directed hand movements. *Exp. Brain Res.* **1995**, *106*, 475–484. [CrossRef]
12. Helsen, W.F.; Elliott, D.; Starkes, J.L.; Ricker, K.L. Temporal and spatial coupling of point of gaze and hand movements in aiming. *J. Motor Behav.* **1998**, *30*, 249–259. [CrossRef]
13. Lunenburger, L.; Kutz, D.F.; Hoffmann, K.P. Influence of arm movements on saccades in humans. *Eur. J. Neurosci.* **2000**, *12*, 4107–4116. [CrossRef]
14. Diaz, G.; Cooper, J.; Rothkopf, C.; Hayhoe, M. Saccades to future ball location reveal memory-based prediction in a virtual-reality interception task. *J. Vis.* **2013**, *13*, 20. [CrossRef]
15. Mann, D.L.; Nakamoto, H.; Logt, N.; Sikkink, L.; Brenner, E. Predictive eye movements when hitting a bouncing ball. *J. Vis.* **2019**, *19*, 28. [CrossRef]
16. Beaubaton, D.; Hay, L. Contribution of Visual Information to Feedforward and Feedback Processes in Rapid Pointing Movements. *Hum. Mov. Sci.* **1986**, *5*, 19–34. [CrossRef]

17. Desmurget, M.; Grafton, S. Forward modeling allows feedback control for fast reaching movements. *Trends Cogn. Sci.* **2000**, *4*, 423–431. [CrossRef]

18. Saunders, J.A.; Knill, D.C. Humans use continuous visual feedback from the hand to control fast reaching movements. *Exp. Brain Res.* **2003**, *152*, 341–352. [CrossRef]

19. Thaler, L.; Goodale, M.A. The role of online visual feedback for the control of target-directed and allocentric hand movements. *J. Neurophysiol.* **2011**, *105*, 846–859. [CrossRef]

20. Tchalenko, J. Eye movements in drawing simple lines. *Perception* **2007**, *36*, 1152–1167. [CrossRef]

21. Ueyama, Y.; Miyashita, E. Devising a Robotic Arm Manipulandum for Normal and Altered Reaching Movements to Investigate Brain Mechanisms of Motor Control. *Instrum. Sci. Technol.* **2013**, *41*, 251–273. [CrossRef]

22. Hinder, M.R.; Tresilian, J.R.; Riek, S.; Carson, R.G. The contribution of visual feedback to visuomotor adaptation: How much and when? *Brain Res.* **2008**, *1197*, 123–134. [CrossRef]

23. Milner, T.E. A model for the generation of movements requiring endpoint precision. *Neuroscience* **1992**, *49*, 487–496. [CrossRef]

24. Plamondon, R.; Alimi, A.M. Speed/accuracy trade-offs in target-directed movements. *Behav. Brain Sci.* **1997**, *20*, 279. [CrossRef]

25. Matthis, J.S.; Yates, J.L.; Hayhoe, M.M. Gaze and the Control of Foot Placement When Walking in Natural Terrain. *Curr. Biol.* **2018**, *28*, 1224–1233. [CrossRef]

26. Todorov, E.; Jordan, M.I. Optimal feedback control as a theory of motor coordination. *Nat. Neurosci.* **2002**, *5*, 1226–1235. [CrossRef]

27. Liu, D.; Todorov, E. Evidence for the flexible sensorimotor strategies predicted by optimal feedback control. *J. Neurosci.* **2007**, *27*, 9354–9368. [CrossRef]

28. Izawa, J.; Rane, T.; Donchin, O.; Shadmehr, R. Motor adaptation as a process of reoptimization. *J. Neurosci.* **2008**, *28*, 2883–2891. [CrossRef]

29. Nagengast, A.J.; Braun, D.A.; Wolpert, D.M. Optimal control predicts human performance on objects with internal degrees of freedom. *PLoS Comput. Biol.* **2009**, *5*, e1000419. [CrossRef]

30. Ueyama, Y.; Miyashita, E. Optimal Feedback Control for Predicting Dynamic Stiffness During Arm Movement. *IEEE Trans. Ind. Electron.* **2014**, *61*, 1044–1052. [CrossRef]

31. Shadmehr, R.; Mussa-Ivaldi, F.A. Adaptive representation of dynamics during learning of a motor task. *J. Neurosci.* **1994**, *14*, 3208–3224. [CrossRef] [PubMed]

32. Krakauer, J.W.; Pine, Z.M.; Ghilardi, M.F.; Ghez, C. Learning of visuomotor transformations for vectorial planning of reaching trajectories. *J. Neurosci.* **2000**, *20*, 8916–8924. [CrossRef] [PubMed]

33. Osu, R.; Burdet, E.; Franklin, D.W.; Milner, T.E.; Kawato, M. Different mechanisms involved in adaptation to stable and unstable dynamics. *J. Neurophysiol.* **2003**, *90*, 3255–3269. [CrossRef] [PubMed]

34. Saijo, N.; Gomi, H. Multiple motor learning strategies in visuomotor rotation. *PLoS ONE* **2010**, *5*, e9399. [CrossRef]

35. Wang, J.X.; Kurth-Nelson, Z.; Kumaran, D.; Tirumala, D.; Soyer, H.; Leibo, J.Z.; Hassabis, D.; Botvinick, M. Prefrontal cortex as a meta-reinforcement learning system. *Nat. Neurosci.* **2018**, *21*, 860–868. [CrossRef]

36. Sugiyama, T.; Schweighofer, N.; Izawa, J. Reinforcement meta-learning optimizes visuomotor learning. *bioRxiv* **2020**. [CrossRef]

MDPI

St. Alban-Anlage 66

4052 Basel

Switzerland

Tel. +41 61 683 77 34

Fax +41 61 302 89 18

www.mdpi.com

Applied Sciences Editorial Office

E-mail: applsci@mdpi.com

www.mdpi.com/journal/applsci